Hasan Padamsee

RF Superconductivity

Related Titles

H. Padamsee, J. Knobloch, T. Hays

RF Superconductivity for Accelerators

2008
ISBN: 978-3-527-40842-9

J. Clarke, A. I. Braginski

The SQUID Handbook
Volume I: Fundamentals and Technology of SQUIDs and SQUID Systems

2004
ISBN: 978-3-527-40229-8

J. Clarke, A. I. Braginski (Eds.)

The SQUID Handbook
Volume II: Applications of SQUIDs and SQUID Systems

2004
ISBN: 978-3-527-40408-7

W. Buckel, R. Kleiner

Superconductivity
Fundamentals and Applications

2004
ISBN: 978-3-527-40349-3

T. P. Wangler

RF Linear Accelerators

2008
ISBN: 978-3-527-40680-7

Hasan Padamsee

RF Superconductivity

Science, Technology, and Applications

WILEY-VCH Verlag GmbH & Co. KGaA

The Author

Hasan Padamsee
Laboratory of Nuclear Studies
Cornell University, Newman Lab
Ithaca, NY, USA
hsp3@cornell.edu

Cover illustration
9-cell cavities for XFEL, ILC and a number of upcoming light sources around the world.
Hasan Padamsee, Cornell University

All books published by **Wiley-VCH** are carefully produced. Nevertheless, authors, editors, and publisher do not warrant the information contained in these books, including this book, to be free of errors. Readers are advised to keep in mind that statements, data, illustrations, procedural details or other items may inadvertently be inaccurate.

Library of Congress Card No.: applied for

British Library Cataloguing-in-Publication Data
A catalogue record for this book is available from the British Library.

Bibliographic information published by the Deutsche Nationalbibliothek
Die Deutsche Nationalbibliothek lists this publication in the Deutsche Nationalbibliografie; detailed bibliographic data are available in the Internet at http://dnb.d-nb.de.

© 2009 WILEY-VCH Verlag GmbH & Co. KGaA, Weinheim

All rights reserved (including those of translation into other languages). No part of this book may be reproduced in any form – by photoprinting, microfilm, or any other means – nor transmitted or translated into a machine language without written permission from the publishers. Registered names, trademarks, etc. used in this book, even when not specifically marked as such, are not to be considered unprotected by law.

Printed in the Federal Republic of Germany
Printed on acid-free paper

Composition K+V Fotosatz GmbH, Beerfelden
Printing betz-druck GmbH, Darmstadt
Bookbinding Litges & Dopf Buchbinderei GmbH, Heppenheim

ISBN 978-3-527-40572-5

Contents

Preface *XI*

List of Abbreviations *XIII*

Part I   **Science**

1 **Introduction** *3*
1.1   RF Superconductivity Benefits *3*
1.2   Overview of Chapters *4*

2 **New Cavity Geometries** *9*
2.1   Classification of Structures *9*
2.2   High-β Cavities *10*
2.2.1   Re-Entrant and Low-Loss Shapes *13*
2.3   Medium-β Cavities *18*
2.3.1   Medium-β Elliptical Cavities *18*
2.3.2   Medium-β Spoke Resonators *22*
2.4   Low-β Resonators *26*
2.4.1   Quarter-Wave Resonators *26*
2.4.2   Half-Wave Resonators *29*
2.5   Very Low-β Cavities *30*
2.5.1   Superconducting RFQ *30*
2.6   Mechanical Aspects *32*
2.6.1   Mechanical Stresses *32*
2.6.2   Vibrations *35*
2.6.3   Lorentz-Force Detuning *36*
2.7   Codes for Cavity Design *38*

3 **Surface Resistance and Critical Field** *41*
3.1   BCS Surface Resistance *41*
3.1.1   Mean Free Path Dependence of BCS Resistance *42*
3.1.2   Oxygen Diffusion *44*
3.2   Nonlinear (Field-Dependent) BCS Resistance *46*

3.3	RF Magnetic Field Dependence of BCS Surface Resistance	48
3.3.1	Low-Field Q-Slope: Experiments and Models	48
3.3.2	Medium-Field Q-slope and Thermal Feedback	50
3.3.2.1	Thermal Feedback with Standard BCS Resistance	51
3.3.3	Medium-Field Q-Slope and Field-Dependent BCS Resistance	60
3.3.3.1	Linear Dependence of the Medium-Field Q-Slope Due to Josephson Fluxons	62
3.4	Residual Resistance	65
3.4.1	Hydrogen-Related Q-Disease	66
3.4.2	Residual Losses from Trapped dc Magnetic Flux	69
3.4.3	Baking Effects	71
3.5	DC Critical Magnetic Fields: H_{c1}, H_{c2}, H_c, and H_{c3}	72
3.5.1	H_{c1}, H_c, and H_{c2}	73
3.5.2	Surface Superconductvity and H_{c3}	77
3.6	RF Critical Magnetic Field, H_{sh} the Superheating Critical Field	79
3.6.1	Theoretical Updates	79
3.6.2	RF Critical Magnetic Field: Experiments	81
4	**Multipacting and Field Emission**	**85**
4.1	Multipacting	85
4.1.1	Two-Point Multipacting in Elliptical Cavities	87
4.1.2	Multipacting in the Beam Pipe	89
4.1.3	Multipacting in Low-β Cavities	92
4.1.4	Codes for Multipacting	92
4.2	Field Emission	92
4.2.1	Field Emission in Cavities: Temperature Maps and X-Rays	95
4.2.1.1	Emitter Activation	95
4.2.1.2	Field-Emission-Induced Thermal Breakdown	96
4.2.1.3	X-ray Diagnostics	96
4.2.2	Microparticles and Cleanliness	98
4.2.2.1	Statistical Models for Cavity Performance with Field Emission	103
4.2.2.2	Field Enhancement Factors	104
4.2.3	Processing Field Emission	105
4.2.3.1	CW RF Processing	106
4.2.3.2	RF Processing of a Planted Particle	109
4.2.3.3	High Pulsed Power Processing	109
4.2.3.4	DC Voltage Breakdown Studies	111
4.2.3.5	DC Breakdown Studies on Deliberately Planted Particles	112
4.2.3.6	The Nature of the Starburst	113
4.2.4	Model for RF Processing and Simulations	116
4.2.5	Model for Helium Processing	125
4.2.6	Summary for Field Emission	127

5	**High-Field Q-Slope and Quench Field** *129*
5.1	RF Measurements and Temperature Maps *129*
5.1.1	General Description of Q-slopes *129*
5.1.2	Magnetic Field Effect *130*
5.1.3	Spatial Nonuniformity of Losses *133*
5.1.4	Comparison of Q-Slopes: Electropolishing and Buffered Chemical Polishing *134*
5.1.5	The Mild Baking Benefit *138*
5.1.5.1	Baking BCP Cavities *141*
5.1.5.2	Baking EP Cavities *142*
5.2	Candidate Mechanisms for the High-Field Q-Slope *145*
5.2.1	Thermal Feedback *145*
5.2.2	Roughness and Q-slope *147*
5.2.2.1	Macroroughness and Microroughness *147*
5.2.2.2	Field Enhancement at Grain Boundary Edges *157*
5.2.3	Oxygen Pollution Layer Model *159*
5.2.3.1	Difficulties with the Pollution Models *166*
5.2.3.2	The Nonrole of Hydrogen *171*
5.2.4	Fluxoids *172*
5.2.4.1	Role of Grain Boundaries *174*
5.2.5	Summary of Key Features of High-Field Q-Slope *180*
5.3	Surface Studies *181*
5.3.1	Nondestructive Methods *181*
5.3.1.1	XPS Studies of the Oxide Structure *181*
5.3.1.2	XPS Studies of Baked Samples *184*
5.3.1.3	Electron Backscatter Diffraction (EBSD) *187*
5.3.2	Destructive Methods *188*
5.3.2.1	Secondary Ion Mass Spectrometry (SIMS) *188*
5.4	Quench Fields *192*

Part II	**Technology**
6	**Cavity Fabrication Advances** *203*
6.1	Overview *203*
6.2	Niobium Material *204*
6.2.1	Specifications *204*
6.2.1.1	Availability of High RRR Nb *205*
6.2.1.2	Sheet Production from Ingot *206*
6.2.2	Eddy Current and SQUID Scanning *208*
6.2.3	Large-Crystal and Single-Crystal Niobium *211*
6.2.3.1	Material Properties *211*
6.2.3.2	Fabrication from Large Grain *213*
6.2.3.3	Single-Crystal Nb *218*
6.3	Advances in Welding *219*

6.4	Low-β Resonators	221
6.5	Seamless Cavities	222
6.5.1	Hydroforming	222
6.5.1.1	Niobium Clad Copper Hydroformed Cavities	224
6.5.2	Spinning	227
6.5.2.1	Tube Forming	229
6.6	Nb–Cu-Sputtered Cavities	230
6.6.1	High-β Cavities	230
6.6.2	Medium- and Low-β Cavities	231
6.6.3	Research for Better Film Quality	234
7	**Cavity Treatment Advances**	**237**
7.1	Overview of Cavity Treatment Procedures	237
7.2	Inspection	239
7.3	Barrel Polishing	240
7.4	Electropolishing	241
7.4.1	A Short History of EP	241
7.4.2	Basics of Electropolishing	242
7.4.3	Horizontal Electropolishing	249
7.4.4	Vertical Geometry Electropolishing	251
7.4.5	Contamination Issues	253
7.5	Heat Treatment and Hydrogen Degassing	255
7.6	Final Cleaning	256
7.6.1	High-Pressure Rinsing	256
7.6.2	Alternate Final Cleaning Methods	259
7.7	Dust-Free Assembly	260
8	**Input Couplers**	**263**
8.1	Overview of Requirements and Design Principles	263
8.2	Power	267
8.3	Waveguide versus Coaxial Couplers	268
8.4	Windows: Warm and Cold	270
8.5	Coaxial Coupler Examples	272
8.5.1	Couplers for Low-β Resonators	280
8.6	Waveguide Coupler Examples	280
8.7	Materials and Fabrication	282
8.8	Multipacting in Couplers	285
8.9	Conditioning	290
9	**Higher Mode Couplers and Absorbers**	**295**
9.1	Overview of HOM Excitation and Damping	295
9.2	Mode Trapping in Multicell Structures	301
9.3	Coaxial Couplers	302
9.3.1	Multipacting in Coaxial Couplers	305
9.3.2	Coaxial Coupler Kicks	306

9.4	Waveguide Couplers 306
9.5	Beam Pipe Couplers and Absorbers 308

10	**Tuners** 313
10.1	Overview 313
10.2	Tuner Examples 315
10.2.1	Saclay TTF Tuner 317
10.2.1.1	Modified Saclay Tuner 317
10.2.2	INFN Coaxial Blade Tuner 318
10.2.3	KEK Coaxial Ball Screw Tuner 320
10.2.4	KEK Slide Jack Tuner 321
10.2.5	JLab Tuners 321
10.3	Fast Tuning 324
10.4	Low-β Cavity Tuners 329

Part III Applications

11	**Applications and Operations** 335
11.1	Storage Rings 335
11.1.1	LEP-II and LHC 335
11.1.2	CESR and KEK-B 338
11.1.3	Storage Ring Light Sources 342
11.1.4	Crab Cavities 348
11.2	Electron Linacs 353
11.2.1	CEBAF and Upgrade 353
11.2.1.1	CEBAF Upgrade 356
11.2.2	Free Electron Lasers (FEL) 357
11.2.3	TTF SASE FEL (FLASH) 361
11.2.4	European XFEL 366
11.3	Heavy-Ion Accelerators 369
11.4	Heavy-Ion Accelerators for Rare Isotope Beams (RIBs) 373
11.4.1	ISAC-II 374
11.4.2	SPIRAL2 376
11.4.3	MSU Reaccelerator 378
11.5	Neutron Source 380
11.5.1	SNS 380

12	**Future Applications** 389
12.1	Overview for Next Generation Light Sources 389
12.1.1	Single-Pass FELs 390
12.1.2	ERL-Based Light Sources 392
12.2	ERL for Nuclear Physics 398
12.3	Nuclear Astrophysics 399
12.3.1	Rare Isotope Accelerator (RIA) 400

12.3.2	EURISOL Driver	*401*
12.4	High-Intensity Proton Linacs	*401*
12.4.1	Japan Proton Accelerator Research Complex (JPARC)	*403*
12.4.2	Proton Driver	*403*
12.4.3	SRF Proton Linac (SPL)	*404*
12.4.4	Transmutation Applications	*404*
12.5	International Linear Collider	*406*
12.6	Muon Collider and Neutrino Factory	*408*
12.7	Concluding Remarks	*410*

References *411*

Subject Index *443*

Preface

It has now been nearly ten years since the successful publication and enthusiastic reception of *RF Superconductivity for Accelerators* [1] (ISBN 0-471-15432-6, 1998). Much has happened in the field since the appearance of this work, which is heavily referenced as a standard text. There has been an explosion in the number of accelerator applications and in the number of laboratories engaged in the field. RF superconductivity has become a major subfield of accelerator science.

The overall aim of this book is to discuss advances in the science, technology, and applications of superconducting rf (SRF) since the 1998 publication. Basics covered in [1] are not repeated here although essentials are summarized for the sake of completeness. The book also explores exciting prospects for future advances in capability and applications. The rapid growth of this technology has alerted the world accelerator community and its supporting agencies to the importance of SRF for frontier instruments in high energy, nuclear, and astrophysics, as well as in the materials and life sciences. I hope that newcomers to the field, of which there are many, will also benefit from a discussion of the progress in the science and technology.

This book is not intended to be a comprehensive review of the progress in SRF technology taking place at many laboratories in a variety of arenas. Examples presented are rather selective for illustration purposes only. Due to space and time limitations the book does not cover the extensive topics of cryomodule and cavity/cryomodule integration, as well as feedback and low level rf (LLRF) controls. Accelerator operations experience is only briefly discussed.

Many review articles [2–12] are now available covering the state-of-the-art in rf superconductivity and its application to particle accelerators. There have been 13 international workshops on rf superconductivity. The proceedings carry a comprehensive coverage of the substantial work going on at many institutions with many excellent tutorials on special subjects. The proceedings will soon be also available in electronic form on the JACOW website. The book draws heavily from papers at Particle Accelerator Conferences, presentations at collaboration meetings, such as the TESLA Technology Collaboration (TTC), and special meetings, such as the Spoke Workshop and Coupler Workshops. For more general reference texts in Solid State Physics and Superconductivity readers should refer to [13–15].

RF Superconductivity: Science, Technology, and Applications. Hasan Padamsee
Copyright © 2009 WILEY-VCH Verlag GmbH & Co. KGaA, Weinheim
ISBN: 978-3-527-40572-5

Many thanks to Sergey Belomestnykh (Cornell), Mathias Liepe (Cornell), and Wolfgang Weingarten (CERN) for reviewing various chapters. My sincere apologies to those laboratories and individuals whose work I missed.

Some General Remarks

In the Appendix the full name for all the many acronyms throughout the book has been spelled out. In most (but not all) cases, the acronyms are also spelled out in the text.

Units: In general we use MKS units, with occasional exceptions. A common variation is the use of different units for the magnetic field. In the MKS system the units for magnetic field (H) have the dimensions of A/m. The magnetic flux density (B) has the dimensions of Vs/m^2. In the cgs system the units for B are gauss and for H are oersted. To convert from MKS to cgs units, use 1 A/m = $4\pi \times 10^3$ Oe. Finally 1 T = 10^4 Oe. Since free space permeability, $\mu = 1$, in the cgs system, the numerical values for B and H are the same. We tend to use B and H interchangeably on occasion according to the original plots used from the references.

List of Abbreviations

ADS	accelerator-driven system
AES	Advanced Energy Systems
AFM	atomic force microscopy
ALPI	A Heavy Ion Superconducting Linear Post-Accelerator at Legnaro
ANL	Argonne National Laboratory
ANSYS	Analysis System
ARC-EN-CIEL	Accelerator-Radiation Complex for Enhanced Coherent Intense Extended Light
ASTM	American Society for Testing and Materials
ATLAS	Argonne Tandelm Linear Accelerator System
ATW	accelerator-based transmutation of waste
BARC	Bhaaba Atomic Research Center
BCP	buffered chemical polishing
BCS	Bardeen, Cooper, Schrieffer
BESSY	Berliner Elektronenspeicherring-Gesellschaft für Synchrotronstrahlung mbh
BNL	Brookhaven National Laboratory
CAD	computer-aided design
CARE	Coordinated Accelerator Research in Europe
CAT	Center for Accelerator Technology (Indore)
CBP	centrifugal barrel polishing
CEA	Commissariat à l'énergie atomique (Saclay France)
CEBAF	Continuous Electron Beam Accelerator Facility
CESR	Cornell Electron Storage Ring
CHESS	Cornell High Energy Synchrotron Source
CLS	Canadian Light Source
CM	cryomodule
CMM	coordinate measuring machine
CR	circulator ring
cw	continuous wave
DC	direct current
DESY	Deutsches Elektronen-Synchrotron
DIAMOND	DIpole And Multipole Output for the Nation at *Daresbury* (now at Rutherford Appleton Lab)
EBIT	electron beam ion trap
EBSD	electron backscatter diffraction
EBW	electron beam welding
ECR	electron cyclotron resonance
EDM	electro discharge machining

RF Superconductivity: Science, Technology, and Applications. Hasan Padamsee
Copyright © 2009 WILEY-VCH Verlag GmbH & Co. KGaA, Weinheim
ISBN: 978-3-527-40572-5

EDX	energy dispersive x-ray
ELETTRA	light source in Trieste, Italy
eLIC	electron-light-ion collider
EP	electropolishing
ERL	energy reocovery linacs
ESRF	European Synchrotron Radiation Facility
ESS	European Spallation Source
EURISOL	European Isotope Separation On-Line (Radioactive Ion Beam Facility)
FE	field emission
FELs	free electron lasers
FESEM	field emission scanning electron microscopy
FLASH	Free-Electron LASer in Hamburg
GANIL	Grand Accélérateur National d'Ions Lourds
GDFIDL	"Gitter drüber – fertig ist die Laube"; could be translated as "put a grid, and ready you are"
GSI	Gesellschaft für Schwerionenforschung
GTI	global thermal breakdown
HEP	high energy physics
HER	high-energy ring
HERA	Hadron Elektron Ring Anlage
HF	hydrofluoric acid
HFSS	high-frequency structures simulator
HGHG	high-gain-harmonic generation
HIP	hot isostatic pressing
HoBiCaT	Horizontal Bi-Cavity Test-facility (BESSY)
HOM	higher order mode
HPP	high pulse power processing
HPPMS	high power pulsed magnetron sputtering
HPR	high-pressure rinsing
HWR	half-wave resonator
ILC	International Linear Collider
INFN	Istituto Nazionale di Fisica Nucleare
IPHI	high-intensity proton injector
ISAC	Isotope Separator & Acceleration (TRIUMF)
ISO	International Organization for Standardization
ISOL	Isotope Separation On-Line
ITRP	International Technology Recommendation Panel
IUAC	Inter-University Accelerator Centre
JACOW	Joint Accelerator Conference Website
JAEA	Japan Atomic Energy Agency
JAERI	Japan Atomic Energy Research Institute
JIS	Japan Industrial Standards
JLab	Jefferson Laboratory
JPARC	Japan Proton Accelerator Research Complex
KEK	Koh Ene Ken (Japanese: National Laboratory for High-Energy Physics)
KOMAC	Korea Multipurpose Accelerator Complex
LANSCE	Los Alamos Neutron Science Center
LCLS	Linac Coherent Light Source (Stanford linear accerator center)
LEP	large electron positron (collider)
LER	low-energy ring
LF	Lorentz force
LHC	large hadron collider
LLRF	low-level rf

LN2	liquid nitrogen
LNL	Laboratori Nazionali di Legnaro (Italy)
LOM	lower order modes
LPSC	Laboratoire de Physique Subatomique et de Cosmologie (Grenoble)
MAFIA	Maxwell's Equations Solved by the Finite Integration Algorithm
MAGIC	MAGnetic Insulation Code (Plasma Pic Code)
MASK	particle-in-cell (PIC) code
MIT	Massachussetts Institute of Technologhy
MP	*multi*ple im*pact* electron amplification
MSU	Michigan State University
MW	megawatt
MWS	Microwave Studio
NERSC	National Energy Research Scientific Computing Center
NSCL	National Superconducting Cyclotron Laboratory
OFE	oxygen free electronic (copper grade)
OOPIC	object-oriented particle-in-cell code
OSCAR	Open Source Cluster Application Resources
PAC	Particle Accelerator Conference
PF	particle fragmentation
PFA	polyperfluoro alkoxyethylene
PIAVE	positive ion accelerator for very-low energy
PTFE	polytetrafluoroethylene (teflon)
PVDF	polyvinylidene fluoride
QWR	quarter-wave resonator
RFQ	radiofrequency quadrupole
RHIC	relativistic heavy-ion collider
RIA	Rare Isotope Accelerator
RIB	rare isotope beams
RIKEN	Rikagaku Kenkyusho (Institute of Physical and Chemical Research, Japan)
RRR	residual resistivity ratio
SAIC	Science Applications International Corporation
SARAF	Soreq Applied Research Accelerator Facility
SASE	self-amplified-spontaneous-emission
SciDAC	Scientific Discovery through Advanced Computing
S-DALINAC	Darmstadt Linac
SEC	secondary emission coefficient
SEM	secondary electron microscopy
SIMS	secondary ion mass spectrometry
SLS	Swiss Light Source (Paul Scherrer Institute, Zurich, Switzerland)
SNS	Spallation Neutron Source at Oak Ridge National Laboratory
SOLEIL	Source optimisée de lumière d'énergie intermédiaire du LURE
SPES	Study & Production of Exotic Species, Legnaro
SPIRAL	Système de Production d'Ions Radioactifs en Ligne (at GANIL, France)
SPL	superconducting proton linac
SQUID	superconducting quantum interference device
SRF	superconducting rf
SRFQ	superconducting rf quadrupole
SRRC	Synchrotron Radiation Research Center (Taiwan)
SS	stainless steel
SuperLANS	calculation of azimuthal-homogeneous modes in axisymetrical cavities and periodic structures
TDR	time domain reflectometry

TE	transverse electric
TEM	transverse electric and magnetic
TeV	tera electron volt
TIFR	Tata Institute of Fundamental Research
TM	transverse magnetic
TRASCO	TRAsmutazione SCOrie
TRISTAN	Transposable Ring Intersecting Storage Accelerator in Nippon (KEK electron-positron collider)
TRIUMF	TRI-University Meson Facility (Vancouver, Canada)
TTC	Tesla Technology Collaboration
TTF	TESLA Test Facility
UHV	ultrahigh vacuum
VDI	Society of German Engineers
VUV	vacuum ultraviolet
W&M	William and Mary
XADS	eXperimental Accelerator Driven System
XFEL	x-ray free electron laser
XPS	x-ray photoelectron spectroscopy

**Part I
Science**

1
Introduction

1.1
RF Superconductivity Benefits

Superconducting rf (SRF) cavities excel in applications requiring continuous wave (cw) or long-pulse accelerating fields above a few million volts per meter (MV/m). We often refer to the accelerating field as the "gradient". Since the ohmic power loss in the walls of a cavity increases as the square of the accelerating voltage, copper cavities become uneconomical when the demand for high cw voltage grows with particle energy. A similar situation prevails in applications that demand long rf pulse length, or high rf duty factor. Here superconductivity brings immense benefits. The surface resistance of a superconducting cavity is of many orders of magnitude less than that of copper. Hence the intrinsic quality factors (Q_0) of superconducting cavities are usually in the 10^9 to 10^{10} range (Q_0 is often abbreviated as Q). Characterizing the wall losses, the Q_0 is a convenient parameter for the number of oscillations it takes the stored energy in a cavity to dissipate to zero. After accounting for the refrigerator power needed to provide the liquid helium operating temperature, a net gain factor of several hundred remains in the overall operating power for superconducting over copper cavities. This gain provides many other advantages.

Copper cavities are limited to gradients near 1 MV/m in cw and long-pulse operation because the capital cost of the rf power and the ac-power related operating cost become prohibitive. For example, several MW/m of rf power would be required to operate a copper cavity at 5 MV/m. There are also practical limits to dissipating high power in the walls of a copper cavity. The surface temperature becomes excessive causing vacuum degradation, stresses, and metal fatigue due to thermal expansion. On the other hand, copper cavities offer much higher accelerating fields (≈ 100 MV/m) for short pulse (μs) and low duty factor ($<0.1\%$) applications. Still, for such applications it is still necessary to provide abundant peak rf power (e.g., 100 MW/m) and to withstand the aftermath of intense voltage breakdown in order to reach the very high fields.

There is another important advantage that SRF cavities bring to accelerators. The presence of accelerating structures has a disruptive effect on the beam, limiting the quality of the beam in aspects such as energy spread, beam halo,

RF Superconductivity: Science, Technology, and Applications. Hasan Padamsee
Copyright © 2009 WILEY-VCH Verlag GmbH & Co. KGaA, Weinheim
ISBN: 978-3-527-40572-5

or even the maximum current. Because of their capability to provide higher voltage, SRF systems can be shorter, and thereby impose less disruption. Due to their high ohmic losses, the geometry of copper cavities must be optimized to provide a high electric field on axis for a given wall dissipation. This requirement tends to push the beam aperture to small values, which disrupts beam quality. By virtue of low wall losses, it is affordable to design an SRF cavity to have a large beam hole, reduce beam disruption and provide high quality beams for physics research.

For low velocity, heavy-ion accelerators, a major advantage of superconducting resonators is that a cw high voltage can be obtained in a short structure. The linac to boost ion energies can be formed as an array of independently phased resonators, making it possible to vary the velocity profile of the machine. The superconducting booster is capable of accelerating a variety of ion species and charge states. An independently phased array forms a system which provides a high degree of operational flexibility and tolerates variations in the performance of individual cavities. Superconducting boosters show excellent transverse and longitudinal phase space properties, and excel in beam transmission and timing characteristics. Because of their intrinsic modularity, there is also the flexibility to increase the output energy by adding higher velocity sections at the output, or to extend the mass range by adding lower velocity resonators at the input.

1.2
Overview of Chapters

This book is divided into three parts.

Part I (Science) will start with a review of fundamental design principles to develop cavity geometries for accelerating velocity-of-light particles ($\beta = v/c = 1$), moving on to corresponding design principles for medium-velocity (medium-β) and low-velocity (low-β) structures. There is an in-depth presentation of several new geometries that have evolved in the last decade, such as the re-entrant and low-loss shapes for high-β cavities, compressed-elliptical cavities for medium-β applications and spoke resonators for intermediate-β, bridging the medium- and low-β regimes. For ultralow velocities, a superconducting radiofrequency quadrupole (RFQ) has been developed to combine acceleration and strong focusing. The last part of Chapter 2 touches upon mechanical aspects of cavity design including Lorentz-force detuning and vibrations. Wherever appropriate, the chapter mentions various electromagnetic field calculation codes available for cavity design.

Chapter 3 delves into recent theories and experiments that address the fundamental aspects of rf surface resistance. Since the main aspects of the Bardeen, Cooper, Schrieffer (BCS) resistance are covered extensively in [1], we concentrate on the low-field and medium-field dependence of surface resistance, referred in the SRF community as medium-field Q-slope and low-field Q-slope. The chapter also updates recent findings about residual resistance. The rest of the chapter addresses developments defining fundamental limits for the maximum expected gradients based on the maximum expected surface magnetic fields. Chapter 4 discusses progress in studying electric field dependent phenomena that determine the maximum gradient and Q_0 of cavities. Multipacting and field emission have been successfully reduced but challenges remain in both arenas. High-pressure water rinsing has played a major role in reducing field emission. There has also been substantial progress in understanding the basic mechanisms involved in rf processing. Chapter 5 discusses progress in studying magnetic field dependent phenomena that determine the maximum gradient and Q_0 of cavities. Techniques such as electropolishing and mild baking (120 °C, 48 h) play a major role in achieving the best cavity performances, raising regularly achievable gradients above 25 MV/m, and the best gradients above 50 MV/m. The phenomena of high-field Q-drop have been extensively studied and various models have emerged, but none with complete success. Surface studies have made major contributions to understanding the origin and reduction of the Q-drop. Several of the prevailing models will be discussed to account for the performance improvements with new preparation techniques.

In **Part II** (Technology), Chapters 6 and 7 cover cavity fabrication and treatment advances. High-performance demands excellent control of niobium material properties, purity, fabrication stages (forming and welding) surface smoothness, and surface cleanliness. Techniques for fabrication and surface treatment have evolved considerably over the last decade to achieve the desired levels of control. Advances are discussed in depth, along with the performance levels achieved. The performance of low- and medium-velocity structures is also covered. New techniques for fabricating seamless cavities, such as monolithic spinning and hydroforming continue to make progress. Previous efforts continue to deposit thin superconducting films onto a copper substrate, and new approaches are underway to improve the quality of films. If successful, the seamless and thin film approaches will allow a cost reduction of future facilities. Although the tantalizing subject of new materials, such as Nb_3Sn, YBCO, and MgB_2, remains of great interest, there has been little progress in cavity results over the last decade. Hence the scope will exclude this topic.

The next three technological chapters (Chapters 8 to 10) deal with input couplers, higher order mode extraction couplers and absorbers, and tuners of both the slow and fast varieties. There have been substantial advances in the average and peak power capability of input couplers. Chapters 8 and 9 discuss electromagnetic, mechanical, and thermal design principles and implementation, culminating in a presentation of several popular coupler designs and variants. Multipacting issues are covered for both types of couplers. Chapter 10 presents the advantages and disadvantages of a remarkable variety of slow mechanical tuner designs. Piezo techniques to control Lorentz force and microphonics detuning are covered.

Part III (Applications) describes SRF systems at major accelerators and briefly discusses the operating experience which demonstrates that SRF is a robust technology. The Argonne Tandem Linear Accelerator System (ATLAS) has been operating for nearly three decades as a national user facility for heavy ion, nuclear and atomic physics research, logging well over 100 000 h of beam-on-target operation. CEBAF at Jefferson Lab (JLab) has been operating for a decade. Jefferson Lab is developing high gradient cavities and cryomodules to upgrade CEBAF's energy to 12 GeV. Doubling the beam energy is an important priority for advancing understanding of the strong force and its manifestation in gluonic matter. The FEL at Jefferson Lab, LEP-II at CERN (now decommissioned for LHC installation), HERA and TTF at DESY, CESR at Cornell, KEK-B factory in Japan, provide a few among many excellent examples of systems and operations.

As a natural outcome of the LEP-II Nb–Cu technology, superconducting cavities and cryomodules are ready to meet the voltage and high current demands of the large hadron collider (LHC) at CERN. The spallation neutron source (SNS) at Oak Ridge National Laboratory has been completed and commissioned. Neutron scattering is an important tool for material science, chemistry, and life science. SNS will provide 1.4 MW of beam power on target to produce a neutron flux comparable to the average flux of the Grenoble reactor, the largest neutron science facility. SNS switched to superconducting technology in 2000 for a shorter linac due to the higher operating gradient possible with superconducting cavities, a large savings in the rf installation, and the potential of reduced activation due to large beam holes of superconducting structures.

Electron storage rings as light sources have an enormous impact on materials and biological science. Superconducting cavity accelerating systems have upgraded light sources, such as CHESS at Cornell, and the Taiwan Light Source. New light sources such as the Canadian Light Source and DIAMOND in UK have adopted SRF. The Swiss Light Source and ELETTRA in Treiste have installed higher frequency (third harmonic) superconducting cavities to improve beam lifetime and stability.

The Jefferson Lab FEL has generated 14 kW of cw laser power in the infrared, and demonstrated energy recovery by recirculating nearly 1 MW of beam power. This is an important milestone toward energy recovery linacs (ERL) for future light sources and electron beam cooling applications. An upgrade to ultraviolet is underway.

High gradient SRF technology developed at the TESLA Test Facility (TTF) at DESY will drive the European XFEL, a linac-based free electron laser to provide Angstrom wavelength x-ray beams of unprecedented brilliance. The brilliance, coherence, and ultrashort pulses will open a wide range of novel experiments not possible with present x-radiation sources. TTF will continue as FLASH to serve users of ultraviolet radiation, as a test bed for the XFEL and continue its vital role as a proving ground for the technology needed for the future International Linear Collider (ILC).

In the low-β arena, ALPI at Legnaro has a new injector PIAVE based on a superconducting RFQ followed by a string of quarter-wave resonators (QWRs). Heavy-ion linacs in New Delhi and Mumbai have come on-line. TRIUMF in

Canada is expanding its radioactive beam facility (ISAC) by adding a superconducting heavy-ion linac supplying more than 40 MV. Among other fundamental questions, radioactive beams will provide basic insight into the origin of the heavy elements. With the SPIRAL2 project, GANIL in France aims to make an intermediate step between existing radio isotope beams (RIB) facilities and future projects like EURISOL.

Chapter 12 moves on to cover new accelerators under construction and planning. Designs for the nuclear astrophysics Rare Isotope Accelerator (RIA) are able to call on advanced preparation techniques that deliver high performance cavities. A future US facility will use superconducting structures suitable for particle velocities ranging from a few percent to about 70% the speed of light. Record radioactive beam intensities will allow the study of a large number of exotic isotopes that will provide quantitative information necessary for theories of stellar evolution and the formation of elements in the cosmos. In Europe, EURISOL and its first stage SPES are similar facilities also under study.

A variety of new applications are under study for linac-based light sources, such as high-power free electron lasers (FEL) and high-brilliance ERL. FEL and ERL studies are flowering around the world. Cornell University, Argonne National Labs, MIT, University of Wisconsin, Lawrence Berkeley Laboratories, BESSY in Berlin, FZ Rossendorf and the Cockroft Institute in Daresbury are all conducting a wide range of activities. High intensity beams of ERL have spurred explorations for electron cooling applications and for electron–ion colliders, for example to upgrade RHIC at Brookhaven.

High-intensity proton linacs will likely fulfill future needs in a variety of arenas: upgrading the injector chains of proton colliders and accelerators, heavy-ion radioactive beams for nuclear physics, medical therapy, industrial applications, high intensity spallation neutron sources, transmutation applications for treatment of radioactive nuclear waste, nuclear energy production using thorium fuel, neutrino beam lines, neutrino factories, and muon colliders. Fermilab is studying 1–2 MW beam power superconducting linac proton driver to upgrade its injector and provide intense neutrino beams. Prototype work is underway at CERN for the superconducting proton linac (SPL), at KEK for upgrading the Japan Proton Accelerator Research Complex (JPARC) facility, at the Korea Multipurpose Accelerator Complex (KOMAC). A consortium of European laboratories is conducting Design Studies for an eXperimental Accelerator Driven System (XADS) for transmutation of nuclear waste. With abundant thorium resources, India is interested in development of ADS systems for nuclear energy production using thorium fuel. A multitask facility, including an ADS and a neutron spallation source, is envisioned at RRCAT Indore.

In August 2004 the International Technology Recommendation Panel (ITRP) recommended the superconducting option for the next linear collider. Complementing the LHC at CERN, the linear collider will start with 500 GeV energy to be upgraded eventually to 1 TeV. The collider will provide new insights into the structure of space-time, matter, and energy. Among the discoveries expected from the multibillion dollar project are new particles to explain the origin of mass, the mystery of dark matter, and the possibility of extra spatial dimensions. Some of the main features that stem from the choice of a superconducting linac are: lower operating power, better conversion efficiency of ac power to beam power, and the ability to use long-wavelength (larger beam opening) structures than possible for the warm structures. The larger beam aperture of the cold structure reduces electromagnetic wakefields which disrupt the beam quality. The chosen design gradient is 31.5 MV/m. As proof-of-principle, individual 1 m long cavities have been successfully operated in cryomodules at 35 MV/m and in FLASH at 31.5 MV/m. Several tens of meters of superconducting cavities in accelerator modules have been operated at the 25 MV/m level.

In conclusion, superconducting cavities have been operating routinely in a variety of accelerators with a range of demanding applications. With continued progress in basic understanding, cavity performance has steadily improved to approach theoretical capabilities. Niobium cavities have become an enabling technology offering upgrade paths for existing facilities, and pushing frontier accelerators for nuclear physics, high energy physics, and materials and life sciences.

2
New Cavity Geometries

2.1
Classification of Structures

There are three major classes of superconducting accelerating structures: high, medium, and low-β (here $\beta = v/c$, where v is the speed of the accelerated particle and c is the speed of light). The high-β structure, based on the TM_{010} resonant cavity, is for acceleration of electrons, positrons, or high-energy protons with $\beta \sim 1$. The cavity gap length is usually $\beta\lambda/2$, where λ is the wavelength corresponding to the frequency choice for the accelerating structure. Section 2.2 will discuss advances in cavity designs for $\beta \sim 1$ applications including the "elliptical" cell shape which has become the basis for most high-β cavities. For a review of high-β design aspects see [17].

Medium (intermediate) velocity structures with β between 0.2 and 0.7 are used for protons with energies less than 1 GeV as well as for ions. At the higher β end, these are "foreshortened" speed-of-light structures with longitudinal dimensions scaled by β. Near $\beta = 0.5$ spoke resonators with single or multigaps are becoming popular. Spoke resonators operate in a TEM mode and are so classified. The overlap between foreshortened elliptical and spoke structures near $\beta = 0.5$ involves several tradeoffs which we will discuss. Elliptical shape cells for $\beta < 0.5$ become mechanically unstable as the accelerating gap shortens and cavity walls become nearly vertical. The choice of a low rf frequency, favored for ion and proton applications, also makes the elliptical cells very large, aggravating the structural weakness.

Low-velocity structures are for particles moving at a small fraction (e.g., 0.01–0.2) of the speed of light, such as heavy ions emerging from a Van de Graff accelerator or an electron cyclotron resonance (ECR) ion source. The accelerating gap is proportional to $\beta\lambda/2$ so that low frequencies are necessary to provide a significant acceleration length. There are a large variety of low-β structures which evolve from the resonant transmission line, either quarter-wavelength or half-wavelength, and operate in a TEM-like mode. Spoke resonators are also based on half-wave transmission lines and serve as a building block for the longer multigap spoke structures. When the number of loading elements is large and rotated by 90° from one to the next, the multigap spoke cavities are sometimes referred to as H-type or ladder resonators. Unlike axially symmetric TM structures, TEM resonators are essentially three-dimensional geometries. For reviews of low- and medium-β design aspects see [18–23].

In subsequent sections of this chapter we will discuss structures of progressively lower β values.

2.2
High-β Cavities

A typical high-β accelerating structure consists of a chain of coupled cells operating in the TM_{010} mode, where the phase of the instantaneous electric field in adjacent cells is shifted by π to preserve acceleration as a charged particle traverses each cell in half an rf period. Figure 2.1 shows a 9-cell accelerating structure [24, 25] developed by the TESLA collaboration and used at the TESLA Test Facility (TTF). The TTF has now become the Free-Electron LASer in Hamburg (FLASH). The TESLA cavity is also intended for use in the European X-ray Free Electron (XFEL) project, and remains a strong candidate for the International Linear Collider (ILC). Single-cell cavities generally used for SRF R&D also find

(a)

(b)

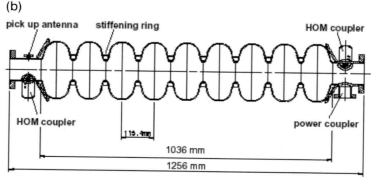

Fig. 2.1 (a) Photograph of a 9-cell TESLA accelerating structure with one input power coupling port at one end and one HOM coupler at each end. (b) Layout of the components for the 9-cell TESLA-style structure (courtesy of DESY and ACCEL).

Fig. 2.2 (a) Single-cell 500 MHz cavity for CESR with a waveguide input coupler and fluted beam tube on one side to remove the first dipole HOMs. The cell length is about 28 cm, a bit shorter than $\lambda/2$ to optimize the cell shunt impedance. (b) Single-cell 508 MHz cavity for KEK-B with the coaxial input coupler port and large beam pipe on one side for propagation of HOMs [28] (courtesy of KEK).

accelerator application, as for example in high current ring colliders, such as CESR, KEK-B, as well as many storage ring light sources. Figure 2.2 shows the single-cell CESR and KEK-B cavities [26–28]. The beam enters and exits the structure from the beam tubes. Input coupler devices attached to ports on the beam tubes bring rf power into the cavity to establish the field and deliver beam power; higher order mode couplers extract and damp the higher order modes (HOMs) excited by the beam, and smaller ports carry pick-up probes to sample the cavity field for regulation and monitoring.

The main figures of merit for an accelerating structure are defined and discussed in [1]. These are: rf frequency, accelerating voltage (V_c), accelerating field (E_{acc}), peak surface electric field (E_{pk}), peak surface magnetic field (H_{pk}), surface resistance (R_s), geometry factor (G), dissipated power (P_c), stored energy (U), Q value, geometric shunt impedance (R_a/Q_0 often mentioned as R/Q for short), cell-to-cell coupling for multicell structures, Lorentz-force (LF) detuning coefficient, input power required for beam power (P_b), coupling strength of an input coupler (Q_{ext}), higher order mode frequencies, shunt impedances, and Q values.

Most $\beta=1$ structures are now based on the elliptical cavity. The elliptic cell shape [29] emerged from the more rounded "spherical" shape [30] which was first developed to eliminate multipacting. Chapter 4 reviews and updates multipacting topics. The tilt of the elliptical cell also increases the stiffness against mechanical deformations and provides a better geometry for acid draining and water rinsing. As Fig. 2.3 shows, the profile of the elliptic cavity consists of several (usually two) elliptic arcs and straight lines between arcs as needed to smoothly merge the arcs [17].

Some of the main choices that need to be made for accelerator structure design are cavity frequency, cell shape, number of cells, beam aperture, operating gradient, operating temperature, input coupler, and HOM coupler types. Two classes of considerations govern structure design: the particular accelerator

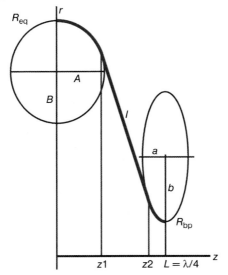

Fig. 2.3 Design of the elliptical cavity half-cell for $\beta=1$ applications. The half-cell length L is a quarter of the wavelength. Four independent parameters, the ellipse dimensions, A, B, a and b, are adjusted for optimization of cavity features, such as the peak surface fields and the shunt impedance. The radius of the beam pipe R_{bp} is set by the accelerator, and the equatorial radius R_{eq} can be adjusted to tune the cavity resonant frequency. The length of the straight segment l is chosen so that it is tangential to the two ellipses [17].

application and superconducting rf properties. Typical accelerator aspects are the velocity of the particle under acceleration, the desired voltage, the duty factor of accelerator operation, beam current, or beam power. Other properties of the beam, such as bunch length, also play a role in the selection of the operating frequency, as these influence the longitudinal and transverse wakefields along with higher order mode impedances. Typical superconducting properties are the microwave surface resistance at the chosen frequency, and the peak surface electric and magnetic fields at the design accelerating field. These properties set the operating field levels, the rf power required, as well as ac operating power, together with the operating temperature. Mechanical properties also play a role in design aspects to ensure stability under atmospheric loading and temperature differentials, to minimize Lorentz-force (LF) detuning, and to keep microphonics detuning under control. Finally, input and output power coupling issues interact with cavity design. In general, there are many tradeoffs between competing requirements. For example, the higher the power capability of the input coupler, the larger the allowed number of cells per structure. But the difficulties of handling long structures set an upper limit to the number of cells. A large number of cells will also increase the probability that some HOMs remain trapped inside the structure, as discussed in Chapter 9. A large aperture will

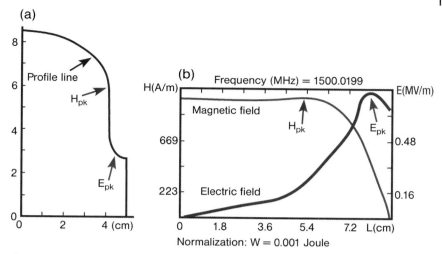

Fig. 2.4 (a) Geometry of an inner half-cell of a multicell cavity.
(b) Electric and magnetic field distributions along the structure profile line for one milli-joule structure stored energy [17].

improve the propagation of HOMs out of the structure, but will increase the peak surface electric and magnetic fields.

Two of the major accelerator structure design considerations are the peak surface electric (E_{pk}) and magnetic (H_{pk}) fields at the operating accelerating field. High surface electric fields can cause field emission, which increases exponentially with E_{pk} (Chapter 4). The high surface magnetic field can cause breakdown of superconductivity, also called quench (discussed in Chapter 5). Figure 2.4 shows the electric and magnetic field distributions along the surface of a cell. The maximum E_{pk} occurs near the iris and the maximum H_{pk} occurs near the equator. At the iris, the elliptic arc reduces the peak surface electric field. Therefore, the larger the minor axis of the iris ellipse, the lower the peak electric field. However the cell-to-cell coupling strength of a multicell structure also decreases, which increases the difficulty of tuning the structure to obtain a flat field profile along the beam axis. A large equatorial radius (R_{eq}) reduces the peak magnetic field by increasing the surface area in the high magnetic field region, thus lowering the current density. Increasing the beam aperture increases both surface fields (for a given accelerating field). The next section discusses many of these design issues with respect to the development of new shapes beyond the commonly used elliptical shape.

2.2.1
Re-Entrant and Low-Loss Shapes

As discussed above, traditional wisdom is to optimize the cavity shape for a minimal E_{pk}/E_{acc} to minimize field emission. Although avoiding intense field emission is an important challenge for clean structure preparation, there is no known fundamental limit to the maximum tolerable surface electric field. Experiments with special superconducting cavities have demonstrated surface electric fields of 140 MV/m, cw [31] and 220 MV/m, pulsed [32], without encountering any fundamental limitation. Many single-cell cavities have reached surface fields E_{pk} above 100 MV/m [33, 34] with low field emission, mainly due to advances in surface cleaning. These cleaning techniques, which will be discussed extensively in Chapter 7, are high-pressure water rinsing (HPR) to eliminate microparticle contaminants and electropolishing (EP) to achieve surface smoothness with an average macroroughness less than 0.5 µm. Residual field emitters can also be eliminated by rf processing, which is often continued for stubborn emitters with high peak power, as discussed in Chapter 4. However, as cavity performance advances, surface magnetic fields approach an impenetrable barrier, the fundamental critical magnetic field limit. The physics of the rf critical magnetic field will be discussed in Chapter 3. To push for higher E_{acc} in the face of the fundamental critical magnetic field limit, new geometries were first proposed [35–38] to reduce H_{pk}/E_{acc} even at the expense of increasing E_{pk}/E_{acc}. The resulting cavity geometries are the "re-entrant" (Ω) shape conceived at Cornell [36–38] and the "low-loss" (LL) shape with reduced aperture proposed by a DESY/JLab/KEK collaboration [39–41]. A slight variation of the LL shape, renamed the ICHIRO shape, is under development by KEK [42, 43]. Figure 2.5 provides a graphical comparison of the new shape cell profiles with the reference TTF profile [44], and Table 2.1 [45, 46] compares the important parameters of the shapes. Figure 2.6 compares the surface electric and magnetic field profiles of the two new shapes with the baseline TTF shape.

The basic idea of the new shapes is to increase the surface area of the equatorial region, where the magnetic field is highest, and hence to lower the cur-

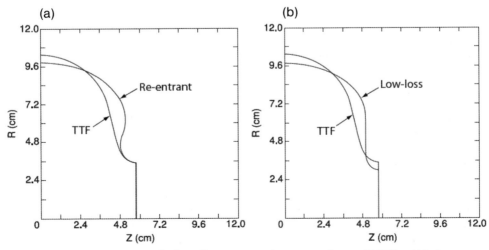

Fig. 2.5 Comparison of cell profiles: (a) TTF and re-entrant shapes; (b) TTF and LL shapes.

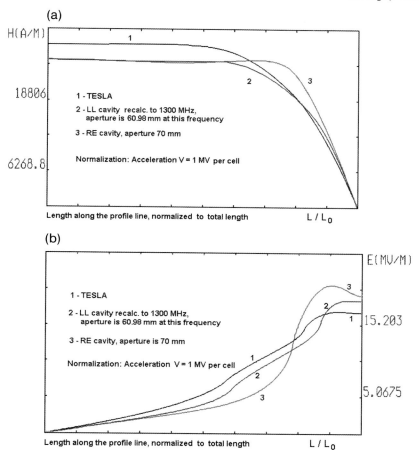

Fig. 2.6 Comparison of the surface field profiles for three cell shapes: TTF-like, re-entrant and LL; (a) magnetic field and (b) electric field. Note that the magnetic field for both new shapes is reduced whereas the electric field for both new shapes is increased.

rent density. Making the geometry re-entrant (Ω-like shape) maximizes this strategy. The LL shape makes reductions in H_{pk} by reducing the aperture. Similarly, a second re-entrant shape cavity design incorporates the benefit of smaller aperture to make further reduction in H_{pk}.

The new shapes offer a lower peak surface magnetic field by 10–15% at the expense of raising the surface electric field by 15–20%. The new shapes also result in a higher shunt impedance (R/Q) and a slightly higher geometry factor (G), which is the product of the quality factor and the surface resistance. The higher value for the product $G \times R/Q$ results in lower power dissipation on the cavity surface for the same cavity voltage, and hence a lower cryogenic loss (the reason for choosing the name "low-loss"). As Table 2.1 shows, compared to the

Table 2.1 Comparison of rf parameters of the new shapes with the TTF shape.

Parameter	Unit	TTF	Re-entrant 70	Low-loss	Re-entrant 60
Aperture	mm	70	70	60	60
H_{pk}/E_{acc}	Oe/MV/m	41.5	37.8	36.1	35.4
E_{pk}/E_{acc}		1.98	2.4	2.36	2.28
GR/Q	Ω^2	30 840	33 762	37 970	41 208
Cell–cell coupling factor	%	1.9	2.38	1.52	1.57
Loss factor (long) $\sigma_z = 1$ mm	V/pC	1.46	1.45	1.72	
Loss factor (trans) $\sigma_z = 1$ mm	V/pC/cm^2	0.23	0.23	0.38	
LF detuning coefficient[a]	Hz/(MV/m)^2	−0.74	−0.81	−0.83	

a) Assuming 2.8 mm wall thickness and optimal stiffening ring location.

TTF shape, the LL shape offers a 23% saving in cryogenic power. The 60 mm aperture re-entrant shape offers a 34% saving in cavity wall losses.

There are some disadvantages to decreasing the aperture from 70 mm to 60 mm. The smaller aperture raises the longitudinal beam-excited wakefields (by 18%) and transverse wakefields (by 65%) [47, 48]. Higher transverse wakes demand more stringent cavity alignment in order to preserve beam emittance in a linac. Hence beam dynamics requirements set a lower limit to the aperture size choice. Table 2.1 shows the longitudinal and transverse loss factors for a bunch length of 1 mm. Since energy couples from cell to cell via the electric field through the irises, a smaller aperture also reduces the cell-to-cell coupling in a multicell structure, increasing the difficulty of achieving field flatness from cell to cell. The coupling for the 70 mm re-entrant shape is higher than that for the TTF shape, and significantly lower for the LL shape. Note that the LF detuning coefficients (to be discussed in Section 2.6) for the new shapes are comparable to the TTF shape as long as the stiffening rings are placed in the optimal position for each shape.

Experimental work pursued at Cornell, KEK, and JLab shows encouraging progress in advancing E_{acc} using the new shapes and H_{pk} using the best treatment procedures discussed in Chapter 7. Using a single-cell re-entrant, niobium cavity at 1.3 GHz with a 70-mm aperture beam pipe (Fig. 2.7) Cornell first demonstrated a record gradient of 47 MV/m ($H_{pk} = 178$ mT) [49]. This exceeded the previous record accelerating field of 42 MV/m for a TTF-shape single-cell cavity [50–56]. Later, a Cornell/KEK collaborative effort [34] reached a new record of 52.7 MV/m ($H_{pk} = 199$ mT) with another 1.3 GHz re-entrant cavity, designed and built at Cornell, and processed and tested at KEK. Using a 60-mm aperture re-entrant cavity (Fig. 2.7) the collaboration [33] reached 59 MV/m ($H_{pk} = 205$ mT). Using several single-cell, 1.3 GHz ICHIRO cavities with a 60 mm aperture (Fig. 2.7), KEK also reached gradients of 45–53 MV/m ($H_{pk} = 162$–188 mT) [34].

2.2 High-β Cavities

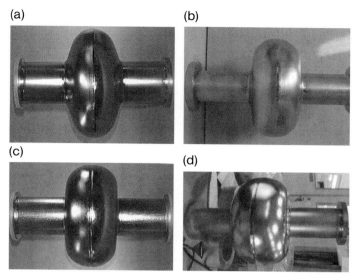

Fig. 2.7 Single-cell, 1.3 GHz niobium cavities of various shapes:
(a) TTF-like; (b) KEK, LL (ICHIRO) 60 mm aperture (courtesy of KEK);
(c) Cornell re-entrant 70 mm aperture; (d) Cornell re-entrant 60 mm aperture.

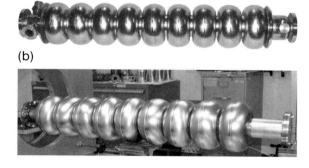

Fig. 2.8. 9-cell, 1.3 GHz niobium cavities:
(a) KEK, LL (ICHIRO) shape;
(b) Cornell re-entrant, 70 mm aperture (courtesy of AES).

These encouraging single-cell results make the new cavity shapes attractive for eventual application to the TeV energy upgrade of ILC, the International Linear Collider (see Chapter 12 for a more detailed discussion of the ILC). Approaching 200 mT at high Q provides the important proof of the principle that the ultimate critical magnetic field for niobium is above H_{c1} and near H_{sh}, as discussed in Chapter 3.

Experimental work on multicell niobium cavities of the new shapes is under active exploration at Cornell, KEK, and JLab. Figure 2.8 shows a 9-cell ICHIRO [43] cavity, and a 9-cell re-entrant cavity under development at KEK and Cornell, respectively.

2.3
Medium-β Cavities

With the growing interest in accelerators for spallation sources, such as for example SNS, elliptical resonators have been extended to high-energy (1 GeV) proton acceleration. Medium-β superconducting cavities are also gaining interest for radio-isotope beam (RIB) production and postacceleration, high-current, high-brightness proton and deuteron linacs for energy production via accelerator-driven reactors, material irradiation, and nuclear waste transmutation. Chapter 11 discusses the SNS experience in more detail and Chapter 12 covers upcoming prospects for RIB and high-energy proton accelerators.

The design of a medium-β structure involves several tradeoffs. The choice of a low frequency increases the voltage gain per cell, the beam energy acceptance, and beam quality, while decreasing rf losses and beam losses. But a low rf frequency increases structure size and microphonics level, making rf control more challenging. Consider next the number of cells. The larger the number the higher the voltage gain per structure, but the narrower the velocity acceptance, and the larger the number of cavity designs needed to optimize the voltage gain with changing particle velocity for heavy ion and proton accelerators. In the medium velocity range, structures must efficiently accelerate particles whose velocity changes along the accelerator. Several structure geometries are therefore needed, each of which is optimized for a particular velocity range. The lower the velocity of the charged particle under acceleration, the faster it will change, and the narrower the velocity range of a particular accelerating structure. This implies that the smaller the β of a cavity, the smaller the number of cavities of that β which can be used in the accelerator. Also, failure of a low-β cavity to achieve its design gradient means that the particle will not be captured by the following accelerating section. As a consequence of their small number, and importance of achieving their design gradient, medium-β cavities need to be designed and operated more conservatively than high-β cavities. As β increases structures can be designed more aggressively with the expectation of achieving the design gradient on average. SNS for example uses two elliptical cavity geometries, one at $\beta=0.6$ between 200 and 600 MeV, and the other at $\beta=0.8$ from 600 MeV to 1 GeV. Table 2.2 shows the parameters of the SNS structures. Chapter 11 will discuss the performance of SNS cavities.

2.3.1
Medium-β Elliptical Cavities

Efficient acceleration for $0.5<\beta<0.8$ is achieved in a straightforward manner by axially compressing the dimensions of standard elliptical resonator geometry while maintaining a constant frequency as shown in Fig. 2.9. As mentioned, the lower limit of usefulness for the compression approach is about $\beta=0.5$, when the vertical flat walls make the structure mechanically unstable. Figure 2.10 shows a collection of example structures down to $\beta=0.5$. An exceptional case at

Table 2.2 Parameters of SNS medium- and high-β structures [57, 58].

Parameter	Cavity type	
	$\beta=0.61$	$\beta=0.81$
Operating gradient (MV/m)	10.2	15.6
Q_0 spec at operating gradient	$\geq 5\times 10^9$	$\geq 5\times 10^9$
E_{peak} (MV/m)	27.6	34.2
H_{peak} (mT)	58.0	73.2
E_{peak}/E_{acc}	2.71	2.19
B_{peak}/E_{acc} (mT/(mV/m))	5.72	4.72
Operating temperature (K)	2.1	2.1

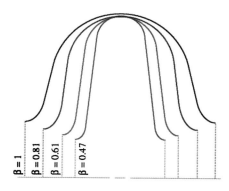

Fig. 2.9 A progression of compressed elliptical cavity shapes at the same rf frequency but for decreasing β values [59] (courtesy of MSU).

$\beta=0.1$ of the re-entrant cavity is also included in this collection because it belongs to the class of TM resonators, not because it is a medium-β cavity.

As the cells compress to low-β geometries, cavity properties exhibit interesting general trends [19]. E_{pk}/E_{acc} increases from its typical value of 2–2.5 for $\beta=1$ up to 3–4 for $\beta=0.5$. The peak electric field remains at the iris region. For the 1 MV/m accelerating field, the peak surface magnetic field near the equator increases from 4.0–4.55 mT for $\beta=1$ to 6–8 mT for $\beta=0.5$. The geometry factor obeys a simple scaling law $G(\Omega)=275\beta$. The geometric shunt impedance per cell decreases roughly quadratically as $R/Q(\Omega)=120\beta^2$. For constant E_{acc}, the stored energy (U) per cell is roughly independent of β. Structure stored energy plays an important role in amplitude and phase control in the presence of microphonics detuning since the rf power required for phase stabilization depends on the product of the energy content and the amount of detuning. A typical value is $U=200$–250 mJ per cell at 1 MV/m.

As illustration for these trends, Table 2.3 [64] gives the properties of three 700 MHz medium-β elliptical cavities (Fig. 2.11) designed for a proton accelerator TRASCO by optimizing the cell geometry. Chapter 7 will discuss the performance of medium-β elliptical prototype cavities.

(a)

(b)
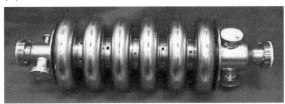

(c)
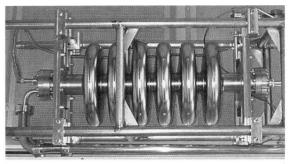

(d)

Fig. 2.10 Examples of TM$_{010}$ cavities with $\beta<1$. From top to bottom: (a) 805 MHz, $\beta=0.82$ [60]; (b) 805 MHz, $\beta=0.62$ [60] (courtesy of JLab); (c) 800 MHz, $\beta=0.5$ MSU/JLab [59, 61] (courtesy of MSU); (d) 352 MHz, $\beta\sim0.1$ re-entrant TM010 cavity [62, 63] (courtesy of INFN).

2.3 Medium-β Cavities

Table 2.3 Geometrical and electromagnetic parameters for three medium-β elliptical resonators designed for TRASCO [64].

Geometrical parameters			
Cavity synchronous beta	0.50	0.68	0.86
Number of cells	5	5	6
Cell geom. length [mm]	100	140	180
Geometrical beta	0.470	0.658	0.846
Full cavity length [mm]	900	1100	1480
Iris diameter [mm]	80	90	100
Tube ∅ at coupler [mm]		130	
Internal wall angle, α [°]	5.5	8.5	8.5
Equator ellipse ratio, R	1.6	1	1
Iris ellipse ratio, r	1.3	1.3	1.4
Full cavity electromagnetic parameters			
Max. E_{peak}/E_{acc}	3.59	2.61	2.36
Max. B_{peak}/E_{acc}	5.87	4.88	4.08
Cell to cell coupling [%]	1.34	1.10	1.28
R/Q [Ohm]	159	315	598

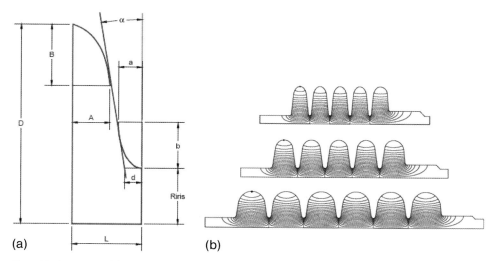

Fig. 2.11 Geometry of three medium-β elliptical structures designed for TRASCO. See Table 2.3 for structure properties. (a) Parameters defining the cell shape. (b) 5-cell optimized resonators profiles [64] (courtesy of INFN).

2.3.2
Medium-β Spoke Resonators

An alternate path to medium velocity structures with β near 0.5 is via multigap spoke resonators. Here each spoke element is a half-wavelength resonant transmission line operating in a TEM mode. Resonant transmission lines have been developed mostly for low-β applications to be discussed further in the next section. A half-wavelength ($\lambda/2$) transmission line, with a short at both ends, has maximum voltage in the middle and behaves as a half-wave resonator. Figure 2.12 shows a simple circuit equivalent along with voltage and current distributions in one of the $\lambda/2$ loading elements of a single-spoke resonator [20]. TEM resonators are essentially three-dimensional geometries with one or several field-free drift tubes (also called loading elements) that hang from the center conductor in the maximum electric field region.

For a multispoke resonator, the lowest frequency mode in the pass-band is the accelerating mode. For this mode, the electric field in adjacent gaps differs in phase by π radians – thus the name π-mode applies. The spoke elements are

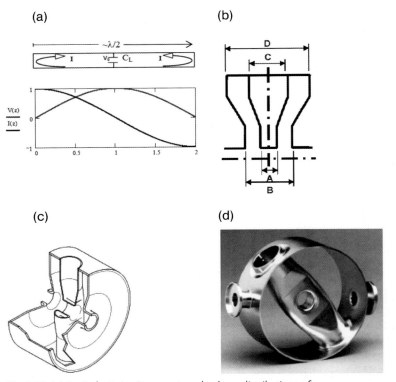

Fig. 2.12 (a) Equivalent circuit, current, and voltage distributions of a single-spoke half-wave resonator [20] (courtesy of INFN); (b) spoke and gap profile [65]; (c) 3D sketch [65]; (d) photograph of the first spoke resonator with $\beta=0.28$, 800 MHz [66] (courtesy of ANL and JLab).

elliptical in cross section in order to minimize the peak surface fields. The major axis of the ellipse is normal to the beam axis in the center of each spoke to minimize the surface electric field and maximize the beam aperture. A typical beam aperture is 4 cm at 345 MHz. In the region of the spokes near the outer cylindrical diameter, the major axis is parallel to the beam axis in order to minimize the peak surface magnetic field. There have been extensive efforts for design optimization by controlling A/B (in Fig. 2.12b) to reduce E_{pk}/E_{acc} and C/D to reduce B_{pk}/E_{acc}. Figure 2.13 shows the electric and magnetic field vectors computed by Microwave Studio (MWS) for a multispoke resonator [67].

In the $\lambda/2$ structure the coupling does not rely on the electric field at the beam holes, as for elliptical cavities, but takes place chiefly via the magnetic field linking cells through the large openings. As a result the coupling is very strong (20–30%) as compared to 2% for $\beta=1$ elliptical cavities, which makes the spoke structures robust and field profiles insensitive to mechanical tolerances. Half end-cells (half-gaps) terminate the structure to derive a flat π mode.

The range of spoke resonator application continues to be extended into the medium-β regime. In principle, there is no clear-cut transition energy from spoke resonators to elliptical ones. In the overlap region of $\beta=0.5$, one striking difference between TM elliptical cavities and the TEM spoke structures is their transverse dimensions [59, 61, 68–70]. Typically TM cavities have an inside diameter of about 0.9λ. Spoke structures are nearly a factor of 2 smaller, with outer diameters below 0.5λ. Thus a spoke cavity can be much smaller at the

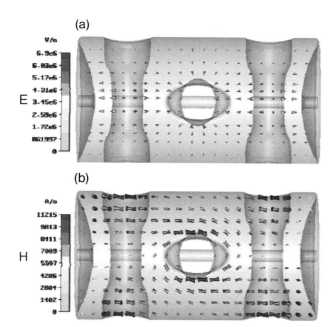

Fig. 2.13 Microwave Studio field calculations for a multispoke resonator: (a) electric fields; (b) magnetic fields [67] (courtesy of ANL).

2 New Cavity Geometries

same frequency, or the spoke structure can be made at half the frequency, for roughly the same dimensions as the elliptical structure, as shown in the comparison of Fig. 2.14. Choosing a lower frequency opens the option of 4.2 K operation, saving refrigerator associated capital and operating costs.

With half the number of cells the spoke structure also provides a broader velocity acceptance for the same overall accelerating length. Due to the change in the field during the finite time of transit across the accelerating gap, a particle of charge q traversing the gap with velocity β has an energy gain lower than qV, where V is the maximum gap voltage. The "transit time factor" integral takes into account the corresponding drop in voltage gain. A typical structure is optimized for the maximum transit time factor at a particular β value, but the transit time factor will drop for other β values.

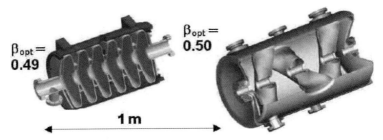

Fig. 2.14 Comparison of $\beta=0.5$ elliptical (805 MHz) and spoke (345 MHz) resonators [41, 68]. Both structures have approximately the same size, but the spoke operates at half the frequency (courtesy of MSU and ANL).

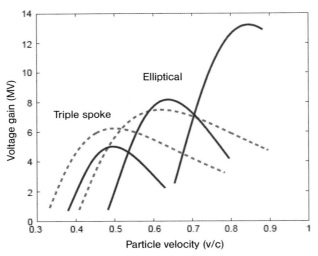

Fig. 2.15 A comparison of the velocity acceptance of the spoke and elliptical resonators over a range of velocities [70] (courtesy of ANL).

Table 2.4 Comparison of the properties of elliptical and spoke cavities for two β values [68].

Cavity type	3-spoke	6-cell	3-spoke	6-cell
Beta geometric	0.50	0.47	0.62	0.61
Frequency (MHz)	345	805	345	805
Length (cm)	65	53	81	68
G (Ω)	86	137	93	179
R/Q (Ω)	494	160	520	279
At an accelerating gradient of 1 MV/m				
RF energy (mJ)	397	341	580	330
Peak E-field (MV/m)	2.9	3.4	3.0	2.7
Peak B-Field (G)	87	69	89	57

For a structure design with broad velocity acceptance, the voltage gain will drop off more gradually for β values lower (and higher) than the optimum. Increased longitudinal acceptance reduces beam loss and control tolerances. Figure 2.15 compares the velocity acceptance of several families of the spoke and elliptical structures over a range of β values [70–72].

Table 2.4 compares the key properties of spoke resonators with elliptical cavities for two β choices. The spoke structure has a higher shunt impedance and so exhibits a higher accelerating field for a given power loss than the elliptical structure. On the other hand, TM structures have lower H_{pk} and larger apertures for lower beam loss. E_{pk} values are comparable. Spokes are mechanically more stable and exhibit a stable field profile due to the high cell-to-cell coupling.

For the RIA project (see Chapter 12), MSU considered 6-cell 805 MHz elliptical cavities with $\beta=0.49$, 0.63, and 0.83 at 15 MV/m and 2 K. ANL explored two types of 3-spoke cavities with $\beta=0.5$ and 0.62, at a gradient of 9.5 MV/m at 4.2 K. These projects will be discussed further in Chapter 12.

Many-gap spokes or ladder structures are under study for frequencies between 170 and 800 MHz, and β from 0.1 to 0.3. Figure 2.16 gives an example of a 19-gap, 1.05 m structure under development. If successful, these would efficiently provide a large energy gain for a compact linac. But the low-velocity acceptance, small aperture, and high peak fields are general disadvantages. The narrow beam velocity profile is not a problem if only one type of beam is used. Compared to the quarter-wave resonator (QWR) in this velocity range (to be discussed below), a many-gap spoke structure puts severe requirements in achieving good alignment, field flatness, tuning, and final surface finishing. At Frankfurt, a 19-gap, $\beta=0.1$ spoke cavity prototype has been demonstrated to reach an accelerating field above 6 MV/m [73–76].

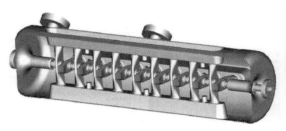

Fig. 2.16 A 19-gap, $\beta=0.1$ multi-spoke cavity prototype under development at Frankfurt [73] (courtesy of Julich).

2.4
Low-β Resonators

Low-β resonators have been in use for heavy ion boosters for more than three decades. The short independently phased cavities provide flexibility for operation and beam delivery. Applications continue to expand toward both the lower-β and the medium-β range, as with spoke resonators discussed above. Low-velocity structures must accelerate a variety of ions with different velocity profiles. Different cavity geometries with many gaps have been developed suitable for different beam energy, beam current, and mass/charge ratios.

2.4.1
Quarter-Wave Resonators

The QWR derives from transmission-line-like elements and belongs to the TEM resonator class. Figure 2.17 shows a coaxial line, $\lambda/4$ in length, shorted at one end to form a resonator with the maximum electric field at $\lambda/4$ where the accelerating gaps are located. Low frequencies, typically 100–200 MHz, must be used because the active and useful length of the structure is proportional to $\beta\lambda$. The low frequency results in a large resonator. The typical structure height is about 1 m. The inner conductor, which is made from niobium, is hollow and filled with liquid helium. 4.2 K operation is usually possible due to the low rf frequency.

The shunt impedance is kept high by use of a small aperture. But the small aperture means low transverse acceptance for the beam. These properties result in cavities with rather high peak-to-accelerating field ratios as compared to high-velocity elliptical cavities. Table 2.5 shows that for the typical low-velocity QWRs, the surface electric and magnetic fields are more than a factor of 2 higher than those for $\beta=1$ elliptical cavities [77–79]. When comparing performance gradients in resonators developed at different laboratories, it is important to keep track of how the cavity lengths are defined. Figure 2.5 shows the different choices used [20].

One of the variants of the QWR is the low-β split ring structure, discussed in [1]. It has been in use for many years before the invention of the more familiar

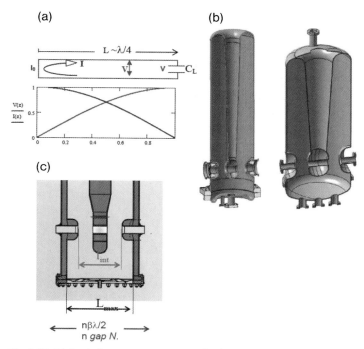

Fig. 2.17 (a) Equivalent circuit, current, and voltage distributions of a quarter-wave resonator [20] (courtesy of INFN). (b) 3D drawings of two quarter-wave resonators for SPIRAL II with different β (0.07 and 0.12) values [77] (courtesy of Saclay). (c) Different definitions used for the effective accelerating gap [20] (courtesy of INFN).

Table 2.5 Main rf parameters of two QWRs at 88 MHz with $\beta=0.07$ and $\beta=0.12$ in Fig. 2.17 (b) [77].

Frequency [MHz]	88.05	88.05
β_{opt}	0.07	0.12
E_{pk}/E_{acc}	5.0	5.6
$B_{pk}/E_{acc}\left(\dfrac{mT}{mV/m}\right)$	8.9	10.2
r/Q (Ω)	632	518
V_{acc} at 6.5 MV/m, β_{opt} (MV)	1.54	2.65
G $[\Omega]$	22.4	38
Beam tube $\varnothing$ (mm)	30	36
Cavity ext. $\varnothing$ (mm)	230	380
Q_{ext}	6.610^5	1.110^6

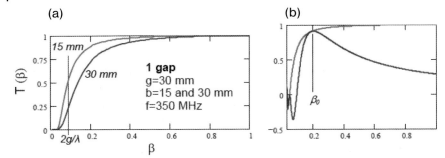

Fig. 2.18 (a) One-gap transit time factor, using two different aperture values 15 mm and 30 mm. Acceleration takes place efficiently above $\beta \sim 2g/\lambda$ and it is maximum at $\beta=1$. Here g is the gap and b is the aperture. (b) 2-gap transit time factor for the β-mode compared to the 1-gap in (a). Acceleration is maximum for a particular value of β. The transit time factor falls off steeply with β on either side of the optimum [20] (courtesy of INFN).

QWR of Fig. 2.17. The split ring arms and the outer conductor wall form a pair of coupled QW transmission lines, with two principal modes. With large energy gain and good efficiency, the frequency and β application range of the split-ring resonators has been $90 \leq f \leq 150$ MHz and $0.05 \leq \beta \leq 0.15$, respectively [80].

The larger the number of gaps in a QWR the larger the energy gain, but the narrower the velocity acceptance. Figure 2.18 shows the transit time factor for 1-gap and 2-gap resonators in the simple approximation of constant field in the gap and zero field outside.

Rules of thumb for good gap (g) and aperture (b) choices are $2gf < \beta_c$ and $2 < g$. Thus, to be efficient at low β, it is necessary to work at low rf frequency, short gap length, and small beam aperture. Figure 2.19 shows the normalized transit time factor for multigap resonators [20].

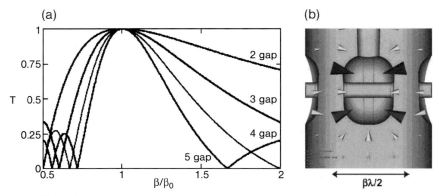

Fig. 2.19 (a) Normalized transit time factor versus normalized velocity β/β_0, for cavities with different number of equal gaps [20] (courtesy of INFN). (b) Electric field in the beam aperture region for the QWR [81] (courtesy of ANL).

Being a nonsymmetric structure, the QWR has nonsymmetric electromagnetic fields in the beam region which produces undesirable beam steering through electric and magnetic dipole field components. Figure 2.19 (b) portrays the electric fields in the beam aperture region. Steering is similar to the rf defocusing effect in misaligned cavities. In resonators with a small aperture/gap aspect ratio, a slight misalignment is usually sufficient to cancel most of the steering problem. In cavities where the required misalignment becomes too large compared to the beam port aperture, compensation can be obtained by gap shaping: the magnetic deflection can be cancelled by enhancement of the electric deflection [82]. Cavities with dipole compensation have been built at MSU [83], LNL [84], and ANL [85].

Cancellation of dipole and quadrupole field components is particularly important in high-current proton–deuteron accelerators with high space charge. Above approximately 160 MHz, the best dipole compensation can be obtained with the more symmetric half-wave resonator (HWR) discussed in the next section.

Quarter-wave structures are sensitive to mechanical vibrations because of their large size and large load capacitance. The related phase stability is an important issue, particularly for the lowest velocities and for a small beam loading, high-Q_{ext} operation. The mechanical stability problems have been solved by electronic fast tuners [86] or by the addition of a mechanical damper in the cavity stem [87, 88].

QWRs have covered a wide overall application range: $48 \leq f \leq 160$ MHz, $0.001 \leq \beta \leq 0.2$, with 2 gaps and 4 gaps. The extensions to the very low-β regime use a tuning fork arrangement [89]. Compact and modular, QWRs have proven to be efficient, high-performance resonators. They can achieve reliably 6 MV/m. Niobium QWRs are operating in accelerators ALPI in LNL (INFN, Legnaro) [102], in NSC (New Delhi) [90, 91], and in ISAC-II at TRIUMF (Canada) [92]. They are foreseen for future projects such as RIA, SPIRAL2, and EURISOL, to be discussed in Chapter 12.

2.4.2
Half-Wave Resonators

An HWR is equivalent to two quarter-waves facing each other providing the same accelerating voltage as a QWR but with almost twice the power dissipation. Figure 2.20 shows an example [93]. The symmetry of the structure cancels steering and opens the use of HWRs at β from 0.1 to 0.5, above the range customary for QWRs. HWRs also show improved mechanical vibration properties over the QWRs.

The peak surface electric field occurs at the center of the loading element. By suitable sizing and shaping of the cross section, the surface to accelerating field ratio of 3.3 can be obtained, independent of β. The maximum H_{pk} occurs where the loading element meets the outer enclosure and is sensitive to the size and shape of the center conductor. Values of 7 mT/MV/m can be obtained by proper shaping. Most structures are designed with somewhat higher surface field values.

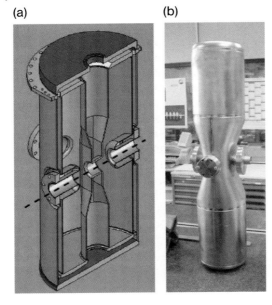

Fig. 2.20 (a) 3D schematic of a HWR. (b) 176 MHz, $\beta=0.09$ HWR built for SARAF by ACCEL [93] (courtesy of ACCEL).

Coaxial 2-gap HWRs have been built with β up to about 0.3 (these are efficient up to about $\beta=0.5$ to 0.6). A 322 MHz, $\beta=0.28$ HWR has been built and tested at MSU [94] for RIA, and a 352 MHz, $\beta=0.3$ HWR at LNL [95].

2.5
Very Low-β Cavities

The very low-β range $(0.01 < \beta < 0.04)$ is covered by 4-gap QWR structures [89] and the superconducting RFQ [96]. At ANL the interdigital, 48 and 73 MHz, 4-gap QWRs replaced the Tandem accelerators in the injector section, and have been operating successfully for many years. These structures are discussed in the reference text [1].

2.5.1
Superconducting RFQ

Radiofrequency quadrupoles (RFQs) have been in use for many decades for the acceleration of very low velocity ions [97, 98]. They combine the strong electric focusing provided by the rf quadrupole with effective acceleration by the modulation of the vanes. These are precisely machined to provide a longitudinal electric field component, synchronous with ion bunches. RFQ structures are ideal for very low ion velocities $\beta < 0.01$. They are typically normal conducting, span a frequency range 50–400 MHz, and provide a 100–200 kV voltage difference between vanes with a quality factor $Q \sim 10^4$. Power consumption limits their duty cycle to less than 20%. The superconducting RFQ offers comparable vane voltage but lower power consumption (at $Q \sim 10^8$–10^9) and cw operation. The combination of focusing and acceleration allows approximately one meter long structures with a large number of cells to provide real-estate gradients of 2–3 MV/m. With an aperture of 1.5 cm, the surface electric field to accelerating field ratios are 7.3 and 10, while the magnetic field ratios are 25 and 30 mT/MV/m. Hence the design values of the accelerating field are limited. The stored energy is acceptable for state-of-the-art phase-amplitude stabilization circuits.

INFN at Legnaro built prototypes (see Fig. 2.21) with 12 cells which reached the design voltage of 280 kV at a surface field of 25 MV/m [99]. Subsequently, they built two full-scale RFQs for the new injector PIAVE of the ALPI accelerator [100–102]. Table 2.6 lists the main parameters for both structures. Two struc-

Fig. 2.21 Superconducting niobium RFQ for cw operation [99] (courtesy of INFN).

Table 2.6 Parameters of two superconducting RFQs [101].

	SRFQ1	SRFQ2
Energy in (MeV)	8.8	83.6
Energy out (MeV)	83.6	139.3
Beta in	0.0089	0.0275
Beta out	0.0275	0.0355
Voltage (kV)	148	280
Length (m)	1.34	0.74
No. of cells	43	13
Diameter (m)	0.8	0.8
Aperture (cm)	1.5	1.5
E_{pk}/E_a	10	7.3
B_{pk}/E_a (mT/MV/m)	25	30
P_{diss} (at design field) (W)	10	10
U (at design field) (J)	2	3
Q	1×10^8	2×10^8

tures delivered 4.7 MV in 2.13 m, resulting in a good real-estate gradient up to $\beta = 0.035$. An ANL-type fast tuner achieved reliable phase lock in these geometrically complex cavities. During beam commissioning they reached the design field [103]. The interelectrode voltages are 148 and 280 kV, respectively, significantly higher than those achieved by typical normal conducting RFQs. The superconducting performance also compares favorably with the ANL module (3 m long) of four interdigital structures of $\beta = 0.01$–0.04, and superconducting focusing quadrupoles delivering a total voltage of 3 MV [59].

2.6
Mechanical Aspects

The design of a superconducting cavity must take into account several mechanical aspects: stresses, vibrations, and Lorentz forces. The cavity must withstand stresses induced by the differential pressure between the beam pipe vacuum and atmospheric pressure. Differential thermal contraction due to cool-down from room temperature to cryogenic temperatures induces stress on the cavity walls. Mechanical vibrations of the cavity and the cavity–cryomodule system (microphonics) form another aspect of cavity mechanical design. External vibrations couple to the cavity and excite mechanical resonances, which modulate the rf resonant frequency inducing pondermotive instabilities [104, 105]. These translate to amplitude and phase modulations of the field becoming especially significant for a narrow rf bandwidth. Lorentz-force (LF) detuning becomes important in cavity designs for high-field pulsed operations [44, 106]. Surface currents interact with the magnetic field to exert a Lorentz force on the cavity wall. This stress causes a small deformation to change the cavity volume and frequency.

2.6.1
Mechanical Stresses

To avoid plastic deformation the cumulative mechanical stress on the cavity walls must not exceed the cavity material yield strength, including some engineering margin. The frequency shifts due to these stresses must be taken into account for targeting the final frequency or tuner settings and tuner range. Stresses due to the operation of the tuner mechanism should not exceed yield strength while cold. The mechanical requirements may be dealt with by proper choice of cavity wall thickness or by adding stiffening rings or ribs at locations of high strain.

We present a few specific examples of these mechanical aspects of cavity design.

Figure 2.22 shows mechanical stress calculation results by ANSYS for the CESR B-cell cavity [107, 108]. The presence of HOM conducting flutes with flat walls on one of the beam pipes increases stresses. The flat waveguide walls also need to be braced to avoid the deformation and consequent change of the waveguide coupling strength.

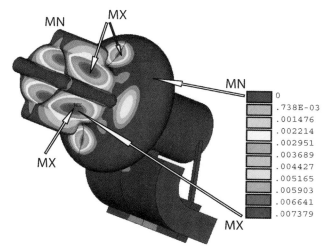

Fig. 2.22 ANSYS stress calculation results for the 500 MHz CESR B-cell cavity [107] (courtesy of SRRC).

Medium-β elliptical cavities are especially vulnerable to mechanical stresses due to the flattening of the wall. As mentioned, medium-β elliptical cavities are especially vulnerable to stresses. The structural behavior of a 700 MHz single-cell elliptical cavity with $\beta = 0.42$ has been studied [109] by finite element structural analysis using COSMOS/M. Figure 2.23 shows that the maximum von Mises stress for a 3 mm wall thickness is 54 MPa. Since the yield stress of niobium after the heat treatment at about 700 °C (to remove dissolved H gas) is 50–100 MPa [110, 111], the engineering safety factor is insufficient. Either the wall thickness must be increased to 5 mm or a stiffener must be added. With a conical stiffener at an optimal location the maximum stress decreases to 11.8 MPa as shown in Figs. 2.23 and 2.24. The resonant frequency shift due to vacuum load was calculated by the SUPERFISH code. The calculated frequency shift of 306 Hz/mbar for a 3 mm Niobium cavity dropped to 74 Hz/mbar for 5 mm. Additional calculations were carried out for stresses induced due to thermal contraction to 4.2 K, and the vacuum load. With a 5 mm wall thickness and both the ends of the beam tubes held fixed the max stress after cool-down is 77 MPa. This is satisfactory since the yield strength of Nb increases by a factor of 5 from room temperature to 4.4 K.

The mechanical stability of spoke resonators is excellent when radial stiffening ribs are used at the ends. Figure 2.25(a) shows the stress distribution of a 4-gap (3-spoke) cavity with the four supports [112]. The peak stress is a safe 15 MPa, much lower than the elastic limit of the niobium. The frequency shift caused by changing the pressure of helium in the cooling jacket was −9.6 kHz/atm. Besides radial stiffening ribs at the ends, additional ribs on the cylindrical wall are also used [113], as exemplified by the FNAL spoke resonator (Fig. 2.25 b).

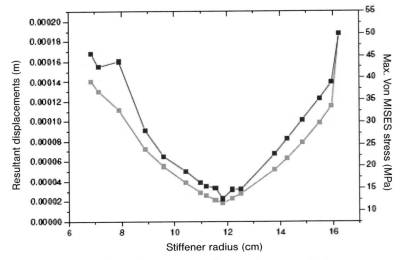

Fig. 2.23 The dependence of the maximum Von MISES stress on the location of the conical stiffener for a $\beta=0.5$, 700 MHz elliptical cavity [109] (courtesy of BARC).

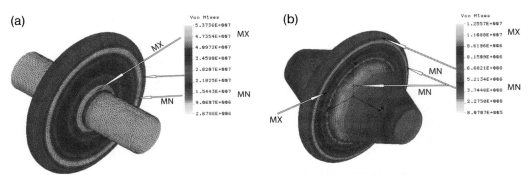

Fig. 2.24 COSMOS stress calculation results for the $\beta=0.5$, 700 MHz elliptical cavity. (a) Without conical stiffener, the maximum stress is 54 MPa. (b) With conical stiffener at the optimum location, the maximum stress drops to 11.8 MPa [110, 111] (courtesy of BARC).

Each end wall of the spoke resonator is reinforced by two systems of ribs: a tubular rib with elliptical section in the end wall outer region and six radial daisy-like ribs in the inner region (nose). These two systems are not connected to facilitate a controlled displacement of the nose area for cavity slow tuning. A third and final system of four ribs is present on the cylindrical portion of the cavity. All ribs are made of reactor grade niobium.

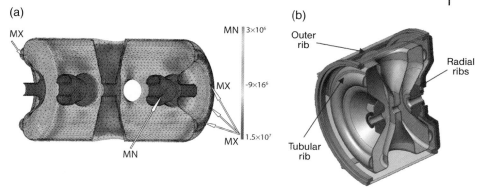

Fig. 2.25 (a) ANSYS stress calculations for the triple-spoke resonator, 350 MHz, $\beta=0.4$. The peak stress is 15 MPa [112] (courtesy of Julich). (b) FNAL single spoke resonator $\beta=0.22$, 30 mm aperture and 325 MHz [113] (courtesy of FNAL).

2.6.2 Vibrations

Stiffeners added at appropriate locations raise the cavity mechanical resonant frequencies so that these no longer couple to the lower frequency external vibration sources. Dampers introduced in the mechanical system of cavity and cryomodule reduce the mechanical Q of the resonances. The rf bandwidth can also be widened by increasing the strength of the input coupler, but this demands higher rf power and lowers the operating efficiency. Mechanical tuners are usually too slow to counteract cavity wall deformations from microphonics. The stored energy per cell plays an important role in amplitude and phase control in the presence of microphonics detuning. When beam loading is negligible, the amount of rf power required for phase stabilization is given by the product of the energy content and the amount of detuning. Fast tuners of the piezoelectric or magnetostrictive type added to the tuning system provide active damping of microphonics together with sophisticated electronic feedback systems. Slow and fast tuners will be covered in Chapter 10.

For mechanical vibrations of the medium-β elliptical cavity example above, the lowest frequency of the 3 mm wall thickness cavity (with one end free) is a dangerous 66 Hz and rises to 141 Hz with the conical stiffener. For the triple-spoke cavity, microphonic frequency fluctuations were measured at 4.2 K with the cavity operating at $E_{acc}=9.7$ MV/m. A 1 Hz rms magnitude of the microphonics was sufficiently small that a tuning window of a few tens of Hertz was adequate for phase control.

The TESLA cavity (Fig. 2.1) has a stiffening ring installed between adjacent cells to reduce the LF detuning, as we will discuss below. These stiffeners also raise the cavity mechanical resonant frequencies. Figure 2.26 shows mechanical resonant modes of a 7-cell elliptical cavity under development for the ERL main linac [17, 114]. Table 2.7 shows the results of a study to increase the resonant frequencies by positioning stiffening rings in various locations.

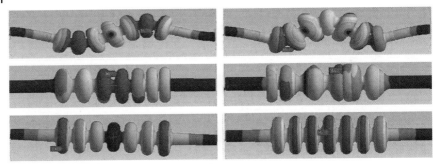

Fig. 2.26 Examples of vibration modes of a 7-cell, 1.3 GHz cavity. The active length of the cells is 80 cm. Modes from top to bottom are: transverse, longitudinal, and breathing (ANSYS) [17, 114].

Table 2.7 Increase of mechanical resonant frequency of a 7-cell, 1.3 GHz cavity with the location of a stiffening ring [114].

Stiffening ring location Mode	0.7 R_{eq} Freq (Hz)	0.65 R_{eq} Freq (Hz)	0.4 R_{eq} Freq (Hz)	No ring Freq (Hz)
1	131	115	85	54
2	131	115	85	54
3	316	268	191	133

2.6.3
Lorentz-Force Detuning

The cavity wall tends to bend inwards at the iris and outward at the equator as shown in Fig. 2.27 (a) [115]. The resonant frequency shifts with the square of the field amplitude distorting the frequency response [116] as shown in Fig. 2.27 (b). Typical detuning coefficients are a few $Hz/(MV/m)^2$. This frequency shift can be compensated by mechanical tuning once the operation field is reached. A fast tuner is necessary to keep the cavity on resonance, especially for pulsed operation. However, a large LF coefficient can generate "pondermotive" oscillations, where small field amplitude errors initially induced by any source (e.g., beam loading), cause cavity detuning through Lorentz force, and start a self-sustained mechanical vibration which makes cavity operation difficult [117]. LF detuning is especially important in pulsed operation, where the dynamics of the detuning play a strong role.

Stiffeners must be added to reduce the LF coefficient, as shown in Fig. 2.28 [118], but these increase the tuning force. For the TESLA-shape 9-cell elliptical structure (see Fig. 2.1) the LF detuning coefficient is about 2–3 $Hz/MV/m^2$ resulting in a frequency shift of several kHz at 35 MV/m, much larger than the cavity bandwidth (300 Hz) chosen for matched beam loading conditions for a

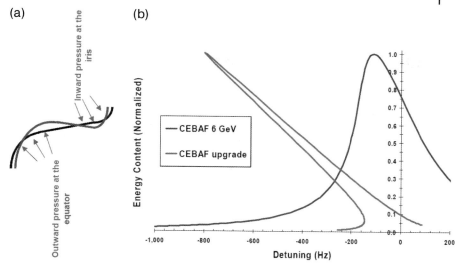

Fig. 2.27 Cavity shape distortions due to LF detuning [115]. (a) Lorentz forces acting on different parts of the cavity wall. Note the rotated orientation of the cavity. (b) Distortion of the frequency response of the cavity response at two field levels [116] (courtesy of JLab).

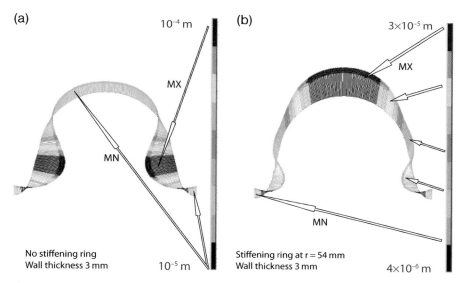

Fig. 2.28 (a) Deformations of the cell wall without stiffening rings. (b) Reduced deformation with stiffening rings [118] (courtesy of FNAL and DESY).

linear collider (or XFEL). Stiffening rings in the 9-cell structure reduce the detuning to about 1 Hz/MV/m^2 at 35 MV/m pulsed operation [119].

Feedforward techniques can further improve field stability [119–123].

In cw operation at a constant field the Lorentz force causes a static detuning which is easily compensated by the tuner feedback, but may nevertheless cause problems during start-up which must be dealt with by feedforward in the rf control system.

The LF detuning coefficients of medium-β elliptical cavities increase with lower-β. 805 MHz 6-cell SNS structures at βs of 0.8 and 0.6 with stiffening rings have $k\sim 0.4$ and 2 Hz/(MV/m)^2 [115, 124], but the LF coefficient rises to $k\sim 7$ for the 5-cell 700 MHz structure with β of 0.5 [125]. The LF detuning coefficient for the medium-β (0.5) elliptical cavity example of Fig. 2.24 is also unacceptable at about 10 Hz/MV/m^2 for the 3 mm wall but drops to a tolerable 3 Hz/MV/m^2 for the 5 mm wall without stiffener [110].

Spoke resonators also have significant LF detuning, but in comparison to the medium-β elliptical cells, these have lower LF detuning coefficients and are less sensitive to microphonics for the same β. A k of 3.7 Hz/(MV/m)^2 was obtained on a single-element spoke cavity with lower frequency (350 MHz) and lower β (0.4) [126]. It is interesting to note that, for the TM cavities, the LF deformation takes place in all the cells, and the coefficient is independent of the number of cells. For spoke structures, a large contribution comes from the deformation of the end plates, and thus LF detuning decreases as the number of loading elements increases. For the 3-spoke resonator of Fig. 2.25, the Lorentz detuning was 7.3 Hz/(MV/m)^2.

2.7
Codes for Cavity Design

Electromagnetic software packages for modeling accelerating cavities have been in existence for decades, first in 2D with SUPERFISH [127], SUPERLANS [128], GDFIDL [129] and more recently in 3D, with standard codes like MAFIA [130], Microwave Studio (MWS) [131] and HFSS [132] now in routine use by the cavity development community. Vector Fields 3D SOPRANO [133] code has been used for QWR design. Codes which provide mesh models to fit curved surfaces better provide and higher accuracy for the same number of mesh elements. Most codes are memory limited as they only run on a single computer while the latest supercomputers consist of thousands of processors with a significantly larger total memory. Under the Scientific Discovery through Advanced Computing (SciDAC) [134, 135] collaboration, SLAC has developed parallel processing finite element electromagnetic codes to obtain gains in accuracy, problem size, and solution speed by harnessing the computing power and exploiting the huge memory of the latest supercomputers. The suite of electromagnetic codes available is based on unstructured grids for high accuracy, and use parallel processing to enable large-scale simulation. The new modeling capability supports meshing, solvers, refinement, optimization, and visualization. The code suite to date includes the eigensolver Omega3P, the S-matrix solver S3P, the time-domain solver T3P, and the particle tracking code Track3P.

With codes such as Omega3P under SciDAC, direct simulations of the entire cavity with input and HOM couplers have been carried out. The high computing power and adjustable mesh densities allow treatment of the disparate length scales of the cavity accurately, from the cell shape to the fine features in the couplers. Eigensolvers have been developed to directly calculate the external Q of modes as a result of power flow out of the cavity through the couplers.

TEM-class drift-tube loaded cavities have been designed using modern 3D simulation codes such as MAFIA, MWS, and ProEngineer/ANSYS [136] for mechanical properties. MWS has been used to optimize the spoke resonators for the minimization of the E_{pk}/E_{acc} and B_{pk}/E_{acc} ratios. Coupled ANSYS-MWS analyses have been carried out on the mechanical properties of single spoke resonators [113].

Codes such as ANSYS [136] or COSMOS [137] determine structural mechanical properties and help reduce cavity wall deformations in the presence of mechanical loads and vibrations by choosing the appropriate wall thickness or location of stiffening rings or ribs. Lorentz-force detuning can be evaluated using a combination of mechanical and rf codes (e.g., SUPERFISH and MWS).

We will discuss various examples from these codes throughout the book.

3
Surface Resistance and Critical Field

3.1
BCS Surface Resistance

As discussed more fully in the reference text [1], the remarkable properties of superconductivity are due to the condensation of charge carriers into Cooper pairs, which move without friction, and hence the zero resistance hallmark of superconductivity. At $T = 0$ K, all charge carriers condense into pairs. At higher temperatures, pairs break up. The fraction of unpaired carriers increases exponentially with temperature, as $e^{-\Delta/kT}$, where 2Δ is the energy gap of the superconductor, i.e., the energy needed to break up the pairs. Above T_c, none of the carriers are paired which brings about the normal conducting state. In this simplified picture, known as the London two-fluid model [138], when a dc field is turned on, the pairs carry all the current, shielding the applied field from the normal (unpaired) electrons. Electrical resistance vanishes since Cooper pairs move without friction. In the case of rf currents, however, dissipation does occur for all $T > 0$ K, albeit very small compared to the normal conducting state. While the Cooper pairs move without friction, they do have inertial mass. For high-frequency currents to flow, forces must be applied to bring about alternating directions of flow. Hence an ac electric field will be present in the skin layer, and it will continually accelerate and decelerate the normal carriers, leading to dissipation proportional to the square of the rf frequency. A simplified form of the temperature dependence of Nb for $T_c/T > 2$ and for frequencies much smaller than $2\Delta/h \approx 10^{12}$ Hz is [139]

$$R_S = A(1/T)f^2 e^{-\Delta(T)/kT} + R_0$$

Here A is a constant that depends on material parameters, as we will discuss. The operating temperature of a superconducting cavity is usually chosen so that the temperature-dependent part of the surface resistance is reduced to an economically tolerable value. R_0, referred to as the residual resistance, is influenced by several factors to be discussed in Section 3.4.

50 years ago, Bardeen et al. [16] put forward the first quantum mechanical theory (referred to as the BCS theory) which has been very successful in ex-

RF Superconductivity: Science, Technology, and Applications. Hasan Padamsee
Copyright © 2009 WILEY-VCH Verlag GmbH & Co. KGaA, Weinheim
ISBN: 978-3-527-40572-5

plaining many properties of superconductivity, including rf surface resistance. The BCS surface resistance decreases exponentially with temperature, as electrons freeze into Cooper pairs, and increases as the square of the rf frequency, except for small deviations at frequencies above 10 GHz [1]. At any given temperature and frequency, the magnitude of the surface resistance also depends on material parameters, such as the London penetration λ_L (36 nm for pure Nb), intrinsic coherence length ξ_0 (64 nm for pure Nb), Fermi velocity, and in particular the electron mean free path (l), which characterizes material purity. From these parameters the BCS surface resistance can be calculated numerically [140–142].

3.1.1
Mean Free Path Dependence of BCS Resistance

The BCS surface resistance shows interesting behavior with changes in the purity (electron mean free path) of Nb. Figure 3.1 shows the mean free path dependence of the BCS resistance for 1.5 GHz at two different temperatures for specular and diffuse electron reflection. For niobium cavities the diffuse reflection case applies to electrons scattering off interstitial impurity sites. In the dirty regime, $l \ll \lambda$ (~ξ_0), the surface resistance is dominated by multiple impurity scattering inside the rf penetration depth. For this case a useful approximation in the two-fluid model is [139, 145]:

$$R_S = \mu_0^2 \omega^2 \sigma_n \lambda^3 \Delta \ln(\Delta/\omega) \frac{e^{-\Delta/T}}{T} \qquad \sigma_n = ne^2 l/p_F \qquad p_F = (3\pi^2 n)^{1/3} \hbar$$

Here σ_n is the residual conductivity in the normal state, p_F is the Fermi momentum, n is the total electron density, and e is the electron charge. Both coherence length and penetration depth also vary with purity and electron mean free path. As per the nonlocal generalization between current and fields introduced in [146] the coherence length of a superconductor changes with the electron mean free path as follows:

$$\frac{1}{\xi} = \frac{1}{\xi_0} + \frac{1}{l}$$

An approximate expression for the penetration depth at 0 K in the "dirty" limit ($l < \xi_0$) is given by

$$\lambda = \lambda_L \sqrt{1 + \frac{\xi_0}{l}}$$

Hence in the dirty regime ($l \ll \xi$), the surface resistance decreases with increasing l because $\sigma_n \lambda^3 \sim l \times l^{-3/2} \sim l^{-1/2}$. The superconductor behaves like a normal metal where the surface resistance decreases with increasing metal purity.

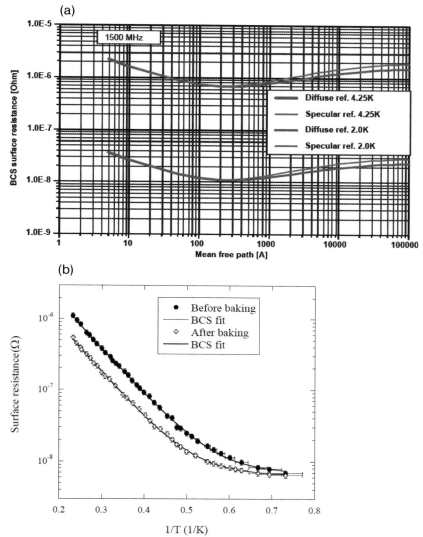

Fig. 3.1 (a) BCS surface resistance versus mean free path for niobium at a frequency of 1.5 GHz and for temperatures 4.2 K and 2 K. [140–142]. (b) Decrease of BCS surface resistance with baking at 120 °C for 48 h. The measurements were carried out with a 1.5 GHz cavity [143, 144]. The residual surface resistance remains unchanged at about 5 nΩ (courtesy of JLab).

When the mean free path becomes larger than the coherence length, the superconducting surface resistance starts to show anomalous behavior, *increasing* with mean free path according to BCS (Fig. 3.1). In the clean rf limit $l \gg \lambda$ ($\sim \xi$), an incident electron which collides with the surface gets reflected to the bulk without impurity scattering in the penetration depth. The BCS surface resistance becomes independent of the impurity scattering, independent of mean free path, and thus independent of the normal state conductivity. Purifying the material further does not change R_s, because electrons get scattered over a different length scale, corresponding to the gradient of the penetrated magnetic field. Hence the London penetration depth λ plays the role of the mean free path [145].

$$R_{s0} \cong \frac{3\Delta}{2T} \mu_0^2 \sigma_{\text{eff}} \omega^2 \lambda^3 \ln \frac{1.2T\Delta\xi^2}{\omega^2\lambda^2} e^{-\Delta/T}$$

Here the Drude conductivity $\sigma_n = ne^2 l/p_F$ is replaced by the effective conductivity $\sigma_{\text{eff}} = ne^2\lambda/p_F$. Hence in the clean limit ($l \gg \xi$), $\lambda \sim \lambda_L$ (constant) and $\sigma_{\text{eff}} \sim \lambda_L$, $\sigma_{\text{eff}}\lambda^3 = \lambda_L^4$ becomes independent of the mean free path. The clean limit is analogous to the anomalous skin effect in normal metals [147].

The clean limit surface resistance value is about a factor of 2 above the minimum which takes place when $l \sim \xi$ (Fig. 3.1) or when the Nb RRR is about 10 and the mean free path about 40 nm. (From the resistivity of niobium at room temperature (15 ×10^8 Ωm) and the constant ρl product (6×10^{16} Ωm^2), the mean free path at room temperature is 4 nm, increasing to 40 nm for 10 RRR Nb near the surface resistance minimum.)

The clean limit applies to high-purity niobium (RRR > 300) commonly used for cavities. As we will discuss in Chapter 5, for best high field performance, cavities are generally baked at 100–120 °C for 48 h to remove the strong Q-drop at high fields. After baking, the physics of the niobium surface changes from the "clean limit" to the "dirty limit" ($\ell < \xi_0$). In addition to reducing the high-field Q-drop baking also lowers the BCS surface resistance as shown in Fig. 3.2 [143]. This is due to a decrease of the electron mean free path, presumably by diffusion of oxygen from near (~10 nm) the surface into the penetration depth layer (~40 nm).

3.1.2
Oxygen Diffusion

The prevailing hypothesis for the drop in BCS resistance due to baking is oxygen diffusion from the oxide or from an oxygen-rich interface layer just below the oxide. In support of this hypothesis, the oxygen concentration versus depth of a baked cavity has been estimated from the BCS resistance, and compared to the oxygen diffusion profile [144, 148]. In a series of experiments, the BCS surface resistance of a high RRR cavity baked at 145 °C for 45 h was measured each time after progressively removing thin surface layers of niobium by several cycles of oxipolishing [144]. During each cycle a layer of niobium was removed by oxidation under an applied voltage (2 nm of oxide growth per volt for anodization) and dissolution of the oxide with a hydrofluoric (HF) acid rinse. After bake, the BCS resistance at 4.3 K dropped by about a factor of 2 from its high RRR value due to the decrease of the mean free path. Several oxipolishing steps restored the higher BCS value. After each cycle, the measured surface resistance

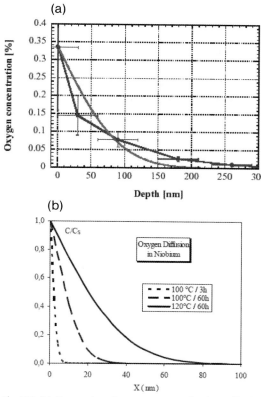

Fig. 3.2 (a) Comparison between oxygen depth profile due to diffusion after baking at 145 °C for 45 h, and depth profile inferred from BCS surface resistance measurements after progressive oxipolishing [149] (courtesy of CERN and JLab). (b) Oxygen diffusion profile for various baking conditions [151] (courtesy of Saclay).

was fit to the BCS prediction by adjusting the mean free path which then yielded the possible oxygen content. Figure 3.2 reports the inferred O concentration versus depth due to the bake. The analysis is consistent with a surface O concentration as 0.33% (atomic), which is close to the solubility limit of oxygen in Nb at 145 °C. The inferred depth profile of oxygen concentration is also consistent with the diffusion coefficient of oxygen in niobium [149]. Oxygen diffuses according to the equation

$$x = \sqrt{2D\exp(-E/RT)t}$$

where x is the averaged diffusion length, $D=0.015$ cm^2/s is the diffusion constant for oxygen in niobium [149] with activation energy $E=112890$ J/mol, $R=8.31$ J/(kmol), T is the temperature, and t is the diffusion time. Figure 3.2(b) shows diffusion profiles for various baking time and temperature conditions.

The high concentration of oxygen at the surface spreads over 150 nm, dropping by a factor of 2 within 50 nm, which is comparable to the penetration depth. The diffusion analysis assumes that during the bake the oxygen concentration at the metal–oxide interface remains at its solubility limit due to oxygen arriving from the dissociating oxide. X-ray photoelectron spectroscopy (XPS) analyses [150] confirm an overall reduction of the oxide layer thickness at 145 °C due to dissociation of the pentoxide and O diffusion into the bulk.

Rather than assuming a constant source of oxygen at the surface, a more advanced analysis takes into account the dissociation of the oxide as the source of oxygen to successfully fit the experimental depth profile with the same diffusion hypothesis [143].

The result described in Fig. 3.2 comes from a cavity experiment which used a rather high baking temperature (145 °C) compared to usual practice for healing the high-field Q-drop to achieve best cavity performance. As we will see in Chapter 5, successful baking temperatures for removing the high-field Q-drop for fine grain Nb lie between 100 and 120 °C (48 h).

3.2
Nonlinear (Field-Dependent) BCS Resistance

The BCS theory predicts that the energy gap does not depend on the rf magnetic field until very high fields approaching the thermodynamic critical field. In the clean limit ($l \gg \xi$) at low temperatures, the gap is independent of the corresponding current density (J) as long as $J < \phi_0/4\pi\lambda^2\xi$, where ϕ_0 is the flux quantum [152]. Hence Δ in clean high-performance Nb cavities at $T < 2$ K is practically independent of field when $\hbar\omega \ll \Delta$ and $H_0 < H_c$.

However, there is another intrinsic correction to R_s which results from the pair-breaking effect of the supercurrent density induced by the rf field. This is often referred to as the nonlinear BCS resistance. The pair-breaking manifests via a change of the electron energy spectrum in a current-carrying superconductor [152, 153], $E(k) = E_0(k) + v_s p_F$, where $v_s = J/en$ is the supercurrent velocity, J is the supercurrent density, n is the number density of superelectrons, $E_0(k)$ is the quasiparticle spectrum at $J=0$, and p_F is the Fermi momentum. The increased density of normal electrons corresponds to a decreased gap

$$\Delta(v_s) = \Delta - p_F |v_s|$$

Solving the kinetic equation for the distribution function of quasiparticles in a superconductor in a strong rf field allows a calculation of the current-induced rf pair-breaking in the clean limit for Type II superconductors. The surface resistance is found to increase quadratically with the rf field:

3.2 Nonlinear (Field-Dependent) BCS Resistance

$$R_s = \left[1 + C\left(\frac{\Delta}{T}\right)^2\left(\frac{H_0}{H_c}\right)^2\right]R_{s0} \qquad C = \frac{\pi^2}{384}\left[1 + \frac{\ln 9}{2\ln(4.1 T\Delta\xi^2/\lambda^2\omega^2)}\right]$$

Here H_0 is the rf field, H_c is the rf critical field, and R_{s0} is the standard BCS resistance. The contribution of the logarithmic term in the brackets for Nb at 2 K and 2 GHz is less than 8%, and can be neglected, which results in a simpler approximate expression with just a quadratic field dependence:

$$R_s \cong \left[1 + \frac{\pi^2}{384}\left(\frac{\Delta}{T}\right)^2\left(\frac{H_0}{H_c}\right)^2\right]R_{BCS}$$

For Nb at 2 K, the factor $\gamma = C(\Delta/T)^2 \approx 2$. Note how the field-dependent contribution increases as the temperature decreases.

The simple quadratic dependence is only valid for small H_0, typically below 40 mT at 2 K. The pair-breaking effect becomes more pronounced when $H_0 > (T/T_c)H_c$ [153]. The full dependence is given by

$$R_s \cong \frac{4R_{BCS}e^{\beta_0}}{\beta_0^3\sqrt{2\pi\beta_0}} \qquad \beta_0 = \frac{v_s p_F}{k_B T} = \frac{\pi}{2^{2/3}}\frac{H_0}{H_c}\frac{\Delta}{k_B T}$$

and shown in Fig 3.3 for H_0 from 0 to ≈ 160 mT for Nb at 2 K. In this case the pair-breaking effect can double R_s at $H_0 \approx 100$ mT as compared to R_{BCS}.

The field-dependent BCS surface resistance discussed so far is for the clean-limit ($l \sim \xi_0$) only. Taking into account impurity scattering is a more complicated problem, but the field dependence in surface resistance is generally expected to decrease in the dirty limit [152, 153]. Section 3.3.3 discusses whether the nonlinear BCS resistance helps to account for the observed medium-field Q-slope.

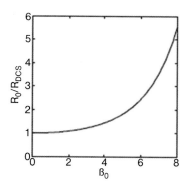

Fig. 3.3 Increase of BCS resistance due to pair-breaking at a high supercurrent density β_0 [152, 153] (courtesy of FSU).

3.3
RF Magnetic Field Dependence of BCS Surface Resistance

The observed Q of a niobium cavity shows several interesting features with increasing field. In the low-field region (between a few mT and 20 mT) the Q *increases* [151,154], a surprising effect, named the "low-field Q-slope." Mild baking (100–120 °C for 48 h) generally enhances the low-field Q-slope. At medium fields (between 20 and 80 mT) the gradual Q fall is called the "medium-field Q-slope" [151, 154, 155] and is a common feature of all niobium cavities. The medium-field Q-slope is generally attributed to a combination of surface heating [156] and nonlinear BCS resistance [152, 153] with varying degrees of success [157, 158]. Besides surface heating and pair-breaking other effects such as "hysteresis losses" due to Josephson fluxons in weak links (oxidation of grain boundaries) are proposed to play a role [159, 160]. Baking generally (but not always) decreases the medium-field Q-slope. Finally, there is a strong Q-drop at the highest field, which is the main subject of Chapter 5. Eventually superconductivity quenches either due to a thermal instability initiated by local heating at a defect (see Chapter 5), or a phase transition at the local rf critical magnetic field (Section 3.5).

3.3.1
Low-Field Q-Slope: Experiments and Models

When raising the rf magnetic field from zero, the Q value *rises*, reaching a maximum near 15–20 mT (Fig. 3.4). The Q rises by about 40% and the maximum is presumably the traditional BCS value. The low-field Q-slope is more pronounced at lower temperatures as Fig. 3.4(b) shows. As expected, the low-field Q-slope manifests more clearly when the residual resistance is low. Two types of experiments shed light on the nature of the low-field Q-slope. Baking either at 100 °C for 48–60 h, or 120 °C for 60 h, enhances the low-field Q-slope (i.e., decreases the starting field Q value) as seen in Fig. 3.4(a). The baking enhancement is observed for surfaces prepared by both chemical etching (Fig. 3.4a) and electropolishing (EP) (Fig. 3.4c). Note some other important effects with increasing baking temperature. The residual resistance increases with baking temperature, an effect we will discuss in Section 3.3.2. Also the quench field decreases with increasing baking temperature, which we will discuss in Chapter 5.

An interesting series of experiments suggests that the low-field Q-slope originates from the metal–oxide layer. Rinsing by HF after baking (Fig. 3.4c) restores the low-field Q-slope to its behavior before baking [151]. Rebaking restores the stronger low-field Q-slope. HF rinsing removes the oxide layer along with some of the oxide–metal interface, and subsequent water rinsing grows a new oxide layer along with a new interface. Since both HF rinsing/re-oxidation and baking modify the metal–oxide interface, it is reasonable to attribute the origin of the low-field Q-slope to the oxide and the interface region.

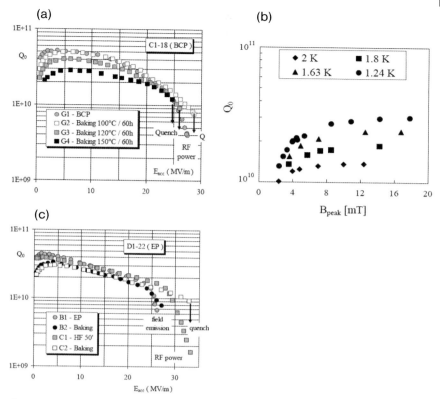

Fig. 3.4 (a) Baking enhances the low-field Q-slope for a cavity prepared by BCP [151] (courtesy of Saclay). (b) Behavior of the low-field Q-slope with decreasing bath temperature [154]. Despite lowering the temperature the Q at fields below a few mT retains a low value, rising to the expected BCS value by 20 mT (courtesy of JLab). (c) Baking drops the low-field Q, and HF rinsing mostly restores the higher BCS value. Rebaking lowers the Q again [151]. These results are from a cavity treated with EP (courtesy of Saclay).

According to one model [161], the Q-slope at low field occurs due to the presence of NbO_x clusters at the oxide–metal interface. Surface analysis studies [150] show that the niobium surface is covered by a few monolayers of hydrocarbon and water followed by about 5 nm of niobium pentoxide Nb_2O_5. Underneath the pentoxide, there are monolayers of niobium suboxides NbO, NbO_2 as well as other suboxide NbO_x ($x<1$) formations, in clusters or channels. Surface studies results will be presented in more detail in Section 3.4 in connection with residual resistance and in Chapter 5 in connection with the high-field Q-drop. Suboxide clusters introduce localized states in the gap region [161], effectively reducing the gap and therefore increasing the surface resistance when excited quasiparticles occupy these states. At low rf fields, quasiparticles which occupy these states remain out of thermal equilibrium due to a mismatch between the photon absorption rate from the rf field and the quasiparticle relaxation rate by phonon scattering. At higher rf fields thermal equilibrium conditions return along with the full energy gap and the standard BCS surface resistance (higher Q). The model predicts that the low-field surface resistance is inversely proportional to the square of the peak magnetic field, which is supported by observations [143, 154]. At lower temperatures, the effect appears more pronounced since the BCS surface resistance (with standard gap) decreases with temperature. After low-temperature baking, oxygen diffusion from the oxide forms more niobium oxide clusters, increasing the density of localized states and enhancing the low-field Q-slope. HF rinsing and regrowing oxide restores the original low cluster density and the prebake milder low-field Q-slope.

An alternate explanation [162] is based on a two-layer superconductor, with a thin layer of poor superconductor on top, such as a contaminated film. The model further assumes that the penetration depth of the poor layer increases with the magnetic field so that the overall rf surface resistance decreases as the fields penetrate more deeply into the better superconductor underneath at higher field values. Baking increases the low-field Q-slope because it makes the overlayer an even poorer superconductor and raises the field for the maximum Q.

3.3.2
Medium-Field Q-slope and Thermal Feedback

The medium-field Q-slope can be quite strong dropping the Q by a factor of 2–3 from 2 to 25 MV/m, and another another factor of 2 out to 40 MV/m. Understanding and controlling the medium-field Q-slope is important to future cw applications, such as the energy recovery linacs (ERL), where cryogenics costs dominate due to cw operation at medium fields (<20 MV/m). These applications will be discussed in Chapters 11 and 12. Analysis of temperature maps (Fig. 3.5) shows

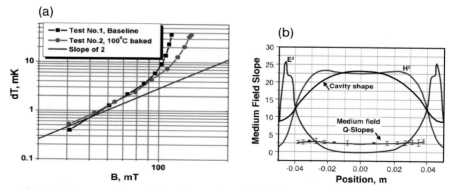

Figure 3.5 Thermometry analysis of the medium-field Q-slope (40<B<80 mT) for a 1.5 GHz single-cell cavity prepared by BCP. (a) Reading of a thermometer near the equator as a function of the peak surface magnetic field. The power law for the field dependence, ΔT versus H, is roughly 2.5. (b) Distribution and spreads of medium-field Q-slopes before baking [163]. H^2 and E^2 are in arbitrary units.

that the medium-field Q-slope is relatively uniform from spot to spot as compared to the high-field Q-drop, which has significant spatial variation, as discussed in Chapter 5. The simplest explanation for the medium-field Q-slope is based on a thermal feedback model with pure BCS resistance. Thermal feedback provides a good account for high-frequency (>2.5 GHz) cavities but otherwise yields a rather weak slope. Inclusion of the pair-breaking BCS resistance (the nonlinear effect) predicts a stronger a medium-field Q-slope, but often exceeds the observed slope. These difficulties stem from the fact that the medium-field Q-slope and thermal feedback models which attempt to fit the data depend on a large number of physical parameters, some of which are not very well known for each cavity, or even for each test: rf frequency, bath temperature, thermal conductivity (especially the magnitude of the phonon peak), Kapitza resistance, wall thickness, electron mean free path (which changes due to mild baking), and residual resistance. Nevertheless, it is useful to study the trends for each parameter variation. Finally, we discuss other proposed mechanisms that may play a role in the medium-field Q-slope, such as entry of Josephson fluxons.

3.3.2.1 Thermal Feedback with Standard BCS Resistance

Thermal effects play an obvious role in the medium-field Q-slope. The surface resistance increases with the rf field because of a thermal feedback process by which the surface temperature increases due to rf heating, while the higher BCS resistance at the higher temperature further increases the rf heating. The exponential temperature dependence of the BCS resistance drives the thermal feedback, but thermal conductivity and heat transfer at the Nb–He interface (e.g., Kapitza conductance below 2.17 K) also play important roles. For low fields, the surface resistance can be expanded in a Taylor series in B^2 and conveniently expressed by [156]

$$R_s = R_{s0}\left(1 + \gamma \frac{H_P^2}{H_C^2}\right) \quad \gamma = H_C^2 R_{BCS} \frac{\Delta}{2kT^2}\left(\frac{d}{\kappa} + R_K\right)$$

where R_{s0} is the surface resistance at about 15 mT [= $R_{BCS0}(T)+R_{res}$], $B_c\sim 200$ mT is the niobium critical field, and T is the He bath temperature. The medium-field Q-slope is customarily represented by an initial slope γ. A simple approximation for γ is derived in terms of the wall thickness d, constant thermal conductivity κ, and constant Kapitza resistance R_k. A full derivation can be found in [158]. The quadratic frequency dependence of γ is subsumed in the BCS resistance, R_{BCS}. As a rough estimate, $\gamma=1$ implies a 25% increase in surface resistance between 15 and 100 mT peak surface magnetic field. The simple formula for γ shows that the medium-field Q-slope increases with increasing wall thickness, Kapitza resistance, bath temperature (here the temperature dependence of R_{BCS} is dominant), rf frequency (again through $R_{BCS}\sim f^2$), and decreases with higher thermal conductivity.

To go beyond the initial slope and derive the Q behavior at higher fields requires a full thermal model feedback calculation [157, 158] using all the temperature-dependent functions, such as BCS surface resistance, thermal conductivity (Fig. 3.6a), and Kapitza conductivity. Starting from a standard set of parameters in Table 3.1, one parameter can be varied at a time to identify the parameters which have a strong effect on the medium-field Q-slope. Heat transfer properties at 2.17 K in superfluid helium (Kapitza conductance) [164] and normal helium (nucleate boiling) [165] are plotted in Fig. 3.6(b). Thermal conductivity can be simply modeled using two parameters, RRR and phonon mean free path (Fig. 3.7a), as suggested by [166]. As-received Nb of RRR = 300 has a grain size of about 50 μm and should have no phonon peak. Mild phonon peaks

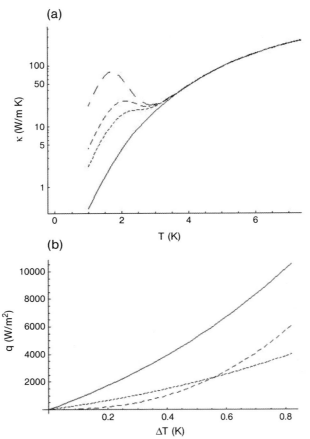

Fig. 3.6 (a) Variation of the thermal conductivity of Nb [166] of RRR = 300 for phonon mean free paths from 0.1, 0.5, 1, and 5 mm in order of increasing conductivity. (b) Heat transfer across the Nb–LHe interface as a function of the temperature difference for annealed niobium (solid line), unannealed niobium (short dashed line), and nucleate boiling (long dashed line).

Table 3.1 Baseline parameters for studying the thermal feedback effect.

RF frequency, f	1.3 GHz
Helium bath temperature, T_b	1.8 K
Residual resistance, R_0	10 nΩ
Wall thickness, d	3 mm
Residual resistivity ratio, RRR	300
Phonon mean free path, l	0.1 mm
Kapitza resistance	Annealed Nb

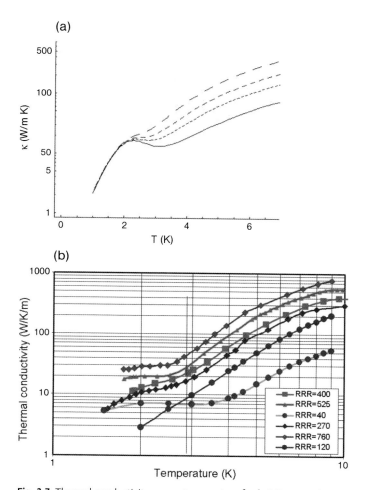

Fig. 3.7 Thermal conductivity versus temperature for $l=0.5$ mm and RRR=100 (solid), 200 (short dashed line), 300 (medium dashed line), and 500 (long dashed line). (b) Measured thermal conductivity of Nb for various RRR [167] (courtesy of DESY).

appear for a grain size of 0.1–0.5 mm and larger peaks for larger grain sizes, as shown in Fig. 3.6 (a). The companion Fig. 3.7 (b) shows a compilation of experimental thermal conductivity measurements on as-received Nb.

It is important to note that the medium-field Q-slope calculations are relevant mostly below 100 mT, since the high-field Q-drop (see Chapter 5) takes over at higher fields. However, the medium-field Q-slope studies discussed in this chapter may also be relevant for electropolished (EP) cavities baked at 100–120 °C since the high-field Q-drop disappears with the mild baking treatment.

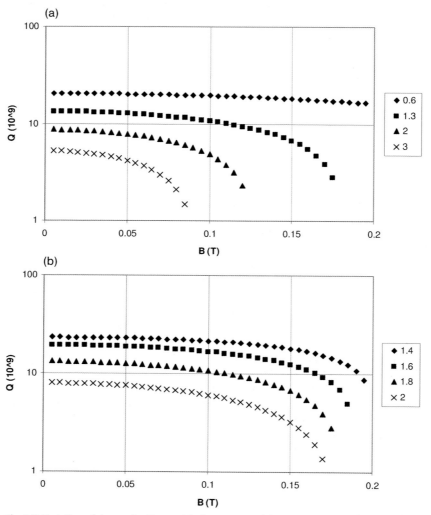

Fig. 3.8 Variation of the quality factor with the magnetic field for (a) various frequencies and (b) bath temperatures. The other parameters are held fixed to the values in Table 3.1. The base choice of 0.1 mm for the phonon mean free path accentuates the thermal feedback effects [158].

3.3 RF Magnetic Field Dependence of BCS Surface Resistance

Starting with a baseline set of parameters of Table 3.1, Figs. 3.8 to 3.12 show the Q-slope trends for systematic variations of frequency, bath temperature, phonon mean free path, and heat transfer at the helium bath. Detailed graphs for the other parameter trends can be found in [158], and will only be discussed briefly here. As one of these parameters varies, the others are held fixed to the values in Table 3.1. Figure 3.9 shows the corresponding trends in γ values for the initial quadratic dependence for various choices of frequency and bath temperature. Computed γ values agree within a few percent with the simple analytical estimate. However, thermal feedback becomes stronger at higher fields as shown by the stronger slope. Note that the baseline thermal conductivity choice has no phonon peak due to the small phonon mean free path choice, appropriate for unannealed Nb. This choice generally accentuates all other thermal feedback effects (Fig. 3.8a).

The first interesting trend is the striking frequency dependence. For high-frequency (>2.5 GHz) cavities, the quadratic frequency dependence of the BCS surface resistance eventually results in a thermal instability well below the rf critical field (Fig. 3.8a). This thermal feedback effect, called the "global thermal instability" (GTI), was first discussed by [1, 168]. The thermal model was applied to the 3 GHz case of interest to predict a strong medium-field Q-slope as well as a thermal instability at high fields. The global thermometer response in a 3 GHz cavity test also confirmed the global nature of the thermal instability. GTI has also been observed for a 2.8 GHz cavity operating in the TE mode [143] as shown in Fig. 3.10 and for the 3.9 GHz, third harmonic cavity [169, 170]. Figure 3.10(b) shows a plot of Q versus B from measurements on a 3.9 GHz along with the results of a thermal model simulation with matching cavity parameters [158]. In the medium-field range, the curves agree fairly well, with the simulation giving Q values only slightly higher than the data.

At low frequencies, however, the calculated thermal feedback effect is much smaller than the observed medium-field Q-slope and does not generally lead to global instability. The absence of GTI at low frequency was one of the original important reasons for selecting a frequency near 1 GHz for high accelerating field applications, such as the linear collider [171].

The thermal feedback effect predicts an overall decrease in γ values with bath temperature from 2 to 1.4 K (Fig. 3.8b). There are several competing effects. Since the BCS surface resistance decreases exponentially with T it dominates the overall trend of a falling γ value with falling temperature, even though the thermal resistance ($1/\kappa$) and the Kapitza resistance (R_k) increase with falling T. Some experimental results [155] show the opposite trend with bath temperature, but there are other possible causes to the medium-field Q-slope which may be relevant, as we discuss later.

A most striking trend is the phonon mean free path dependence. Figure 3.11(a) shows calculated trends in the Q-slope and γ values due to variations in the phonon mean free path. γ values fall from 1.35 to 0.2 as the phonon peak develops and thermal conductivity improves. Thermal feedback therefore predicts a low γ value for postpurified cavities which are heated to 1400 °C with

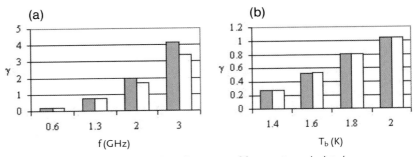

Fig. 3.9 Medium-field Q-slope γ values for varying rf frequencies, calculated numerically (shaded) and from the approximate formula [156] (unshaded). (b) Medium-field Q-slope values for varying bath temperature T_b, calculated numerically (shaded) and from the approximate formula (unshaded) [158].

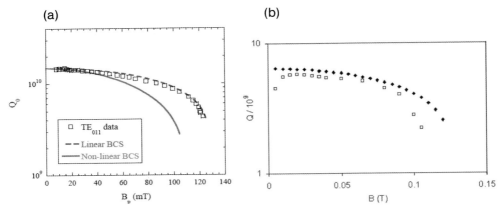

Fig. 3.10 Comparison of the observed and calculated medium-field Q-slopes for a 2.8 GHz cavity operating in the TE_{011} mode using the thermal feedback model. (a) The dashed curve is the calculated result with the standard BCS resistance and the solid curve the nonlinear resistance [143] (courtesy of JLab). (b) Same comparison between measured and calculated [158] results for the TM_{010} mode 3.9 GHz third-harmonic cavity [169, 170].

grain size growth to several mm as well as for large grain annealed cavities. Measurements with postpurified cavities and large grain cavities with higher thermal conductivity generally show reduced slopes [155] most likely due to the appearance of a stronger phonon peak in the thermal conductivity. In general, a detailed knowledge of the real phonon mean free path is essential for matching the observed medium-field Q-slope to calculations.

The heat transfer coefficient at the helium bath also has a strong effect on the medium-field Q-slope as seen in Fig. 3.12(a) for calculated and observed cases. A large medium-field Q-slope develops when the helium bath crosses from superfluid to normal fluid at 2.17 K. This aspect of observed cavity behav-

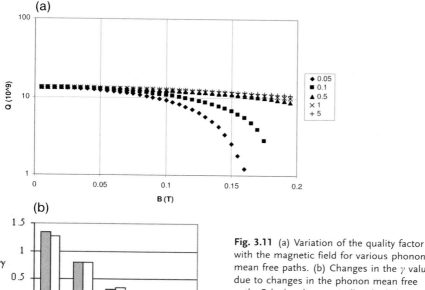

Fig. 3.11 (a) Variation of the quality factor with the magnetic field for various phonon mean free paths. (b) Changes in the γ value due to changes in the phonon mean free path. Calculated numerically (shaded) approximate formula (unshaded).

ior also confirms that thermal feedback plays a clear role in the medium-field Q-slope. For example, Fig. 3.12(b) shows that the medium-field Q-slope increases by about one order of magnitude above the lambda point when niobium is cooled by He-I instead of superfluid He-II.

It is interesting to note that the other parameters, such as wall thickness, residual resistance, and RRR, do *not* have as strong an effect as the frequency, bath temperature, and phonon mean free path on the medium-field Q-slope out to 100 mT. For example, γ changes from 0.4 to 1 for the wall thickness between 1 and 4 mm; γ holds at about 0.8 for the residual resistance between 3 and 20 nΩ; and again γ holds at about 0.8 for the RRR between 100 and 500. The latter trend is because the phonon contribution of the thermal conductivity dominates at 2 K, and this was chosen to be constant at 0.1 mm phonon mean free path as the RRR was varied.

A general observation when comparing the analytical (γ values) and numerical thermal model results with experimental data is that the thermal feedback model with standard BCS resistance generally underestimates the medium-field Q-slope for frequencies below 2.5 GHz. Figure 3.13 summarizes the results of several measurements [144, 151]. The large set of cavities represented here lies well within the range of cavity parameters considered in the previous section, yet these experimental γ values hover around 2 to 4, while those calculated from

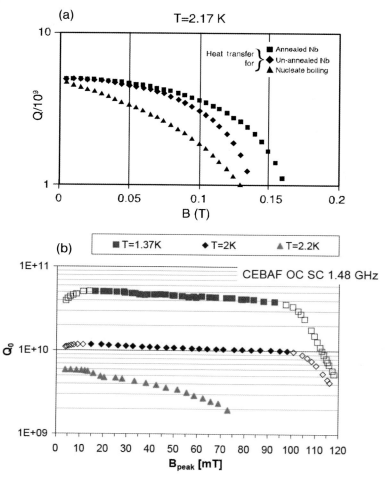

Fig. 3.12 Calculated variation [158] of the quality factor with the magnetic field for various heat transfer coefficients of unannealed niobium [164], annealed niobium [164], and nucleate boiling helium [165] near $T=2.17$ K. (b) Measured Q versus B curves for a post-purified cavity at different bath temperatures. The γ value increases by an order of magnitude from about 1 at 2 K to 14 at 2.2 K (above the lambda point) (courtesy of JLab).

the model are mostly less than 1. There are several possible causes for the strong disagreement. The Kapitza resistance is generally not known accurately. Another difficulty comes from the uncertainty in the strength of the phonon peak in the thermal conductivity since the exact annealing conditions vary among cavities. Another likely cause is the contribution to BCS resistance from pair-breaking (the nonlinear BCS case) as discussed in the next section.

One interesting remark is appropriate here concerning the observed slight reduction of the medium-field Q-slope due to baking in Fig. 3.13(b). This could

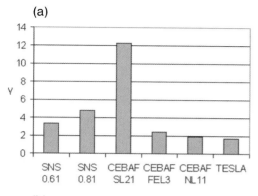
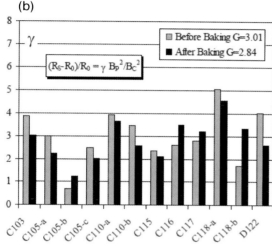

Fig. 3.13 (a) Values of the medium-field Q-slope compiled from measurements [154]. Each value shown here is the average of a set of values obtained from measurements on a set of similar cavities. The cavities have frequencies 804 MHz (SNS), 1.5 GHz (CEBAF), and 1.3 GHz (TESLA). (courtesy of JLab). (b) γ values for 1.3 GHz single-cell cavities at Saclay [151]. Gray bars are after baking (courtesy of Saclay).

primarily be due to the lowering of the BCS resistance with the electron mean free path (Section 3.1.1), most likely due to diffusion of oxygen. Most cavities show a decrease in the medium-field Q-slope after baking, but some show an increase, as Fig. 3.13(b) reports for a large set of data on single-cell 1.3 GHz cavities. One general difficulty to properly model the thermal feedback stems from the different baking conditions at different laboratories. Some use hot air, others hot nitrogen or hot helium. In many cases the outside surface of a cavity is etched. As a result, the Kapitza resistance is generally not known accurately.

3.3.3
Medium-Field Q-Slope and Field-Dependent BCS Resistance

In general, the thermal feedback effect with the standard BCS resistance is not strong enough to account for the observed medium-field Q-slope for low-frequency (<2.5 GHz) cavities, as mentioned. Some additional mechanism plays a role. Unlike the low-field Q-slope, the oxide interface has been eliminated as a possible candidate via HF rinsing experiments. Rinsing a cavity with HF and regrowing a fresh layer of oxide during rinsing does not affect the medium-field Q-slope [151]. This strengthens explanations which rely on more general loss mechanisms such as the field-dependent (nonlinear) BCS theory, or losses which originate from deeper in the rf layer, such as from the weak links at grain boundaries.

As with standard BCS resistance it is useful to study general trends for the medium-field Q-slope with nonlinear BCS. Note that the baseline set of parameters (Table 3.2) for the calculations reported here is somewhat different from the first set of baseline parameters used for the linear case above in order to provide some new information. For example, the baseline phonon mean free path chosen here is 1.3 mm, which will correspond to a small phonon peak in the thermal conductivity, as opposed to the previous baseline case of the 0.1 mm phonon mean free path with no phonon peak. As a result, the medium-field Q-slopes for the standard BCS cases here are much smaller than in Fig. 3.8.

Figure 3.14 compares the thermal feedback predictions for the BCS and nonlinear BCS cases for varying rf frequency and for varying phonon mean free path [172]. Figure 3.15 shows the corresponding substantial changes in γ values due to inclusion of the nonlinear BCS resistance. Several improvements are seen over thermal feedback calculations with the standard BCS resistance. The γ value is now about 2 (instead of 0.7) for the 0.1 mm phonon mean free path and 1.3 GHz, toward better agreement with the γ data of Fig. 3.13. There is also a stronger overall change in Q vs. H in contrast to the standard BCS thermal feedback behavior (Fig. 3.8 b). The γ value (Fig. 3.16 b) which is indicative of the initial slope still increases with bath temperature.

Table 3.2 Baseline parameters for the medium-field Q-slope calculations using the field-dependent BCS theory.

RF frequency, f	1.3 GHz
Cavity wall thickness, d	3.0 mm
Residual resistance, R_0	5 nΩ
Kapitza conductance, H_k	annealed Nb
Residual resistivity ratio, RRR	300
Helium bath temperature, T_b	1.8 K
Phonon mean free path, l	1.3 mm

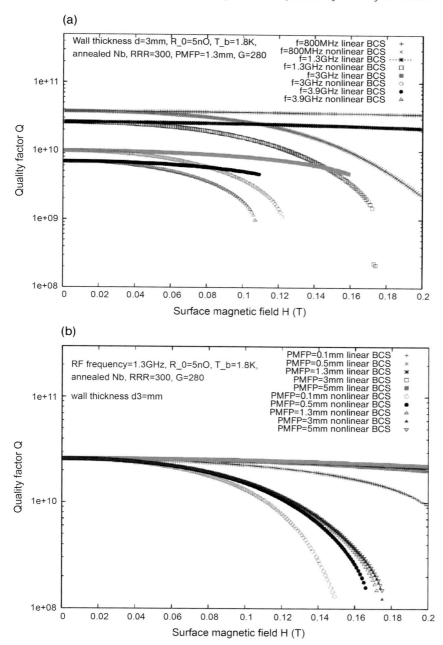

Fig. 3.14 Variation of the cavity Q with the rf surface magnetic field (a) for rf frequencies between 800 MHz and 3900 MHz and (b) for phonon mean free paths between 0.1 and 5 mm. In each case Q versus H curves are given for BCS and nonlinear BCS cases [172]. Residual resistance R_0 is in nΩ.

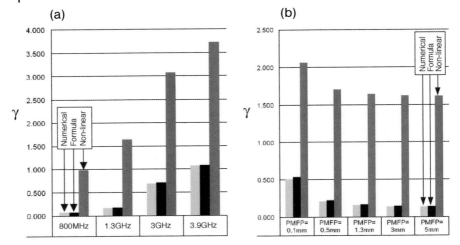

Fig. 3.15 γ values for BCS and nonlinear BCS for (a) various rf frequencies and (b) for various phonon mean free paths [172].

Comparisons of the nonlinear BCS thermal model predictions with experimental Q curves give mixed results. Figure 3.17 shows an attempt to fit the Q behavior of a 1.5 GHz TM$_{010}$ cavity by including just the nonlinear BCS quadratic contribution, but not the full nonlinear effect [143]. This improves the fit to the medium-field Q-slope for the 1.5 GHz cavity. However, inclusion of the stronger field-dependent surface resistance (Fig. 3.3) would clearly overestimate the medium-field Q-slope as compared to the data. Similarly, for the 2.8 GHz case (Fig. 3.10a), the nonlinear BCS quadratic term alone yields too strong a Q-slope as compared to the data. Another comparison is attempted using the full nonlinear BCS contribution [172] for the 1.3 GHz re-entrant cavity data [170] at 1.6 K and 1.9 K. The cavity was prepared by EP, followed by a mild bake. For both temperatures the calculated curves show a stronger field dependence than the data. To be fair, the nonlinear BCS prediction has only been worked out for the clean limit (no baking). In general, the inclusion of the nonlinear BCS resistance comes closer to the data, but often makes the medium-field Q-slope stronger than the observed Q-slope. This was also found in other comparison attempts [157].

3.3.3.1 Linear Dependence of the Medium-Field Q-Slope Due to Josephson Fluxons

Postpurified cavities and single-grain cavities generally show a lower medium-field Q-slope, and a linear plus quadratic magnetic field dependence of the Q-slope [155]. A possible cause for the lower γ is a higher phonon peak, with reduction of grain boundaries. Figure 3.18 shows the transformation from quadratic to more linear behavior in the medium-field Q-slope after postpurification

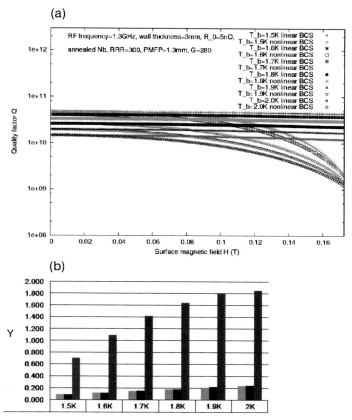

Fig. 3.16 Variation of the cavity Q with the rf surface magnetic field for (a) bath temperatures between 1.5 and 2 K and (b) corresponding changes in γ values. In each case Q versus H curves and γ values are given for BCS and nonlinear BCS cases [172].

[155]. A linear dependence of R_s versus B_{pk} is explained [160] by losses from Josephson fluxons which penetrate weak links formed at oxide channels especially along grain boundaries. Above 90 Oe Josephson fluxons start to penetrate the rf layer. Different from Abrikosov fluxons (encountered in the Abrikosov fluxoid phase), these fluxons nucleate rapidly without any surface barrier, and do not have a normal conducting core. The size is related to the dimension of the weak region rather than the coherence length. Pinning of Josephson fluxons leads to hysteresis and losses in rf fields. Hysteresis losses are enforced by baking, most probably due to oxygen diffusion increasing the length and quantity of links.

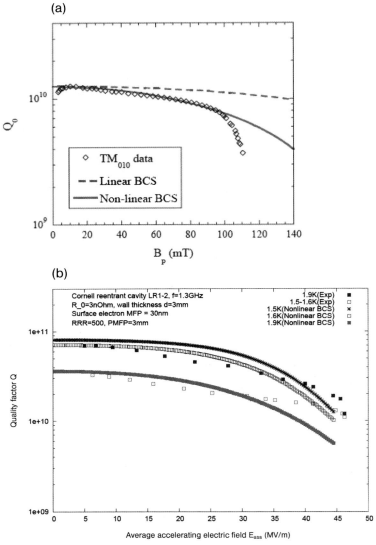

Fig. 3.17 (a) Comparison of the observed medium-field Q-slope for a 1.5 GHz cavity operating in the TM_{010} mode using the thermal feedback model. The dashed curve is the calculated result with the linear BCS resistance and the solid curve the nonlinear resistance (quadratic term only) [143] (courtesy of JLab). (b) Similar comparison for the 1.3 GHz re-entrant cavity test [49, 172].

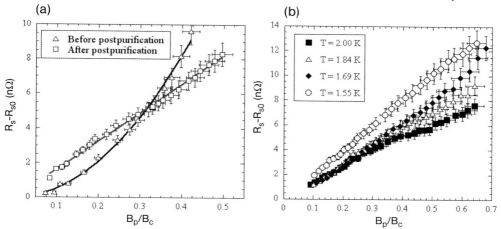

Fig. 3.18 Change from the quadratic to linear field dependence after postpurification for (a) a 1.5 GHz polycrystalline single-cell cavity and (b) temperature dependence of the medium-field Q-slope for a single-crystal postpurified cavity. Note that the slope increases with decreasing temperature [143, 155] (courtesy of JLab).

3.4
Residual Resistance

According to the BCS theory, the surface resistance should fall exponentially with temperature as $e^{-\Delta/kT}$. However measurements at $T<0.2T_c$ generally show that the surface resistance reaches a residual value called the residual surface resistance (R_0). The lowest surface resistance measured in a superconducting cavity is less than 1.5 nΩ for a 1.3 GHz Nb cavity which reached a Q_0 of 2×10^{11} at 1.6 K (Fig. 3.19). The record result yielded a residual resistance of 0.5 nΩ [173]. More typically the residual resistance for Nb cavities is of the order of 5–10 nΩ.

There are several familiar mechanisms for residual resistance described in [1]: insufficient shielding of the ambient dc magnetic field, and the hydrogen-related Q-disease. For the first case the loss mechanism is the normal conducting cores of trapped dc magnetic flux lines. In the second case, niobium-hydride precipitates form at 100–150 K when there is a significant amount of H dissolved in the Nb, as for example if the acid gets hot ($T>20\,^\circ$C) during etching, or if too much H arrives at the cavity surface during electropolishing. The Nb–H effect is often referred to as the H-related Q-disease. Condensed gases also increase residual losses. One example [174] will be discussed in Chapter 4 where condensed gases removed by electron bombardment during multipacting reduced the local low-field residual losses by about 10 nΩ, and recondensation of gas by cycling to 300 K restored the original value. Another mechanism for residual losses is baking at 120 $^\circ$C and above, which increases residual resistance by a few nΩ, presumably due to conversion of pentoxides into suboxides.

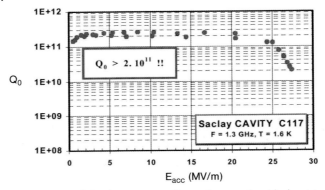

Fig. 3.19 Record Q value corresponding to the record residual resistance of 0.5 nΩ [173] (courtesy of Saclay).

XPS studies show the formation of suboxides at baking temperatures of 120 °C and above [150, 175]. We discuss a few aspects of the better understood mechanisms not previously covered.

3.4.1
Hydrogen-Related Q-Disease

An important residual loss mechanism arises when hydrogen dissolved in the bulk niobium precipitates as a lossy hydride at the rf surface [176]. This residual loss (commonly referred to as the "Q-disease") is a subtle effect that depends on the quantity of dissolved H, the rate of cool down to helium temperatures, and the amount of other interstitial impurities or atomic size defects present in the niobium. Below 10 at.% H will remain in the disordered α-phase at high temperatures. Nb–H phases harmful to superconductivity can form between 100 and 150 K, when the H concentration is greater than 100–200 at. ppm, as shown the Nb–H phase diagram (Fig. 3.20) [177]. The longer the time spent in the dangerous temperature zone the larger the hydride regions grow at nucleation centers due to significant mobility of H at these temperatures (300 μm in 1 h at 120 K). Below 77 K, H mobility is small enough to arrest the growth of the hydride regions. For example, a cavity contaminated with H had a Q greater than 10^{10} on fast cool down, but the Q dropped to 10^9 on slow cool down due to excess time spent in the dangerous 100–150 K zone. Figure 3.20(b) [178] shows the change in Q from 10^{10} to 10^8. The shape of the Q versus E curve with H disease indicates that the Nb–H phase is superconducting at low fields.

Temperature mapping is an effective tool to locate regions of high loss on the cavity surface responsible for residual resistance, such as for example Nb–H precipitates or trapping of the earth's dc magnetic field. The temperature map of a cavity strongly limited by the H-disease shows high losses over the entire surface (Fig. 3.21) [179].

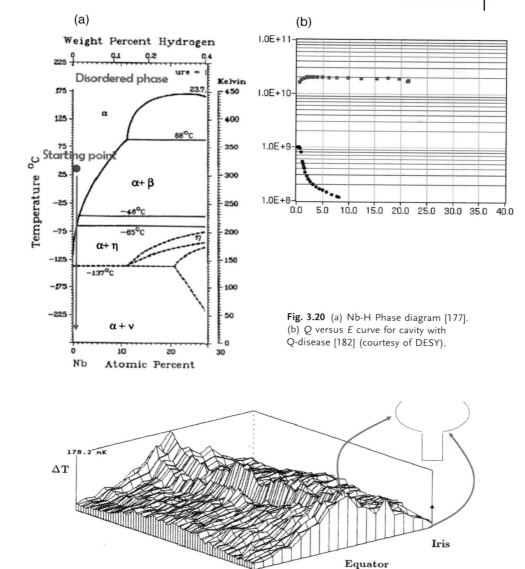

Fig. 3.20 (a) Nb-H Phase diagram [177]. (b) Q versus E curve for cavity with Q-disease [182] (courtesy of DESY).

Fig. 3.21 Temperature map showing uniform heating in the magnetic field region due to the H-related Q-disease [179] (courtesy of Wuppertal).

Past experience has shown that niobium with fewer interstitial impurities (i.e., a higher RRR) is far more susceptible to the Q-virus [176]. The interstitial impurities in low RRR Nb serve as trapping centers for hydrogen, thereby preventing hydride precipitation and Q-disease. Similarly, vacancies and grain boundaries are also

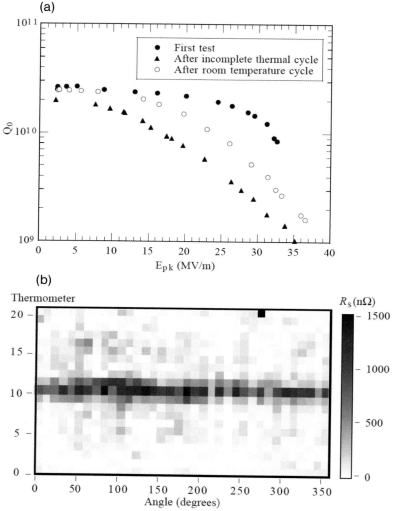

Fig. 3.22 (a) Mild Q-disease manifest in a 1.5 GHz cavity. During the incomplete thermal cycle the cavity warmed up to the dangerous temperature between 100 and 150 K. The Q recovered to the first test values after heat treatment at 900 °C for 2 h (not shown). (b) Temperature map after incomplete thermal cycle, showing strong heating primarily at the equator weld region. There are 36 boards, one every 10° in longitude. Each board carries 20 thermometers from iris to iris. Thermometer 10 is at the equator [181].

known to be effective traps for hydrogen by forming Cottrell clouds [180]. Since the equator weld region has large grains (many mm) it is likely more susceptible to the Q-virus due to the absence of trapping centers for H.

When a "milder" case of the Q-disease occurs it is likely due to Nb–H precipitation primarily in the equator weld region. Figure 3.22 shows a gradual cavity

Q degradation with field [181]. The accompanying temperature map shows a strong temperature rise at the equator weld. The local surface resistance at the equator weld (derived from the temperature map) increases linearly with field to reach nearly 1 µΩ at 60 mT. If the resistance of the entire cavity surface area increased by a comparable amount the Q would be 4×10^8 as would be true for a cavity generally afflicted with a strong H-Q disease.

Since one of the methods whereby niobium becomes contaminated with hydrogen is during chemical etching, it is important to keep the acid temperature below 15 °C. A cavity which is contaminated with H can be fully cured by degassing in a vacuum of better than 10^{-6} Torr at 800 °C for 1–2 h, or at 600 °C for 10–12 h (preferred to avoid grain growth). Fast cool down is one way to circumvent the H-related Q-disease if the H reaches an excess amount despite precautions. For the cavity with the mild Q-disease, (Fig. 3.22) heating at 900 °C for 2 h removed the H dissolved in the Nb, restoring the equator to low resistance and removing the gradual Q-slope.

3.4.2
Residual Losses from Trapped dc Magnetic Flux

A well understood and controllable source of residual loss is trapped dc magnetic flux from insufficient shielding of the earth's magnetic field or other dc magnetic fields in the vicinity of the cavity, such as stray fields from accelerator magnets. Under ideal conditions, if the external field is less than H_{c1}, the dc flux will be expelled from the bulk of a superconductor due to the Meissner effect. However, if there are lattice defects or other inhomogeneities in the material, the dc flux lines can be "pinned," and trapped within the material. Surprisingly, experimental studies show that when niobium cavities are cooled down in the presence of a small dc magnetic field (e.g., less than 1 Oe) all the flux within the volume of the cavity is trapped [183]. It is suspected that the oxide layer on the niobium surface serves as a source for flux pinning sites. The flux comes through the cavity wall in current vortices which contain single quanta of magnetic flux. The normal cores of size ξ cause residual resistance. In rf fields the flux lines oscillate, causing additional losses. Typically, at 1 GHz, R_0 is 1 µΩ/Oe. These losses increase as $\sqrt{f}$, where f is the rf frequency, as to be expected from losses due to the normal cores. To obtain the highest Q, a superconducting cavity must therefore be well shielded from the earth's magnetic field. If a 1 GHz superconducting cavity is not shielded from the earth's field, the maximum Q will be limited to below 10^9.

Figure 3.23 shows an early temperature map of a cavity in a dc magnetic field perpendicular to the equator [184]. The temperature shows the expected sinusoidal heating pattern due to trapped magnetic flux.

Another mechanism for trapping the dc magnetic flux is through thermal currents. Occasionally the Q of a cavity drops after a local quench (by as much as a factor of 2) but can be recovered by warm-up above T_c. This effect has been attributed to the magnetic flux generated by thermoelectric currents (the

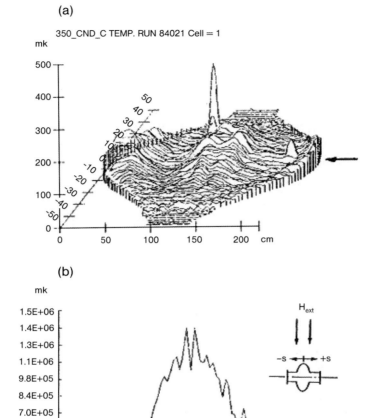

Fig. 3.23 (a) Temperature map at 2.2 K (in subcooled He) showing residual losses due to trapped dc magnetic flux from incomplete shielding. (b) The temperature profile along the equator agrees with the pattern of magnetic flux trapping expected when the dc magnetic field is perpendicular to the equator [184] (courtesy of CERN).

Seebeck effect) during the large temperature differentials generated during a quench. The flux becomes trapped with persistent currents as the cavity cools down rapidly after the quench. Figure 3.24 (a) shows the strong heating during the quench and the companion Figure 3.24 (b) shows the increase in residual resistance in the region of quench heating [185].

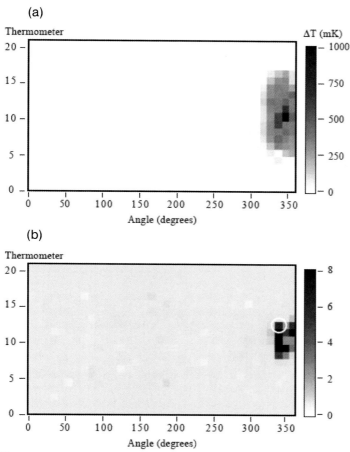

Fig. 3.24 (a) Temperature map during quench initiated by a local defect. (b) Ratio of the surface resistance after several breakdown events to that before breakdown. Dark regions indicate that the surface resistance increased by a factor of about 4 due to trapped magnetic flux generated by thermocurrents [185].

3.4.3
Baking Effects

Baking at 120 °C for 48 h has now become standard practice to heal the high-field Q-drop, as discussed in Chapter 5. Figure 3.25(a) [151] shows an increase in residual resistance of a few nΩ due to baking. After rinsing the surface in HF and water, the residual resistance returns to its original value before baking. The HF rinsing removes the oxide layer and some of the niobium–oxide interface. The subsequent water rinsing grows a new layer by converting about 2 nm of Nb into 5 nm of pentoxide. Higher temperature (150 °C) baking results also

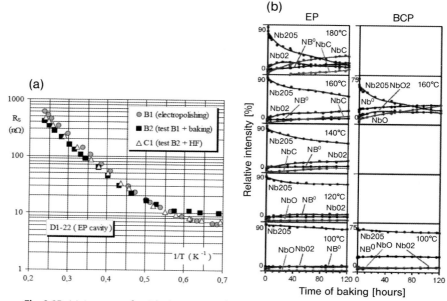

Fig. 3.25 (a) Increase of residual resistance due to baking, and restoration of the lower residual resistance due to HF rinsing [151] (courtesy of Saclay). (b) Growth of suboxides detected by XPS studies [175] (courtesy of UST).

show that the residual resistance increases by a factor of 2 with increasing baking temperature from 120 °C to 150 °C (see Fig. 3.4a).

One reasonable interpretation for the increase of residual resistance is that baking modifies the oxide layer, the more so with increasing temperature. Another explanation is that during baking excess oxygen below the oxide layer diffuses into the penetration depth and increases the residual resistance, although the same effect helps to drop the BCS resistance. XPS studies [175] show an increase in the suboxides during the bake (Fig. 3.25b).

3.5
DC Critical Magnetic Fields: H_{c1}, H_{c2}, H_c, and H_{c3}

Niobium is a Type II superconductor; hence its behavior at high rf surface fields can be related to one or more of several critical fields, H_{c1} – the lower critical field, H_c – the thermodynamic critical field, H_{sh} – the superheating field, H_{c2} – the upper critical field, and H_{c3} – the surface critical field. Table 3.3 summarizes the measured or estimated values of H_{c1}, H_{c2}, H_{c3}, H_c, and H_{sh} to be discussed in the next section.

Table 3.3 DC critical fields for Nb at zero temperature [189–191].

B_c (mT)	200	[189, 191]
B_{c1} (mT)	174	[189]
B_{c1} (mT)	190	[183]
B_{c2} (mT)	390	[191]
B_{c2} (mT)	400	[189]
B_{c2} (mT)	410	[188]
B_{c2} (mT)	450	[183]

3.5.1
H_{c1}, H_c, and H_{c2}

A brief explanation of Type I and Type II superconductors along with their various critical fields is in order.

Besides zero resistance to dc currents, the other hallmark property of superconductors is the Meissner effect, which requires that there be zero magnetic field in the bulk of the superconductor. The Meissner condition prevails until the externally applied field reaches a critical field, when magnetic flux abruptly enters the superconductor. Thermodynamics determines the value of this critical field. When electrons condense into Cooper pairs the free energy of the superconductor decreases by an amount referred to as the condensation energy. In the presence of an external field, the supercurrents which flow at the surface of the superconductor to shield the bulk from the external field increase the free energy. Work must be done to establish the currents to exclude the magnetic field. At the field where the increase in magnetic free energy completely balances the decrease in condensation energy of the Cooper pairs, there is an abrupt transition to the normal state and all the external flux enters the superconductor. This field is known as the thermodynamic critical field, H_c. Superconductors which show an abrupt transition at the thermodynamic critical field are called Type I superconductors.

From a thermodynamic point of view it is possible for flux to enter a superconductor below H_c if the superconductor divides into alternating thin normal and superconducting regions, provided the thickness of the normal region is smaller than the penetration depth. Since the magnetic field does not drop to zero at the centre of the normal sheet, less magnetic energy is needed to expel the magnetic flux. Hence superconductivity can persist in the layer with the penetrated magnetic field. However, since there is a surface energy associated with creating a normal-superconducting boundary, this type of partial flux penetration is only possible for negative surface energy superconductors, also known as Type II superconductors.

The relationship between the intrinsic coherence length, and the London penetration depth λ_L determines the response of a superconductor to an external magnetic field. If $\xi_0 > \lambda_L$ there is a positive surface energy at the boundary.

If $\xi_0 < \lambda_L$, the boundary energy is negative. Ginzburg and Landau determined that the exact crossover between positive and negative surface energies happens for $\kappa = \lambda/\xi = 1/\sqrt{2}$. Materials with $\kappa < 1/\sqrt{2}$ are Type I superconductors, characterized by an abrupt transition to the normal state at H_c. For materials with $\kappa > 1/\sqrt{2}$ (Type II superconductors), there exists a critical field, called the lower critical field (H_{c1}), above which the magnetic flux penetrates the superconductor in the form of a regular triangular array [186] of flux tubes ("fluxoids") (Fig. 3.26), each carrying an elementary quantum of flux $\Phi_0 = hc/2e$. This condition of partial flux penetration is called the "fluxoid state" or the "Abrikosov state." The superconducting order parameter $\psi(r)$ goes to zero at the axis of each flux tube over a distance of the order of the coherence length. These normal conducting "cores" are surrounded by shielding currents, which is the

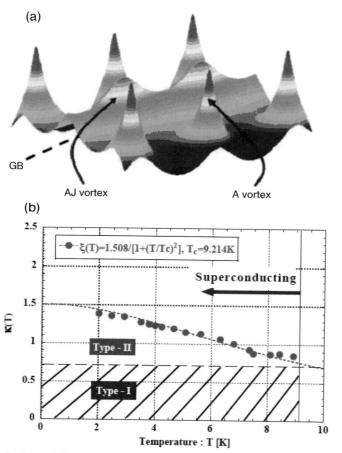

Fig. 3.26 (a) Flux pattern [187] determined by Abrikosov from GL equations (courtesy of FSU). (b) Attempts to measure $\kappa(T)$ from dc magnetization curves [188] (courtesy of KEK).

Cooper-pair vortex current. The center of a fluxoid is normal conducting and covers an area of roughly $\pi\xi^2/2$. As the strength of the applied magnetic field increases above H_{c1} in a Type II superconductor, the fluxoids keep moving into the specimen and pack closer together increasing the average flux density in the superconductor. At a sufficiently high value of the applied magnetic field, called the upper critical field (H_{c2}), the order parameter $\psi(r)$ goes to zero. At this point bulk superconductivity breaks down.

The GL theory expresses H_{c1} and H_{c2} in terms of the thermodynamic critical field H_c and of the Ginzburg–Landau parameter κ. In particular,

$$H_{c2} = \sqrt{2}\kappa H_c$$

The microscopic BCS theory yields H_{c1}, H_c, and H_{c2} in terms of microscopic parameters:

$$H_{c1} = \frac{\phi_0}{4\pi\lambda^2}\left(\ln\frac{\lambda}{\xi} + 0.5\right) \qquad H_c = \frac{\phi_0}{2\sqrt{2}\pi\lambda\xi} \qquad H_{c2} = \frac{\phi_0}{2\pi\xi^2}$$

Attempts to determine $H_{c1}(T)$, $H_{c2}(T)$, and $\kappa(T)$ for Nb have been made through measurements of the isothermal magnetization curves. $\kappa(T)$ is then determined from the GL relationship between H_{c2} and H_c and κ. Figure 3.26 shows the result for $\kappa(T)$ for a high-purity sample [188]. Niobium is confirmed to be a Type II superconductor.

If the magnetization curve is reversible, the area under the curve correctly gives the thermodynamic critical field $H_c(T)$. However, magnetization curves for Nb are notorious for their irreversiblity due to the presence of many types of pinning centers, such as the oxide, suboxides at the oxide–metal interface, interstitial impurities, grain boundaries, and dislocations. Magnetic flux which enters is captured at pinning centers. When the field is reduced again these flux lines remain bound, and the specimen retains a frozen-in magnetization even for a vanishing external field. One has to invert the field polarity to achieve zero magnetization, but the initial state ($H=0$ and no captured flux in the bulk) can only be recovered by warming up the specimen to destroy superconductivity and release all pinned flux quanta, and by cooling down again. Another effect of pinning sites is to inhibit the entry of flux at H_{c1}, thereby leading to false estimates of H_{c1} from dc magnetization curves, as shown in Fig. 3.27 (a) [192]. The apparent values for H_{c1} for Nb decrease with etching, 800 °C annealing, and 1400 °C titanium postpurification treatment, as these treatments anneal defects and remove interstitial impurities. The effect of mild baking (100–120 °C) on the magnetization curves has not yet been studied. As discussed above this treatment gives a large improvement in the Q-drop. Hence it should be interesting to determine whether there are any noticeable effects of mild baking on the dc critical fields.

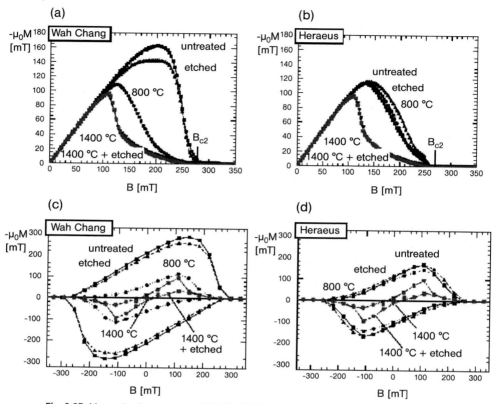

Fig. 3.27 Magnetization curves at 4.2 K for Nb from (a) Wah Chang and (b) Heraeus. Each panel shows magnetization curves for as-received, BCP, 800 °C annealed, and 1400 °C titanium postpurification. Similarly, the lower panels show the magnetization hysteresis curves [192] (courtesy of DESY).

The reversibility improves after removing the surface damage layer with chemistry and after removing stresses by heat treatment [193]. Nevertheless, as Fig. 3.27 [192] shows, the best magnetization curves always end up with a residual magnetic moment due to trapped flux at pinning sites after reducing the external field to zero. Hence the area under the magnetization curve (increasing fields) overestimates H_c. One empirical approach to circumvent this problem has been to take the average of the area under the increasing and decreasing magnetization curves [190]. The best approach is of course to determine H_c via specific heat measurements [194]. An integral of the specific heat difference between the normal and superconducting states yields the entropy difference, and a second integral of the entropy difference gives the free energy difference. The latter is simply related to the square of the thermodynamic critical field.

3.5.2
Surface Superconductvity and H_{c3}

Superconductivity can persist in a surface layer of thickness $\sim \xi$, even in a magnetic field whose strength is sufficient to drive the bulk material normal [195]. This happens up to a critical field called H_{c3}. The value of H_{c3} depends on the angle the applied field makes to the surface and is maximum when the applied field is parallel to the surface. In this case $H_{c3} = 1.695 H_{c2}$ and it is equal to H_{c2} when the field is perpendicular to the surface.

Since surface superconductivity above H_{c2} resides in a layer of thickness ξ (comparable to λ for Nb), measurements of H_{c3} are sensitive to various chemical and electrochemical treatments and to the temperature and duration of the bake-out used for cavity preparation. We will return to this topic in Chapter 5 in connection with the high-field Q-slope. To determine H_{c3}, the ac (10 Hz) susceptibility of Nb samples has been measured at 4.2 K under a dc background field between 0 and 1 T [196]. Figure 3.28 shows that below H_{c2} the real part of the susceptibility is -1, corresponding to perfect diamagnetism as the superconducting surface sheath shields the sample from the probing ac magnetic field. The first deviation from perfect diamagnetism occurs above the upper critical field $H_{c2} = 280$ mT at 4.2 K of the bulk niobium. This is in rough agreement

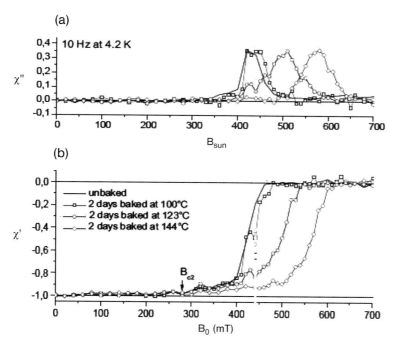

Fig. 3.28 AC susceptibility of Nb samples prepared by BCP and baking at various temperatures (a) Imaginary part and (b) real part [196] (courtesy of DESY).

with H_{c2} values measured from magnetization curves at 4.2 K (see Fig. 3.27). With rising dc field the critical shielding current of the superconducting sheath decreases and is no longer capable of shielding the ac field completely. Hence the real part of the susceptibility rises steeply and reaches zero (no shielding) at the surface critical field H_{c3}. The steep rise of the real part is accompanied by a maximum in the imaginary part (see the lower part of Fig. 3.28). The corresponding energy dissipation is caused by the magnetic flux moving out of the superconducting surface layer.

For all susceptibility measurements done under various chemical treatment and baking conditions, the ratio $H_{c3}/H_{c2}=r_{32}$ always comes out to be greater than 1.69, i.e., H_{c3} *apparently* increases more than H_{c2} [191, 197]. A simple model to explain why the observed r_{32} is larger than the theoretical prediction of 1.69 is that baking decreases the coherence length of the surface layer (to depth ξ) and increases the surface H_{c3} and surface H_{c2}, but does not increase the bulk H_{c2} due to excess oxygen atoms present only near the metal–oxide interface. Since at 10 Hz, the ac susceptibility method samples the bulk H_{c2} and surface H_{c3}, it leads to an apparent increase of the ratio r_{32}. This explanation is consistent with surface studies [150, 175], which show that baking has a tendency to break up the oxide layer and thereby add oxygen to the metal layer underneath the oxide. A more sophisticated explanation invokes percolation theory and will not be discussed here [198].

Measurements of H_{c3} on Nb samples treated by cavity preparation methods BCP and EP, combined with low temperature baking show systematic increases of H_{c3} and for r_{32} (Fig. 3.29). BCP plus bake gives a noteworthy increase which disappears with etching a few μm. The largest increase is for EP plus bake.

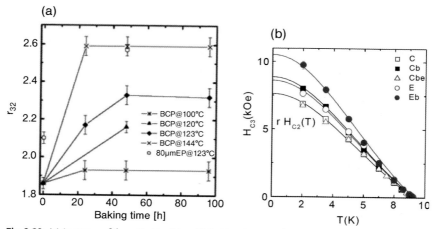

Fig. 3.29 (a) Increase of the ratio H_{c3}/H_{c2} with baking times and temperatures [197]. (b) Measured values of H_{c3} versus temperature for various sample preparations. C = BCP, Cb = BCP+Bake, Cbe = BCP+Bake+BCP, E = electropolished, Eb = electropolished plus bake [191] (courtesy of U. Hamburg).

Remarkably, these H_{c3} increases all correlate rather well with high field performance increases for cavities as we will discuss in Chapter 5 on the topic of a high-field Q-drop.

3.6
RF Critical Magnetic Field, H_{sh} the Superheating Critical Field

3.6.1
Theoretical Updates

Just as a supersaturated solution forms no precipitates, or a superheated liquid stays in the liquid rather than transition to the gaseous state, a superconductor can stay in the Meissner phase if no nucleation sites for fluxoids are present. The maximum field above which this metastability disappears is called the superheating critical field, H_{sh}. It is possible for the "superheated" superconducting state to persist metastably at $H > H_c$, for Type I superconductors, and for $H > H_{c1}$ for Type II superconductors. The time it takes to nucleate fluxoids was measured to be long compared to the rf period of superconducting cavities in the gigahertz range [199]. Therefore, in rf cavities there is a strong tendency for the metastable superconducting state to persist up to H_{sh}. Because the fields change very rapidly (ns), the prevailing definition for the rf critical field is the superheating field.

The superheating field depends on the material properties via the Ginzburg–Landau parameter κ, which is the ratio of the penetration depth (λ) to the coherence length (ξ). Both quantities depend on mean free path and temperature. Figure 3.30 shows the dependence of the superheating field on κ_{GL} by solving the GL equations for the one-dimensional case where half of the space is occupied by a superconductor and the magnetic field is applied parallel to the surface [200].

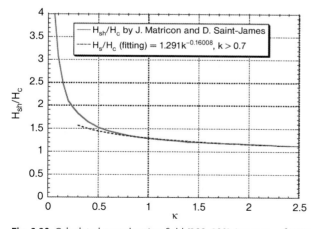

Fig. 3.30 Calculated superheating field [188, 200] (courtesy of KEK).

According to this result from the GL theory, under various limits, H_{sh} is given by:

$$H_{sh} \approx \frac{0.89}{\sqrt{\kappa_{GL}}} H_c, \quad \kappa_{GL} \ll 1$$

$$H_{sh} \approx 1.2 H_c, \quad \kappa_{GL} \approx 1$$

$$H_{sh} \approx 0.75 H_c, \quad \kappa_{GL} \ll 1$$

In principle, the GL theory is valid only near T_c. But cavities operate at $T/T_c \ll 1$. Therefore, a better theory is needed. The correct calculation for H_{sh} in terms of microsocopic BCS still needs to be carried out for the temperature-dependent H_{sh} in the absence of nucleation centers.

Table 3.3 gives the values for various critical fields for Nb, the most popular superconductor for cavities. Since κ is about 1, $H_{sh} = 1.2$ H_{sh} is about 240 mT at zero temperature. This translates to a maximum accelerating field of 54 MV/m (220 mT) at 2 K for a TESLA-shape niobium accelerating structure for velocity-of-light particles. The exact value for maximum allowed accelerating field depends on the detail structure geometry. New geometries discussed in Chapter 2 allow up to 63 MV/m for Nb.

For N_3Sn, which is a high-κ superconductor, T_c is 18 K, H_c is 530 mT [201], and H_{sh} is 400 mT, leading to a basic GL theory prediction of 110 MV/m for an advanced geometry structure.

A common error in the literature is to estimate the superheating field by using a thermodynamic surface energy balance argument. In the one-dimensional formulation [202, 203], this yields a κ dependence similar to that predicted from the one-dimensional solution of the GL equations [200].

$$H_{sh} = \sqrt{\frac{\xi_{GL}}{\lambda_{GL}}} H_c = \frac{1}{\sqrt{\kappa_{GL}}} H_c$$

Although the κ dependence is encouragingly close to theoretical, this analysis is incorrect. The energy balance approach has also been extended to other dimensional forms of nucleation such as line nucleation [188, 202, 203] to yield:

$$H_{sh} = \frac{\xi_{GL}}{\lambda_{GL}} H_c = \frac{1}{\kappa_{GL}} H_c$$

As energy-balance arguments, such approximations, give an upper bound on the *equilibrium* critical field for vortex penetration, which is related to H_{c1}, and not H_{sh}. Nothing in the energy balance argument discusses metastability, which is the key aspect for H_{sh}. Therefore, it is incorrect to use H_{sh} estimates based on energy balance arguments [204].

The line nucleation model is useful in the context of nucleation on inhomogeneities on the scale of the coherence length, but not as a fundamental limit for uniform, flat, pure superconductors. The theoretical H_{sh} needs to be determined in the absence of such nucleation centers. The degree to which real cavities approach the ideal of course depends on sufficiently careful materials preparation, and may not be realizable with present technology, but it may be eventually approachable.

3.6.2
RF Critical Magnetic Field: Experiments

The fundamental rf critical magnetic field has been measured for lead-on-copper, bulk niobium, and Nb_3Sn on niobium using high peak pulsed power to raise the cavity fields well above the cw limits [205–207]. By raising the fields

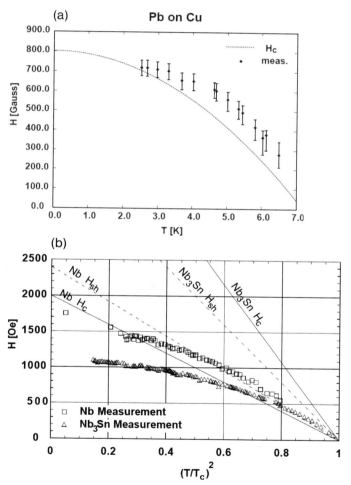

Fig. 3.31 (a) RF critical field measurements of a lead-on-copper 1.3 GHz cavity using a high power pulsed technique. The measured values are well above H_c and agree with the prediction of H_{sh} for lead. (b) RF critical field measurements for Nb and Nb_3Sn [205, 206].

fast (few μs to 100 μs), high magnetic fields can be reached at the superconducting surface before local defects create large normal conducting regions. The Q of the cavity is measured during the pulse to be sure that a significant fraction of the cavity surface is still superconducting at a given rf field level. The rf critical magnetic field is measured as a function of temperature up to T_c for 1.3 GHz cavities. Figure 3.31 shows that niobium and lead measurements are

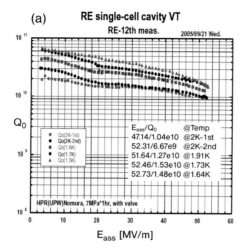

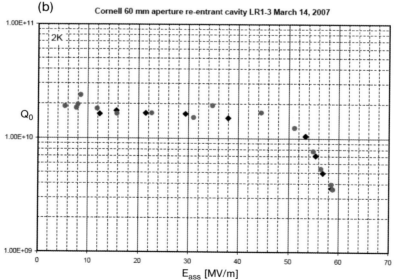

Fig. 3.32 (a) Single-cell re-entrant shape cavity, 70 mm aperture, which reached $E_{acc}=53$ MV/m ($H_{pk}=201$ mT) [34] (courtesy of KEK). (b) Single-cell re-entrant shape, 60 mm aperture, which reached $E_{acc}=59$ MV/m ($H_{pk}=210$ mT) [33].

consistent with the superheating critical field values at $T/T_c > 0.7$, whereas the Nb_3Sn results fall short of that prediction by a considerable amount. Niobium results also fall short of H_{sh} below $T/T_c = 0.7$ for rf magnetic fields above 120 mT. However, the Nb cavity studied here was prepared by the existing chemical etching (BCP) treatment which was later found to give strong Q-drops above 100 mT due to the high-field Q-drop effect discussed in Chapter 5. Similar measurements were conducted earlier to show that Nb_3Sn falls far short of H_{sh}, while Nb comes close to H_{sh} predictions.

With the arrival of better surface preparation techniques, such as electropolishing and baking (discussed in Chapters 6 and 7) and with high-purity annealed niobium, the maximum magnetic field in single-cell cavities is slowly rising, presently reaching 205 mT at 2 K which is clearly higher than H_{c1} and also higher than H_c. Figure 3.32 shows cw measurement results for two 1.3 GHz which reached surface magnetic fields of 200 to 210 mT. The more impressive results of 53 and 59 MV/m for the corresponding accelerating field are also due, of course, to the new shapes, which reduce the surface magnetic field to accelerating field ratios.

In summary, the prospects for accelerating fields of 50 MV/m are well supported by theoretical predictions, while the technology of SRF cavity preparation continues to advance toward reaching the theoretical limits.

4
Multipacting and Field Emission

4.1
Multipacting

The basics of multipacting – *mult*iple im*pact* electron amplification – (MP) in cavities is covered extensively in [1]. Here we give a short review and update findings. For recent reviews of MP see [208–210]. Besides cavities, MP also takes place in input coupler and higher order mode coupler devices. These topics will be addressed in Chapters 8 and 9.

Multipacting is a resonant process in which an electron avalanche builds up within a small region of the cavity surface due to a confluence of several circumstances. Electrons in the high magnetic field region of the cavity travel in circular orbit segments returning to the rf surface near to their point of emission, and at about the same phase of the rf period as for their emission. Secondary electrons generated upon impact travel similar orbits. If the secondary emission yield for the electron impact energy is greater than unity, the number of electrons increase exponentially to absorb large amounts of rf power and deposit it as heat to lower the cavity Q. This form of MP is named "one surface" or "one-point" MP. Depending on the cleanliness of the surface, the secondary emission coefficient of niobium surfaces prepared by cavity treatment methods is larger than unity for electron impact energies between 50 and 1000 eV.

In one-surface MP the travel time for electrons to return is an integral number of rf periods. Before 1980 this phenomenon was the dominant limitation in cavity performance. With the invention of the round wall (spherical) cavity shape [30], one-surface MP is no longer a significant problem for velocity-of-light structures. The essential idea is to gradually curve the outer wall of the cavity. Electron trajectories drift toward the equator of the cavity in a few generations (Fig. 4.1a). At the equator the electric fields are sufficiently low so that energy gain falls well below 50 eV and regeneration stops because the secondary emission coefficient is less than unity for electron impact energies far below 50 eV. The same suppression effect is achieved in the elliptical cavity shape which is generally preferred over the spherical shape due to added mechanical strength and better geometry for rinsing liquids [29]. One-surface MP in spherical/elliptical cavities can be well simulated by several codes [208–214].

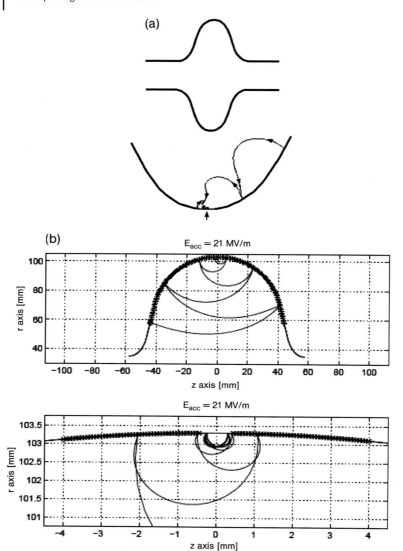

Fig. 4.1 (a) Elimination of one-surface MP with the spherical (elliptical) cell shape. When the cavity shape is rounded, electrons drift toward the zero electric field region at the equator. Here the electric field is so low that the secondary cannot gain enough energy to regenerate. (b) Two-point MP in a single-cell 1.3 GHz TESLA-shape cavity near $E_{acc} = 20$ MV/m. Note resonant trajectories in the lower half [208].

Multipacting levels are suppressed when electron bombardment during MP decreases the secondary yield, most likely by gas desorption. Multipacting can be enhanced when the secondary emission yield increases due to adsorbates or condensed gas. Levels that have been successfully processed can recur for short periods if the cavity is temperature cycled and gases recondense on the surface. Rinsing a cavity with acetone contaminated with trace amounts of oil was observed to substantially enhance MP. The technique was used to facilitate identification of barriers [215]. Occurrence of MP is often accompanied by x-rays. Some MP electrons escape from the MP region and get accelerated in the high electric field region of the cavity to produce high energy x-rays, which penetrate the cavity wall and are detected outside the cryostat.

4.1.1
Two-Point Multipacting in Elliptical Cavities

Multipacting conditions also exist when electrons travel to the opposite surface in half an rf period (or in odd-integer multiples of half an rf period). This form of MP is called two-surface or two-point MP (Fig. 4.1b). Again the round-wall geometry suppresses MP in the cells by pushing succeeding generations of trajectories toward the equator where the electric fields are low. However, two-point MP does survive near the equator of the elliptical cavity because the electron energies remain between 30 and 50 eV near the unity crossover of secondary yield. For analytic approximations to the fields in the equator region and resulting two-point MP there see [216].

Conditioning times can range from fractions of a minute in single-cell cavities to an hour in multicells. During conditioning, MP can grow sufficiently intense to induce local thermal breakdown of superconductivity, as shown in the temperature maps of Fig. 4.2 [217]. The location of intense MP migrates as the secondary emission coefficient of the surface drops in one place due to electron bombardment. Both MP and its associated breakdown events disappear after conditioning, but trapped dc magnetic flux generated by thermoelectric currents during the breakdown events drop the Q values, sometimes by as much as a factor of 2. Figure 4.3 shows local regions of increased surface resistance due to trapped magnetic flux. Warming up to 10 K is necessary to remove the trapped magnetic flux, as discussed in Section 3.3 for breakdown events due to quench from a defect.

Desorption of surface adsorbates by electron bombardment and possibly even a local gas discharge lower the SEC, thereby arresting MP. Figure 4.4 shows the resistance ratios after just a *few* MP-induced breakdown events. Regions with reduced rf losses can be clearly identified. Cycling to 10 K removed the trapped flux and restored the higher resistance regions close to their prebreakdown values. However, it did not affect the lower resistance regions since gas does not migrate at 10 K. Cycling to room temperature increased the resistance of the processed area to preprocessing levels most likely due to recondensation of gas. Gas adsorption also explains why thermal cycling between room temperature and 2 K reactivates MP by redistributing condensed gases on the cavity surface.

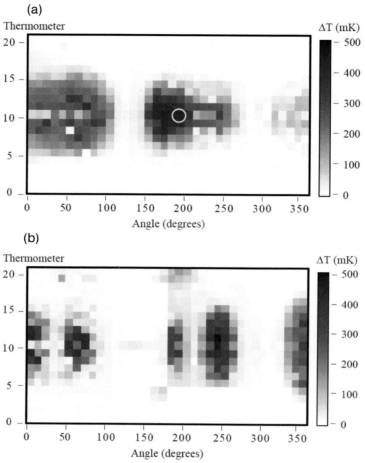

Fig. 4.2 (a) Temperature maps taken on a 1.5 GHz single-cell cavity during breakdown induced by one-point MP. The quench spots move from (a) one location to (b) another as the surface processes due to electron bombardment [217]. There are 36 boards, one every 10° in longitude. Each board carries 20 thermometers from iris to iris. All thermometers in row number 10 are at the equator.

We know that such a redistribution of gases does take place, because the regions which reduced their losses during MP reverted back to their original surface resistance after a complete temperature cycle, whereas losses in the same regions were unaffected by intermediate thermal cycles. The contribution of adsorbates to the surface resistance appears to be on the order of 10 nΩ or less.

Two-point MP has been recognized in $\beta=1$ elliptical cavities for a wide frequency range from 200 MHz to 3000 GHz through numerical simulations supported by experimental evidence [208]. The peak magnetic field levels of the first order two-point MP at various frequencies follow the scaling law:

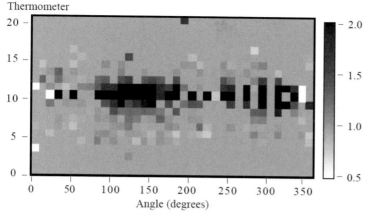

Fig. 4.3 Ratio map of surface resistance (T after processing/T before processing at the same field) due to many breakdown events through MP and processing. Most areas show increased losses (ratio >1) from trapped flux induced by thermocurrents generated by large temperature changes. But some areas show reduced losses (ratio <1) due to processing (see Fig. 4.4) [217].

$$H(\mathrm{mT}) = 5 + 55 f(\mathrm{GHz})$$

This corresponds to MP at 76 mT or $E_{\mathrm{acc}} = 18$ MV/m for the TESLA-shape cavity. In 9-cell tests [218] MP is found between 17 and 20 MV/m confirming the simulations.

Further simple rules give the associated magnetic field for each order as [219]:

One point: $f/N = eB/2\pi m$
Two point: $2f/(2N-1) = eB/2\pi m$

where N is the order of MP, e and m are the charge and mass of the electron, respectively, and B is the local magnetic field at the surface.

4.1.2
Multipacting in the Beam Pipe

Several surprise occurrences of MP in new cavity geometries emphasize that during the cavity design stage MP simulations should be carried out for all the regions of the cavity. One example is the case for steps in the beam pipe, which are often enlarged for better HOM propagation. In the first version of the ICHIRO cavity shape (see Fig. 4.5) the beam pipe at one end was enlarged in anticipation of joining two structures into a "superstructure" [221]. After difficulties with cavity performance, simulations carried out at SLAC showed that MP takes place near $E_{\mathrm{acc}} = 30$ MV/m [222], as shown in Fig. 4.5 (b). Due to many possible resonant trajectories allowed by the geometry of the step region the impact

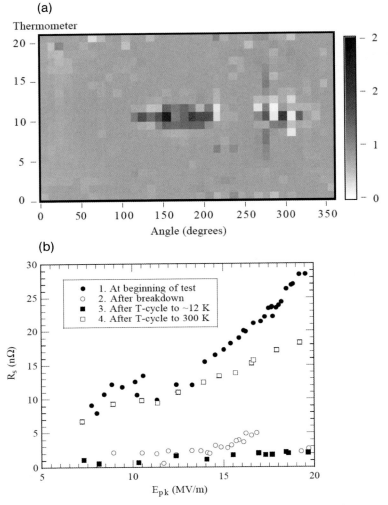

Fig. 4.4 (a) Ratio map showing reduction of residual resistance near the equator at 300°, most likely due to removal of condensed gas by electron bombardment during MP. (b) Tracking changes of residual resistance with processing, and after various thermal cycles. The low-field residual resistance drops by nearly 10 nΩ due to processing, most likely due to removal of condensed gas. Warm up to 12 K has no effect on the reduced resistance, but warm up to 300 K restores the low field resistance to preprocessing values, most likely due to recondensation of gas [217].

energy ranges from 100 eV to several 100 eV, all dangerous because of high secondary yields at these energies.

A similar surprise occurred during the development of the 2-cell 1.3 GHz cavity for the ERL injector. Figure 4.6 shows the MP trajectories at the corner of the beam pipe enlargement segment determined from the code MULTPAC

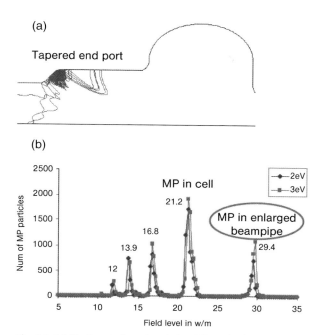

Fig. 4.5 (a) Electron trajectories during MP in the beam pipe of the Low-Loss (ICHIRO) cavity as simulated at SLAC [222]. (b) Simulated MP resonances in the cell and the beam pipe region (courtesy of KEK and SLAC).

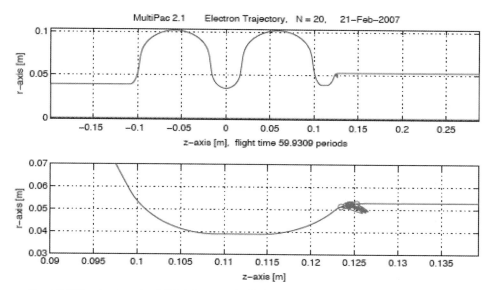

Fig. 4.6 MP region in a 2-cell ERL injector cavity [223].

[211]. At 6.5 MV/m the electron impact energy is between 175 and 195 eV [223]. Various MP levels were observed between 4.5 and 13 MV/m (first order), by detecting electrons on a biased probe and by correlated temperature rises with thermometers. All MP levels processed rapidly. Due to the complicated geometry of the trajectories in the corner region the resonant conditions are not very stable. Such MP levels can be avoided by shaping the transitions so that there is no local minimum in the electric field [220].

4.1.3
Multipacting in Low-β Cavities

Both one- and two-surface MP are incipient in the intricate shapes for low and intermediate-β cavities. Conditioning times can run from a few hours to days. Compared to the elliptical cavities, there has not been as much simulation effort to identify the location of trajectories. One example for both types of MP is shown in Fig. 4.7 for a HWR geometry [20, 93]. For one-surface MP in the high magnetic field region, the impact energies between $E_{pk}=8$–12 MV/m range between 50 and 150 eV for the first 40 impacts. For two-surface MP in the high E field region the impact energy for low fields ranges between 700 and 1200 eV. Inclining the inner (or outer) wall by $5°$ eliminated stable trajectories.

4.1.4
Codes for Multipacting

There exist several codes (2D and 3D) for MP simulation. For a review see [210]. Some codes use external field solvers such as MAFIA, OSCAR2D, SUPERFISH, and SuperLANS, and some use integrated internal solvers. MultiPac [211] is a simulation package for analyzing electron MP in axisymmetric rf structures with the TM_{0nl} mode, such as rf cavities, coaxial input couplers, and ceramic windows. The code has been especially useful for analyzing MP in the cylindrically symmetric parts of input couplers. A multipurpose electron tracking code (FishPact) [224] was developed based on the Poisson/Superfish Field solvers. Analyst (Omega3P) is a powerful tool for MP investigation for complex rf devices in real 3D fields [222] and is helpful to understand the possible roles of nonaxisymmetric features such as the power and HOM couplers. Multipacting in these components is discussed in Chapters 8 and 9.

4.2
Field Emission

Field emission is a general difficulty in reaching high surface electric fields in accelerating structures. The topic of field emission has been discussed at length in [1]. We confine our coverage to a brief review before discussing new findings.

(a)

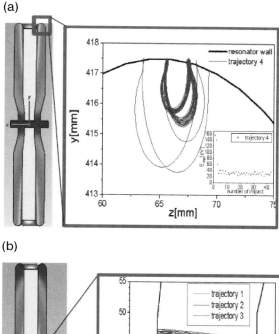

(b)

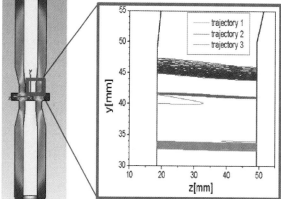

Fig. 4.7 MP in a half–wave resonator. (a) One-surface MP with "horseshoe" shape trajectories in the high magnetic field region. (b) Two-surface MP in a high E field [20, 93] (courtesy of ACCEL).

At the onset of field emission, the Q of a niobium cavity typically starts to fall steeply because of exponentially increasing electron currents emerging from particular emitting spots on the surface (Fig. 4.8). The rf field pick-up probe on the cavity beam pipe detects electron current, and radiation detectors outside the cryostat detect exponentially increasing x-ray intensity due to brehmstrahlung when the field emitted electrons collide with cavity walls. A detailed temperature map shows line heating (Fig. 4.8) along the longitude at the location of the emitter (due to the cylindrical symmetry of the fields in the accelerating mode.)

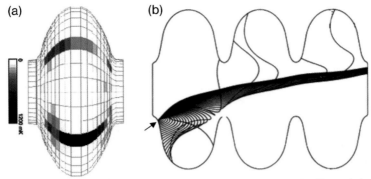

Fig. 4.8 (a) Temperature map from the heating of impacting field emitted electrons. There are two emission sites each producing a line-heating profile. (b) Calculated field emitted electron trajectories in a three-cell 1.5 GHz cavity operating at $E_{pk}=50$ MV/m. Here the field emitter is located (at the arrow) in an end cell, just below the iris, where the surface electric field is 44 MV/m. Note that a significant number of electrons emitted in the part of the rf cycle when the magnetic field is strong bend back and strike the wall near the emitter.

The exponential nature of field emitted current is a characteristic of the quantum mechanical tunneling process first calculated by Fowler and Nordheim (FN) [227]. The observed emission current follows the modified FN law, where the electric field at the emission site, E_{em}, is replaced by a locally enhanced field, $\beta_{FN} E_{em}$, where β_{FN} is the FN field enhancement factor. Since the FN theory calculates the current density, an effective emitter area, A_{FN}, is required to determine the field emission current, I_{FN}. For niobium, the field emitted current (in amperes) is given by the modified FN equation:

$$I_{FN} = 3.85 \times 10^{-7} A/V^2 A_{FN} \frac{(\beta_{FN} E_{em})^2}{t^2(y)} \times \exp\left(-5.464 \times 10^{10} V/m \frac{v(y)}{\beta_{FN} E_{em}}\right)$$

and

$$y = 9.48 \times 10^{-6} (m/V)^{1/2} \sqrt{\beta_{FN} E_{em}}$$

Here $v(y)$ and $t(y)$ are tabulated functions [228] that account for the image charge effect. If the image charge is ignored, these functions are unity. For most reasonable values of E_{em} these functions are on the order of unity and slowly varying. A plot of $\ln(I_{FN}/E^2)$ versus $1/E$ yields a straight line, widely known as a Fowler–Nordheim plot. The slope of the line determines β_{FN}. Both β_{FN} and A_{FN} are empirical parameters, helpful for characterizing field emission, but their physical meaning is debatable as we will discuss.

Although field emission is often found to limit the performance of a significant number of cavity tests, there has been sufficient progress in understanding and avoiding emitters so that many structures do not show any field emission at all up to their maximum field level of $E_{acc}=40$ MV/m. Indeed, field emission does not present any ultimate fundamental theoretical limit to the maximum surface electric field, in contrast to the surface magnetic field which has a limit set by the superheating critical magnetic field (see Section 3.5). The record surface electric field in an SRF cavity is 145 MV/m in cw operation [31] and 220 MV/m in pulsed operation [32]. Recent experimental work with single-cell cavities of new shapes has demonstrated cw rf fields of 120 MV/m over a broad surface area with little or no field emission.

The SRF community has devoted considerable resources to understanding the origins of field emission, avoiding field emission, and dealing with residual emission in cavities. Research reviewed in [1] shows that microparticle contaminants are the dominant emission sources [229, 230]. Increased vigilance in cleanliness during final surface preparation and assembly procedures is important to keep particulate contamination and associated emission under control. New approaches – in particular high-pressure rinsing (HPR) and class 100 clean room assembly – have achieved high levels of cleanliness in cavity surface preparation, leading to fewer emission sites and major improvements in cavity performance [231, 232].

The cleanliness approach reduces field emission substantially. DC field emission studies confirm a substantial reduction in density of emitters after HPR [233]. In large-area structures there is always a probability of dust occasionally falling into the cavity upon installation of power-coupling devices, or during the installation of a cavity into the accelerator. There exist techniques to eliminate emitters *in situ*. These are rf processing with cw power [234, 235], high pulse power processing (HPP) [236–238], and helium processing [239, 240]. The first two methods work by the same principles.

Systematic studies discussed in [1] and in Section 4.2.3 show that when an emitter processes at the field levels accessible with cw power (typically 10–20 W/cell) the emission current destroys the emitter in an rf spark. High pulsed power processing is a continuation of cw rf processing using pulsed high power to reach higher electric fields necessary to destroy emitters in similar rf spark activity. The processing mechanism during HPP is *exactly the same* as for cw rf processing. The essential idea of HPP is to further raise the surface electric field, even if for a very short time (s) so that the exponentially rising electric current can destroy more troublesome emitters that cannot be destroyed at the field levels generally reachable by cw power levels (10–20 W/cell). An important benefit is that HPP can recover cavities that are accidentally contaminated, as for example in a minor vacuum mishap [241].

4.2.1
Field Emission in Cavities: Temperature Maps and X-Rays

4.2.1.1 **Emitter Activation**
Occasionally field emission in a cavity activates irreversibly at a certain field level. For example in the CEBAF accelerator a few cavities per year suddenly show higher levels of emission, higher intensity of x-rays, and cause a higher trip rate

[242]. The available voltage drops by 1–2% per year. CEBAF cavities were prepared in the mid-1990s before HPR became a routine method for surface cleaning. Emitter activation has been attributed to sudden arrival of particles in the high electric field region. The worst affected cryomodules are being removed and reprocessed. If such paths were not available, rf processing and HPP rf processing are good options.

A careful experiment with a single-cell cavity equipped with high sensitivity thermometry has confirmed the phenomenon of field emission triggered by particle arrival [243]. Figure 4.9 shows a sudden Q-drop at $E_{pk} = 33$ MV/m ($E_{acc} \sim 17$ MV/m) due to emitter activation in a 1.5 GHz single-cell cavity. This cavity was also prepared without HPR. The temperature map after activation (Fig. 4.10) shows the development of a clear line heating pattern typical for field emission, whereas the map at the same field level below activation shows no line heating. By carefully fitting the heating pattern with a simulated field emitter heating profile, the location of the emitter was determined to within a few mm^2. In confirmation of particle arrival, the thermometer at the location of the emitter also showed an increase in the ohmic heating signal as shown in Fig. 4.9(b). The cavity was then cleanly dissected at the equator, and the half-cell examined in the SEM to reveal the particle shown in inset of Fig. 4.9(b).

4.2.1.2
Field-Emission-Induced Thermal Breakdown

When emission from an individual field emitter becomes intense it is possible for the power deposited by the impacting electrons to induce thermal breakdown, especially when the impact heating becomes intense in regions of high magnetic field (Fig. 4.11). Note that the center of the breakdown heating in Fig. 4.11(a) does *not* coincide with the maximum heating for the actual location of the emitter in Fig. 4.11(b). This is due to the fact that field-emission-induced breakdown takes place in a region of both significant electron bombardment and significant magnetic field.

4.2.1.3
X-ray Diagnostics

The locations of the field emitter source(s) can also be determined through x-rays using a collimated sodium iodide detector moving along the cavity outside the cryostat [244]. The detector also measures the energy spectrum, with the maximum energy equal to the kinetic energy of impacting electrons. Figure 4.12 shows the x-ray intensity measured in the π mode, superposed with a multicell cavity profile. Simulation codes track the field emitted electrons to infer the emitter locations. In this case, a total of four emitters were required to successfully simulate the x-ray profile and spectra. Additional x-ray measurements from different passband modes of the cavity are necessary to check the predicted emitter locations.

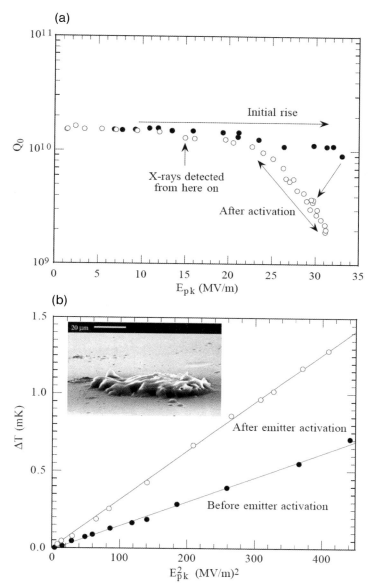

Fig. 4.9 (a) Q versus E_{pk} for a 1.5 GHz single-cell cavity before and after activation of field emission. (b) Temperature rise versus E_{pk}^2 before and after activation showing ohmic heating. Inset: SEM micrograph of the particle found after cavity dissection at the emitter location determined from temperature map analysis below [243].

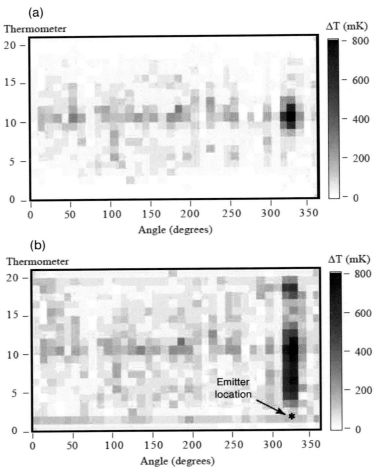

Fig. 4.10 Temperature map of a 1.5 GHz single-cell cavity at $E_{pk}=33.5$ MV/m (a) prior to any significant field emission and (b) at the same field after emission heating activated. The hotspot at the equator is not field emission related. Comparison of the heating pattern with simulations yields the location of the emitter, which was subsequently identified after cavity dissection as shown in Fig. 4.9 (b) [243].

4.2.2
Microparticles and Cleanliness

Niobium samples treated by the best cavity preparation methods have been scanned with dc field emission probes to count the emitter density with rising electric field [233]. As Fig. 4.13 shows, the density of emitters increases rapidly with field level for smooth samples prepared by electropolishing. There is a factor of 2–3 reduction in emitter density after HPR with ultrapure water, showing that even better treatments are necessary if stronger reductions are desired with

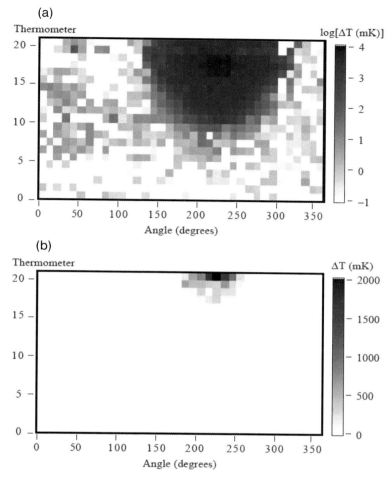

Fig. 4.11 (a) Temperature map due to field-emission-induced thermal breakdown in a 1.5 GHz single-cell cavity. The peak electric field was 28.4 MV/m. Note that the logarithm of T is plotted to demonstrate the extended nature of the breakdown area. (b) High-field temperature map before thermal breakdown occurred in (a). Strong field-emission-induced heating ($T\sim 1$ K) at 220 °C is responsible for the thermal breakdown in (a) [243].

the push for higher gradients. The results of Fig. 4.13 (a) are in rough agreement with previous studies summarized in [1], which showed an emitter density increasing exponentially from $0.3/\text{cm}^2$ at 60 MV/m to about $5/\text{cm}^2$ at 100 MV/m. Studies for the effectiveness of HPR showed a factor of 10 decrease in particle density over a 10 cm^2 sample area [245]. Even after HPR, the remaining density of emitters at $E_{pk} = 100$ MV/m can present problems for cavity performance above $E_{acc} = 40$ MV/m.

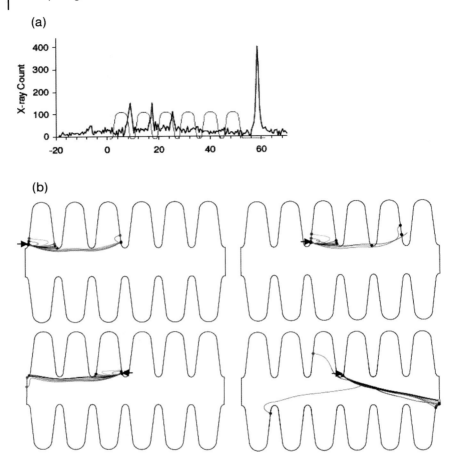

Fig. 4.12 (a) Total flux of x-rays for a horizontal scan in the π mode at $E_{pk} = 22$ MV/m from a 805 MHz $\beta = 0.47$ cavity. X-ray energies for the five peaks ranged from 0.72 to 1.2 MeV. (b) Four emission sites (marked by arrows) in the cavity and simulated electron trajectories for the β mode at $E_{pk} = 22$ MV/m. Electrons emitted at the lowest electric field travel farthest through the structure as the electric field rises through the half rf period. Electrons emitted at the highest field bend back as the magnetic field rises and electric field falls. The FN β value ranges between 140 and 150 with emissive areas of about 1×10^{-11} cm^2 [244] (courtesy of MSU).

The field emission current from the scanned emitters has been characterized by the customary FN field enhancement factor (β_{FN}) and apparent emissive area (A_{FN}) using FN plots for emission current (Fig. 4.13 b). Both β_{FN} and A_{FN} are empirical parameters, helpful for characterizing field emission, but their physical meaning is debatable. For emitters which turn on between 20 and 200 MV/m, β_{FN} values range between 10 and >300, emissive areas between 10^{-7} and 10^{-15} cm^2. However fitted emissive areas can be as high as 1–100 cm^2, which is

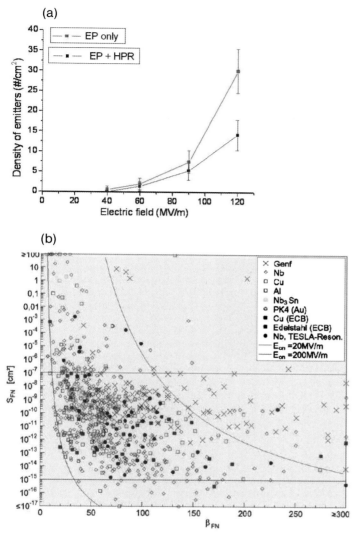

Fig. 4.13 (a) Density of field emitters located in a dc field emission scanning apparatus [233]. (b) β_{FN} and S ($=A_{FN}$) parameters for individual field emitters [246] (courtesy of Univ. Wuppertal).

unphysical. Similarly geometric field enhancement of particles falls in the 1–10 range as compared to β_{FN} values between 10 and >300.

Surface analysis studies (SEM, EDX, and Auger) reveal that dc field emission sites are usually microparticle contaminants, and sometimes surface scratches with sharp features. Figure 4.14(a) shows an example of Al particle found from dc studies [233].

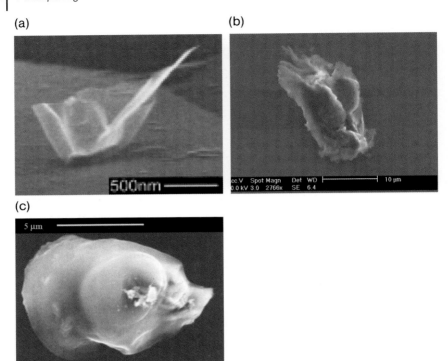

Fig. 4.14 (a) Sub-micron field emitting particle found on sample prepared with 9-cell cavity [247] (courtesy of U. Wuppertal). (b) Al particle found at a field emission site in the dc field emission scanning apparatus and subsequently analyzed with the SEM [233] (courtesy of U. Wuppertal). (c) Field emitting particle found with thermometry followed by dissection of a 1.5 GHz cavity. Carbon, oxygen, iron, chromium, and nickel were among the foreign elements detected [243]. Note the sharp features on all the particles.

Emission sites located in superconducting cavities by using thermometry are also found to be micron-size particles (Fig. 4.14 b). Hence the cleaning of rf cavities by HPR and assembly under Class 10–100 clean room conditions has become a standard practice in SRF cavity preparation, greatly reducing the occurrence of field emission to reach surface fields greater than 80 MV/m in 9-cell, meter-long accelerating structures, and surface electric fields higher than 120 MV/m in 1-cell cavities. Cavities with maximum gradient limited by field emission have been treated with HPR again to lower the field emission and increase the maximum gradient reached.

Combining the field enhancement factor of physical projections from particles (e.g., Fig. 4.14 a) is likely to yield β_{FN} of about 100, in contradiction to observed β_{FN} values in dc scanning experiments of 300 and in rf tests of 600 and higher. Therefore other factors, such as condensed gas, are likely to contribute to β_{FN}, to be discussed a bit later.

4.2.2.1 Statistical Models for Cavity Performance with Field Emission

A statistical model [248] for cavity performance successfully described the yield for single and multicell cavities limited by field emission by using a maximum emitter density (0.3/cm^2), distribution and A_{FN} distribution as below:

$$N(\beta_{FN}) \propto e^{-0.01\beta_{FN}}$$

$$N(\log A_e) \propto e^{-\left(\frac{\log A_e + 13.262}{2.175}\right)^2}$$

The β_{FN} values range from 40 to 600, and β_{FN} distribution is modeled so that the number of emitters decreases exponentially as β_{FN} rises. The log-area values are distributed according to a Gaussian with values ranging from −18 to −8. The simulation runs as follows. A cavity is split up into a number of regions (typically 20), and each region is given a random emitter density between zero and the maximum density specified. A_{FN} and β_{FN} values are distributed so the entire emitter population has the distributions given above. There is no correlation made between the FN area and β values

Both experiments and simulations showed (Fig. 4.15a) that without the HPR treatment single-cell 1.5 GHz cavities reached a 80% success yield at $E_{pk} = 25$ MV/m, whereas 5-cell cavities showed a success yield of 80% at a much lower field of $E_{pk} = 15$ MV/m. A 5-cell (1-cell) test was taken to be successful if the power deposited in field emission was less than 100 (10) W. As expected, the single cell results have a far greater yield due to the smaller surface area.

Since 2000, HPR has improved the gradient yield considerably. Figure 4.15 (b) shows that the yield [249] of 9-cell cavities (prepared by EP, HPR, and bake) reaches 80% success around $E_{pk} = 50$ MV/m, an impressive improvement over the 5-cell results without HPR. The gradient limits due to field emission are likely to continue to improve. The data of Fig. 4.15 (b) does not include even newer treatments that are developing to increase the yield at $E_{pk} = 70$ MV/m. These are ethanol rinsing [250] ultrasonic degreasing [251] and dry ice cleaning [252, 253].

The present gradient yield remains less than 20% at $E_{pk} = 70$ MV/m, and needs to improve to 80% to fulfill the needs of high gradient accelerators, such as the International Linear Collider (ILC). The statistical model simulates the desired yield by lowering the maximum emitter density from 0.3 (without HPR) to 0.1 emitters/cm^2 (with HPR) and modifying the β_{FN} distribution to change the exponent from 0.01 to 0.045, which means that there are must be far fewer emitters with high β_{FN}. The average emitter density used is 0.05 emitters/cm^2. With these improved emitter properties the model also predicts a yield of 80% at $E_{pk} = 70$ MV/m for 9-cell 1.3 GHz cavities. It remains to be seen whether new treatments, such as ethanol rinsing or ultrasonic degreasing can achieve the needed reductions in emitter density.

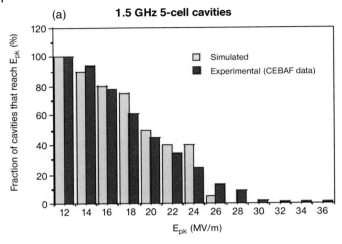

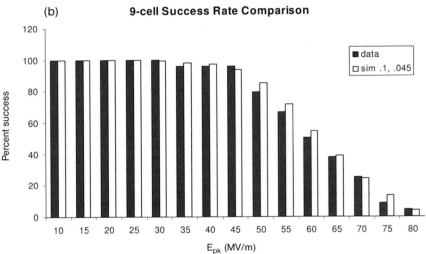

Fig. 4.15 (a) Measured and simulated yield of CEBAF 5-cell cavities prepared in the mid-1990s without HPR. (b) Percent success at E_{pk} in 24 tests on ten 9-cell DESY cavities prepared after 2000, by EP, HPR, and baking [249].

4.2.2.2 Field Enhancement Factors

Studies discussed in [1] have shown that geometric field enhancement does play a role in the value of β_{FN}. Particles with sharp features show stronger emission and voltage breakdown at lower fields, as compared to particles with more rounded surfaces [254]. Figure 4.16 compares two types of vanadium particles introduced into a dc spark gap [255]. The ones with rather sharp features suffered dc breakdown at 40 MV/m compared to the others which could withstand 100 MV/m before breakdown.

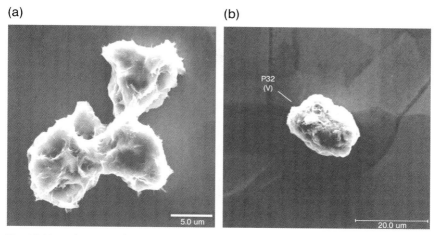

Fig. 4.16 Vanadium particles introduced into a dc spark gap. (a) Particles with sharp features breakdown at much lower fields than (b) particles without sharp features [256].

Other factors such as adsorbed gas, the interface between particle and substrate, also determine β_{FN}, as discussed in [1]. The noise in the field emission current (usually large on raising the field for the first time) and the switching of emitters also suggest nongeometric phenomena at work. The effective surface area of the emission usually does not correlate well with the area of emitting particles (typically 10^{-8} cm^2). In many cases emitting areas take on unphysical values of mm^2 to tens of cm^2, as seen from Fig. 4.13(b)!

4.2.3
Processing Field Emission

RF processing plays an important role to reach the highest accelerating fields. In the mid-1990s during the operating period of LEP-II at CERN, the average gradient of 350 MHz, 4-cell Nb–Cu rose from 6 MV/m (design) to 7.5 MV/m by rf conditioning at high power [117, 257]. Both cw and pulse conditioning with pulse length varying between 10 ms and 100 ms and duty cycle between 1 and 10% were successfully used to reduce field emission. In the 9-cell tests discussed in the previous section, 60% of the tests showed field emission onset at $E_{acc} \leq 25$ MV/m [249]. Many of these emitters had to be rf processed to reach higher fields. The matching statistical model shows that with the present surface preparation technology (EP, HPR, and mild bake), the average number that must be processed to reach $E_{pk} = 70$ MV/m in 9-cell cavities is several emitters per test.

It is therefore important to understand the physics of rf processing. Both rf and dc high voltage emitter processing studies have provided valuable information on this front. RF experiments have been carried out with 1.5 GHz [243, 258], 3 GHz [241, 259], and 6 GHz [234] cavities. Studies of high-voltage breakdown under dc conditions have also been extremely valuable in identifying common features with rf processing.

The following sections will show that rf processing of a field emitter is similar to the dc voltage breakdown. In both the cases there is an explosive event which vaporizes the emitting particle leaving behind a micron-size crater in the metal, as well as a larger (100 μm) surrounding trace of a plasma, which forms during the processing event. The experimental results which lead to this understanding are followed by detailed numerical simulations for field emission and plasma formation.

4.2.3.1 CW RF Processing

When field emission dominates a high gradient cavity, or high voltage dc gap, there frequently occurs a "conditioning event" or "processing event" as the surface electric field rises. After such an event the field emission current drops substantially, as indicated by sharp drops in rf losses and x-ray intensity. It is possible to raise the surface field further. Figure 4.17(a) shows a temperature map of a 1.5 GHz cavity with an active emitter which subsequently processes [243, 258, 260]. Figure 4.17(b) shows the drop in the temperature accompanying the drop in emission current after the conditioning event. A similar processing event was reported in [1] with a 3 GHz single-cell cavity [236]. In both cases it was possible to increase the cavity field after the processing event. On dissecting the cavities, craters with melted rims, and a surrounding dark region (usually with a starburst shape) were found during SEM examination of the suspect emitter location. The 3 GHz processing event showed a clear, dark starburst region surrounding a melted Nb crater with submicron copper residue (Fig. 4.18). A copper microparticle was the likely emitter which processed. In the 1.3 GHz case (Fig. 4.19), the dark region does not have the full starburst pattern possibly due to a long air-exposure of the surface before SEM examination, as we will discuss in more detail. Similar examples (Fig. 4.20) of processing events have been reported from a 6 GHz demountable cavity [234], where emitters processed to $E_{pk} \sim 75$ MV/m were analyzed with the SEM to find craters and starbursts.

The SEM analyses show that a processing event destroys the emitting particle leaving behind a micron-size carter in its place, sometimes with a trace of the original emitter. After such an event the field level in the cavity can be raised further. Due to their small area, such micron-size craters do not increase rf losses significantly and Q values are found to remain above 10^{10} after the processing events.

Auger analysis of single craters found in SRF cavities after single (or few) processing events almost always reveal a thin layer (1–10 nm) of foreign elements, such as copper, titanium, indium, iron, etc. [261]. These thin traces are presumably the residue left behind after the arc (during the rf or dc voltage breakdown) vaporizes the original contaminant particle responsible for field emission.

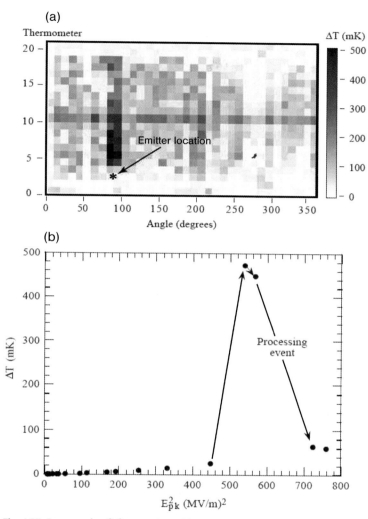

Fig. 4.17 An example of rf processing with cw power. (a) Temperature map showing longitudinal heating due to the presence of an emitter at the marked location determined by fitting the simulated and observed heating patterns. (b) Heating reduction after processing the emitter with cw rf power [243].

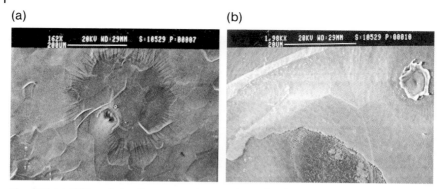

Fig. 4.18 (a) SEM analysis of a processed emitter located with thermometry in a 3 GHz single-cell cavity. (b) Expanded view shows melted Nb crater with isolated copper particles [236].

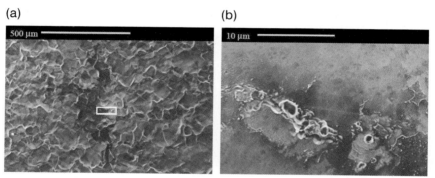

Fig. 4.19 (a) SEM examination of processed emitter site shows a dark region surrounding a crater in the boxed region. (b) The boxed region is enlarged to show crater details. EDX analysis shows only Nb in the micron-size crater [260].

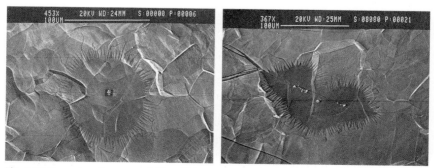

Fig. 4.20 Examples of processed ($E_{pk} \sim 75$ MV/m) emitters found in a 6 GHz demountable cavity. Note each processed site has a starburst feature surrounding central craters of melted Nb [234].

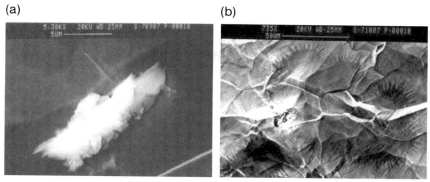

Fig. 4.21 (a) 5 μm silicon dioxide particle introduced into the peak field region of a 6 GHz mushroom cavity. (b) SEM micrograph of the region of the particle after the cavity was tested to $E_{pk} = 75$ MV/m [262].

4.2.3.2 RF Processing of a Planted Particle

A deliberately planted emitter experiment was conducted with an rf cavity. Figure 4.21 shows a 5 μm silicon dioxide particle placed in the high electric field region of a demountable 6 GHz niobium cavity, which was subsequently tested to reach 75 MV/m surface field at the particle site. After the test the cavity was taken apart and the particle site re-examined. SEM analysis (Fig. 4.21b) showed a main crater (at the particle location) surrounded by multiple starbursts. Most of the particle was vaporized during the spark since EDX analysis showed no silicon at the main crater. Auger analysis necessary to find silicon traces was not carried out.

4.2.3.3 High Pulsed Power Processing

SRF cavities are frequently processed for field emission with the available cw power, usually below 200 W. At these cw (or long pulse length) power levels, it is not always possible to process all the limiting field emitters, especially because the emission current grows exponentially with field. With the best techniques of cleaning, such as HPR, field emission often shows up in 1.3 GHz structures above $E_{acc} = 20$ MV/m ($E_{pk} = 40$ MV/m). As the Q of the cavity drops with the growing field emission current, the available cw power is insufficient to raise the electric field high enough to trigger the processing event necessary to destroy the emitter. To reach the high electric field conditions needed for continued processing, it is necessary to use higher power. But to minimize the power dissipation into liquid helium the cavity-fill time and rf pulse length have to be short. As the field emission simulations below will show, it only takes a μs or less to arrive at the conditions necessary for processing, but the peak power required to fill the cavity in 1 μs is prohibitively large (100 MW). However, with a more relaxed fill time of few hundred μs, about 1–2 MW of power is necessary to reach field levels of 100 MV/m in a 9-cell 1.3 GHz structure.

Further rf processing of field emitters can continue after cw rf processing has reached its limits. SEM examination of the processed emitter areas after HPP show the exact same starburst and crater features visible with cw rf processing at lower power. The local processing mechanism with HPP is same for both the cases. HPP merely allows access to high electric field levels and so destroys emitters that cannot be processed by cw low power rf. HPP does *not* introduce any more violent an event than cw rf processing.

The success rate of HPP is quite remarkable. Several 5-cell 1.3 GHz structures prepared by BCP but without HPR were successfully processed with 1 MW and 150 μs pulse length [238, 263]. The field emission limited gradient in each of the three 5-cell structures was raised from 10–22 MV/m before HPP to 26–28 MV/m after HPP. At these field levels the Q versus E curve was then limited by the high-field Q-slope with some residual field emission. With 1 MW the pulsed field accessible was 90 MV/m. In all cases the low-field Q value remained at 10^{10}, showing that the craters introduced during processing did not introduce significant losses, since the areas are a few μm^2. To operate field emission free at a given field level the processing field during the pulsed stage needs to reach 1.7× the desired operating field because many emitters are active. Certainly the ratio 1.7 is discouraging for an eventual goal of $E_{acc}=35$ MV/m ($E_{pk}=70$ MV/m) since the accompanying surface magnetic field will run into thermal or magnetic breakdown. But this experience was with HPP on dirty cavities (not rinsed by HPR). The typical processing field for cavities cleaned by HPR has not yet been studied, although one anticipates that the HPP process will be much more effective since the number of emitters is reduced by factors of 3 or more by HPR [233].

Many 9-cell 1.3 GHz cavities today are prepared sufficiently clean to reach $E_{acc}=30$–40 MV/m without field emission. But the spread in gradients is rather large (±25%), with the successful yield for 35 MV/m about 20% (Fig. 4.15). A significant part of this low yield is due to field emission. In addition, even with the cleanest prepared cavities there remains the risk of contaminants entering later, for example during attachment of power couplers, or installation of the cavity into the beam line. In the long run, HPP will be a useful technique to recover cavities which may become contaminated during assembly or during a mild vacuum accident in the accelerator. Such an experiment was conducted with a 9-cell 3 GHz cavity which had a best performance of $E_{acc}=18$ MV/m, without HPR [241]. A mild vacuum accident (few torr exposure) dropped E_{acc} to 15 MV/m with heavy field emission due to particle contamination. Using HPP with the available 90 kW of power it was possible to recover the original best 18 MV/m performance. Higher power HPP would be necessary to evaluate recovery at higher fields. Reduction of field emission in 9-cell cavities in cryomodules using HPP has been successfully carried out at 10 MV/m at Rossendorf [264] and at 20 MV/m at TTF [265].

The HPP technique does face a fundamental limit above $E_{pk}=100$ MV/m when the surface magnetic field approaches the rf critical field. If any region of a cavity quenches, the fill time of 100 μs is too slow in comparison to the growth rate of the quench region (few μs) [236]. As mentioned, the power demand for μs fill times becomes prohibitive, so that HPP is no longer a useful approach for emitter processing if thermal or magnetic breakdown of superconductivity takes over as the limit.

4.2.3.4 DC Voltage Breakdown Studies

Emitter processing in dc voltage breakdown experiments [256] shows remarkable similarities to rf processing events. DC experiments are conducted at 100–150 MV/m using 10–15 kV across a 100 μm gap. During the breakdown event the current briefly rises to 10–100 A and the gap voltage falls in less than 10 ns. The high current arc lasts several 100 ns during which the voltage remains low and the breakdown current tapers to zero. The vacuum pressure spikes during the breakdown event, which is usually accompanied by a brief flash of light. Residual gas analysis generally shows increased presence of $CO(N_2)$, and CO_2 during the gas burst accompanying the breakdown event. The high current (amps) measured during the dc arc also indicates that ions must be present and involved in the processing event.

Abrupt drops in the field emission current occur after a voltage breakdown event, often identified and localized by a burst of light (the spark). Figure 4.22 [256] shows the increase in surface field as well as the reduction in emission current after a spark event, until the occurrence of the next spark event. After a breakdown event, there was a significant rise in the field for the emission onset current (defined as 10 nA for these experiments.) Subsequent SEM examination of the spark location reveals micron-size craters surrounded by a 100 μm explosive-like pattern (the starburst), with many similarities to those found during rf processing. Follow-up AES examination shows thin layers of foreign elements,

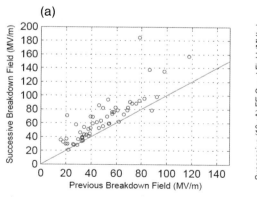
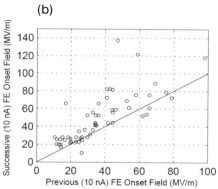

Fig. 4.22 Processing events in a dc high voltage gap. (a) Each data point shows the breakdown fields of two consecutive breakdown events at the same spot, with the earlier breakdown field on the horizontal axis, and the later on the vertical. Successful processing events are above the line $y = x$. Processing increases the breakdown field in (a) and decreases field emission in (b) [256].

residues of the original contaminant particles responsible for the strong field emission, as for rf processing events.

Any new sources of breakdown the arc may create, such as surface features at the rims of craters, are found to be active only at higher fields as evident from Fig. 4.22(b), where the second breakdown event was always at a higher field after the first breakdown event.

4.2.3.5 DC Breakdown Studies on Deliberately Planted Particles

To confirm the strong link between rf and dc processing, microparticles were deliberately injected into a dc high voltage gap to show field emission followed by voltage breakdown [256]. SEM examination of the surface after breakdown reveals micron-size craters surrounded by a 100 µm explosive-like pattern at the location of the particles introduced. Figure 4.23 shows a 5 µm vanadium particle introduced on a niobium surface which was subsequently sparked in a dc gap at about 100 MV/m. Auger analysis revealed traces of the original vanadium contaminant coating on the central crater surface [255]. Occasionally there are satellite craters in the vicinity of the main crater and inside the starburst region.

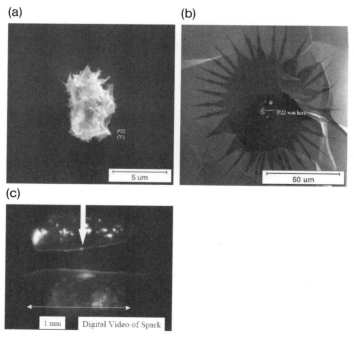

Fig. 4.23 (a) SEM picture of a V particle on a Nb surface which was sparked at 100 MV/m in a dc spark-gap. (b) SEM micrograph of the region of the particle after the dc spark. Note the crater at the location of the particle and nearby satellite craters all within a much larger starburst-shape region. (c) Flash of light captured during the dc spark of this emitter [256].

No foreign elements are generally found in satellite craters. It is suspected that satellite craters form when secondary arcs strike the clean surrounding Nb surface.

Similarly a 1 μm carbon particle introduced into a dc niobium spark gap showed a micron crater and 50 μm starburst after a spark at about 100 MV/m (Fig. 4.24).

4.2.3.6 The Nature of the Starburst

SEM and Auger analyses of the 100-micron size starburst region surrounding craters reveal interesting information most likely related to the nature of the plasma which forms during the breakdown (or rf processing) event. The starburst shape suggests the occurrence of an explosive event. The striking shape only appears in the SEM and is invisible in an optical microscope, although the micron-size craters inside are optically visible. The SEM contrast shows that the dark region inside the starburst has a much lower secondary electron emission coefficient than the surrounding region outside.

The 2D Auger map of the breakdown area due to the carbon particle in Fig. 4.24 (c) shows the starburst to be a region of missing fluorine. Fluorine is always present at a few percent concentration in the oxide layer of niobium that

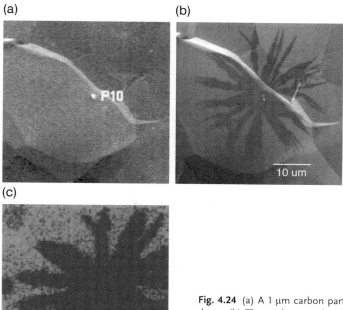

Fig. 4.24 (a) A 1 μm carbon particle before breakdown. (b) The starburst and central crater after breakdown located exactly at the C particle site as referenced to the geometry of the nearby grain boundary. (c) An Auger map of F, showing low F content (dark region) inside the starburst [256].

has been etched in hydrofluoric acid (as niobium cavities always are) [261, 266]. As we see below, explosive activity during the spark removed both the oxide and the fluorine from the surface, resulting in a cleaner region of lower secondary emission coefficient, which appeared dark in the SEM.

More thorough Auger maps showed that not only is F missing, but also C is missing. The Auger map (Fig. 4.25) of the breakdown area due to a 5 μm vanadium particle shows the starburst region to be depleted of C and F. Carbon (in the form of hydrocarbons) is normally found in one to few monolayers on almost all surfaces. Surface C is well known to raise the secondary electron emission, so that its removal explains the dark (fewer secondary electrons) region inside the starburst, visible only in the SEM. Unlike F, C is not completely absent within the starburst region, but the C concentration is noticeably reduced. Therefore Auger analyses of breakdown regions show that explosive activity within any rf or dc processing event (i.e., spark) removes both C and F to show up as a contrast in the SEM image.

Fig. 4.25 (a) A vanadium particle introduced in a dc spark gap. (b) A collection of four AES maps that show (as labeled) the concentrations of Nb (the substrate), C and V. Lighter areas represent higher concentrations. The V residue was detectable in Auger, but not in EDX. The starburst pattern in the SEM contrast is due to removal of F and C from the surface during the arc [256, 267].

The dark starburst shape and Auger analysis show that the arc created during voltage breakdown "cleans" the cathode surface in a starburst-shaped pattern by removing surface C and F. Removal of surface C is also consistent with the observation of CO and CO_2 from residual gas analysis of pressure bursts during the arc. Ion bombardment during the spark event is a likely mechanism. Recall that the high current present during the arc indicates the presence of ions.

Further experiments show that the intensity of the arc is strong enough to remove the entire oxide layer. A Nb sample anodized to build a 50 nm thick oxide layer was subjected to voltage breakdown, and subsequently exposed to air for transportation to the SEM. AES analysis showed the presence of only the thinner natural (5 nm) oxide thickness within the starburst region of the breakdown site. The intensity of the arc was strong enough to remove the entire 50 nm thick oxide layer. A later discussion on breakdown simulations estimates the ion intensity for this depth of material removal.

The starburst contrast has been reported [234] to fade away with long exposure to atmosphere presumably due to reaccumulation of carbon on the surface. When the Nb sample is exposed to air after the rf breakdown event, (i.e., for transport to the SEM), the oxide immediately reforms, but this time without any fluorine contamination, hence the SEM contrast. If the exposure is short, a thin layer of carbon from the atmosphere readheres to the newly formed oxide inside the starburst, but not enough to spoil the SEM contrast. Only after a long exposure does enough C build up.

Since Nb is usually treated with HF, starbursts can be recognized with ease in the SEM. However, surface carbon also contributes to the SEM contrast, so that starbursts should also appear on non-niobium cathodes, not usually treated with HF because all metals tend to have one or more monolayers of carbon contamination from the atmosphere. Indeed, dc breakdown experiments on copper cathodes also reveal a 50–100 µm disc-like region of carbon depletion, although without the striking starburst streamers found in Nb cathodes (Fig. 4.26). Note in some of these cases that the region surrounding the craters is lighter than the surrounding area. An exposure experiment showed that an electron beam (e.g., 10 kV, 1 µA) can turn a starburst from regular (darker than surroundings) to inverse (lighter than surroundings) [256].

Starbursts found in rf cavities are larger than dc starbursts' and larger still in low-frequency cavities. For example, while dc starbursts are generally smaller than 100 µm in diameter, starbursts in 6 GHz cavities average around 200 µm, while starbursts in 1.5 GHz cavities average about 700 µm diameter. This may be attributed largely to the higher stored energy in cavities than in the dc spark gap, and to increased stored energy with lower frequency cavities.

To summarize the experimental status, both rf and dc results show that processing field emitters via the mechanism of voltage breakdown allows higher surface electric fields by vaporizing microparticles which are the most likely sources of field emission. The reduction in emission current and the higher accessible field show that the arc during the breakdown destroys the source. Systematic processing by sequential voltage breakdown events eliminates trouble-

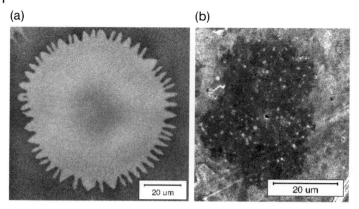

Fig. 4.26 (a) A starburst on diamond-machined copper, (b) electropolished copper. Note that (a) is an inverse starburst, lighter than the surrounding area [256].

some sources of emission and allows high fields to be reached both in dc gaps and in rf cavities.

4.2.4
Model for RF Processing and Simulations

Particle-in-cell plasma simulation codes MASK [268] developed by Science Applications International Corporation (SAIC) and OOPIC-PRO [269] have been used to explore the initiation of voltage breakdown at a strong field emitter upon the release of neutral gas atoms from the emitter and its immediate vicinity. Empirical parameters, A_{FN} and β_{FN}, make the simulated field emitters resemble those found in cavity and dc experiments.

In the simple cylindrical simulation geometry (Fig. 4.27) an ac or dc voltage is applied across a gap between two circular disks. The emitter is placed at the center of the disk. Since the simulation does not use the electrodynamic field solver, but merely the electrostatic field solver, it is more accurately denoted by ac, rather than rf.

The high current (10–100 A) observed during a dc voltage breakdown suggests the presence of a high gas density involved in the arc process. The simulation assumes the presence of a neutral atom density corresponding to melting (e.g., 3000 K) at the tip of a microparticle emitter, for example. In support of this model, a few instances of emitters showing partial melting have been captured in dc and rf tests (Fig. 4.28). The possibility of evaporated metal vapors is therefore a plausible starting source for the gas involved in the high current discharge event. Vapor from electrodes has been detected in microsecond before breakdown (by absorption of light tuned to an atomic transition of the electrode material) in experiments with a 1 mm gap [271]. This study is particularly important because it detected vapor before any significant ionization occurred.

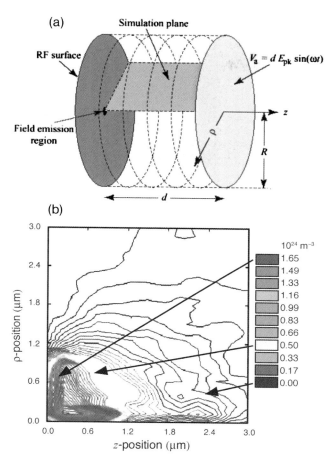

Fig. 4.27 (a) Parallel plate capacitor arrangement used to simulate field emission. (b) Density of gas evolving from the 1 μm region centered on axis at the left surface. The flux was 10^{27} m^2 s^1 and temperature of the gas was 2000 K. Figure only shows the immediate vicinity of the field emitter [243, 260].

Another example which shows that gas plays a role in triggering breakdown comes from a copper rf cavity, which did not go through the standard heat treatment for degassing. Figure 4.29 shows many crater tracks found along grain boundaries, suggesting that gas emerging from the grain boundaries played a major role in the formation of secondary arcs and craters. The correlation between the location of craters and grain boundaries disappeared with vacuum-fired copper cavities as shown in Fig. 4.30 (c).

Figure 4.27 shows the density contours of gas evolving from the emission site. A flux of 10^{27}/m^2/s neutrals is posited with a velocity distribution corresponding to a temperature of 0.25 eV (about 3000 K). The density needs to be

118 | 4 Multipacting and Field Emission

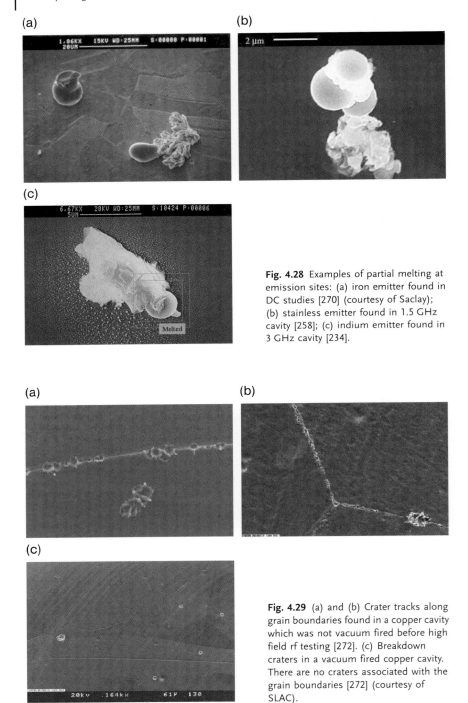

Fig. 4.28 Examples of partial melting at emission sites: (a) iron emitter found in DC studies [270] (courtesy of Saclay); (b) stainless emitter found in 1.5 GHz cavity [258]; (c) indium emitter found in 3 GHz cavity [234].

Fig. 4.29 (a) and (b) Crater tracks along grain boundaries found in a copper cavity which was not vacuum fired before high field rf testing [272]. (c) Breakdown craters in a vacuum fired copper cavity. There are no craters associated with the grain boundaries [272] (courtesy of SLAC).

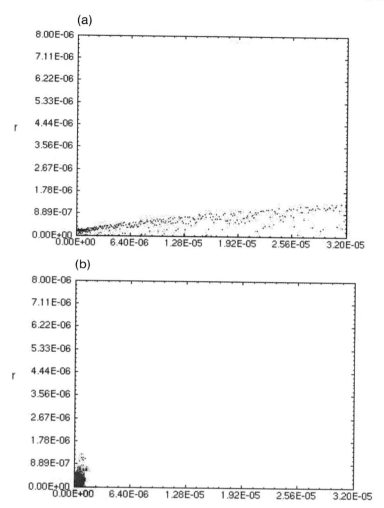

Fig. 4.30 DC field emission simulation (using OOPIC). (a) Electron beam near the emitter. (b) The ion cloud which forms and grows near the emitter [256].

sufficiently high only near the field emitter. More than several microns away from the field emitter, ionization ceases because electrons have too much energy and the ionization cross-section at those energies becomes small.

Among the likely gas candidates for triggering the arc, niobium has mass 93, a typical foreign particle (iron) has mass 56, a typical desorbed gas molecule, carbon monoxide is 28, and water has mass 18. Most simulations were therefore carried out with mass 40 atoms. The simulation is not very dependent on the details of the gas species since the ionization cross-sections fall within an order of magnitude of each other for different species. However, we will see below that light atoms (such as H) are unlikely to be involved, since these atoms move too fast to form the necessary ion cloud above the field emitter. For mass 40 neutrals the flux of $10^{27}/m^2/s$ at 3000 K produces a pressure of nearly 500 torr (more than a half atmosphere) near the emitter, and the number density of atoms reaches about $10^{24}/m^3$.

A typical dc simulation begins with field emission, at the selected field (e.g., 30 MV/m) and emitter properties (β_{FN} and A_{FN}). Neutral gas from a molten region of the emitter expands slowly (relative to electron motion) from the area around the field emitter. The electron beam travels almost perpendicular to the cathode, widening slightly because of transverse velocity spread and space-charge repulsion (Fig. 4.30). Where neutrals collide with electrons at the 15 V equipotential, electrons have gained sufficient energy to ionize the gas atoms, creating positive ions that slowly move toward the cathode, where they are absorbed, while the electrons continue to move away much faster from the cathode. Because ions have low velocity compared to electrons, a cloud with net positive charge builds up in front of the cathode (Fig. 4.30b). The ion cloud increases the surface electric field at the emitter, which increases the electron current. This is the onset of the processing event. More electrons create more ions thus further enhancing the field and leading to a runaway current in a short time.

Due to the resistivity of the field emitter, the emission current causes Joule losses. Since the emission current increases rapidly as the ion cloud develops, the time averaged Joule losses also increase. For the simulation under discussion the dissipated power increases more than 2×10^4 due to the presence of the gas which raises the emitter temperature resulting in the efflux of larger quantities of neutrals. Hence the ionization process is part of a positive feedback mechanism. The simulation shows that an explosion will take place within a few rf cycles once triggered. At lower gas densities, the time until an explosion takes place is longer, but still of the order of several rf cycles.

Simulations in ac closely resemble dc except at the beginning where the ac simulation differs because electrons emerge from the field emitter only in bursts when the field nears its maximum. When neutrals evaporating from the cathode travel far enough away to collide with emitted electrons that have gained enough energy to ionize the atoms, ions are created in the same bursts when electrons are emitted. Figure 4.31 shows the position and energy of electrons and ions after 1.25 rf periods from the MASK simulation. During the rest of the rf period when the field is low or negative and there is no field emission, the ions disperse; self-repulsion pushes the ion cloud apart. As the ion cloud grows in subsequent rf periods, the ac simulation begins to resemble the dc simulation. The electric field due to the ion cloud overwhelms the applied rf field (Fig. 4.32a). As in dc simulations the ion cloud continues to grow and at some point the cloud starts to expand rapidly to form the ion cloud explosion.

In about 2.5 rf periods the positively charged ion cloud (Fig. 4.32a) enhances the local electric field at the emitter from 30 MV/m to more than 100 MV/m. Once injected, the neutrals move without collisions in a straight line with their

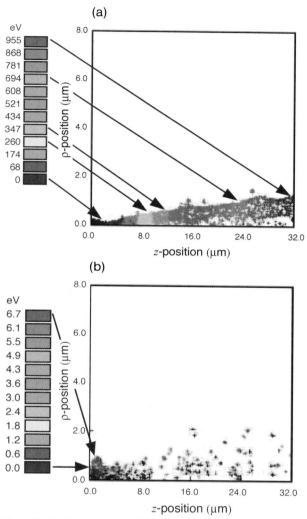

Fig. 4.31. (a) Position plots of the electrons 1 1/4 rf periods into the simulation. Note that the aspect ratio is not 1:1. (b) Position plots of the ions after 1 1/4 rf periods. Each cross represents 10 ions. Electron and ion energies are indicated [243, 260].

initial velocity. Breakdown occurs so quickly that the neutrals hardly get more than a micron or two away from the cathode.

Figure 4.32(b) shows the development of the ion position and energy distributions in the vicinity of the emitter after 2.5 rf periods. MASK ran into a current instability 2.5 rf periods after the current began, making further tracking impossible. Another simulation program, OOPIC, reproduced the results from MASK and carried the simulation to the continued development of the arc.

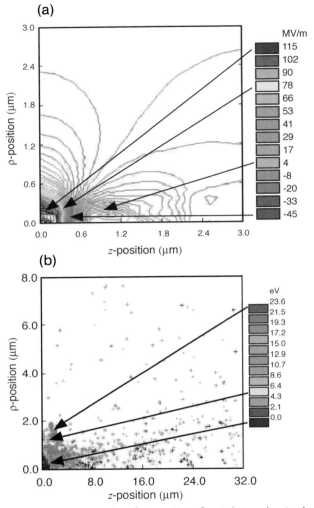

Fig. 4.32 (a) Contour plot of $-E_z$ at 21/4 rf periods into the simulation showing the field enhancement due to the plasma near the rf surface. (b) Position plots of the ions after two rf periods. Each cross represents 10 ions. The ion energy is indicated [243, 260].

As the current continues to rise due to the enhanced electric field (Fig. 4.33), the ion cloud expands further. Presumably the ion bombardment of the surface produces more neutrals to increase the gas density near the emitter, but the simulations do not take this into account.

The simulations show other interesting features of the developing breakdown event. Even though the ion cloud traps some electrons that lose energy through collisions, most high mobility electrons leave the cloud making it positively

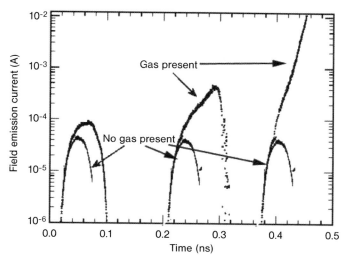

Fig. 4.33 Emitted electron current versus time for a simulation including a neutral gas and the same simulation with no neutrals present. In just over two rf cycles the peak electron current rises from 44 µA to over 10 mA. The fraction of each rf cycle during which field emission is active also increases, because the ions tend to provide a dc bias to the applied rf field. Hence the average current rises even faster than the peak current [243, 260].

charged in the center. The net positive charge repels ions, expanding the cloud of ions which will eventually blow up to leave behind the starburst. The growing ion cloud scatters electrons parallel to the cathode, until they leave the ion cloud and the applied electric field bends them back away from the cathode (Fig. 4.34). As the electron current increases from 100 µA to a few mA (Fig. 4.33), the ratio of the ion cloud field to the applied field increases, and the beam bends from almost perpendicular to almost parallel with respect to the cathode. As the electron current continues to increase through another order of magnitude, the ion cloud, the surface field, and the field emission all increase exponentially. Within a few picoseconds (dc case), with electron current around 100 mA, the ion cloud expands explosively, yet remains nearly spherical (Fig. 4.34b). Before the ion cloud expands 10 µm from the cathode, the electron current grows to 1 A. From the onset of field emission it takes about 300 ps for the current to reach 1 A. The current continues to grow rapidly to about 10 A. Beyond this stage, the ion cloud expands past the boundaries of the simulations and the numerical results are no longer tenable. The simulation is not able to continue to the formation of the starburst.

These simulations demonstrate how field emission plus neutral vapor emission can lead very quickly to voltage breakdown (with currents of many amps). During the period of rapid ion cloud expansion, the radius of the ion cloud increases from about 2 µm to 8 µm in about 25 ps. The close similarity of the dc

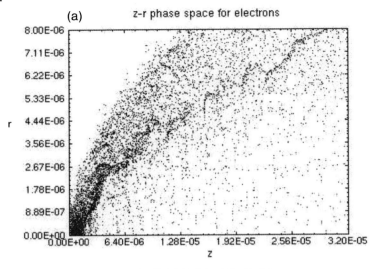

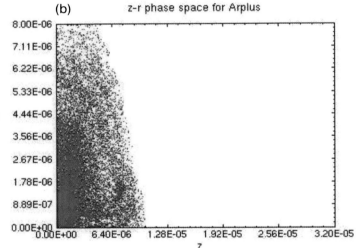

Fig. 4.34 (a) Position plots of electrons near the end of the OOPIC simulation. The aspect ratio in this picture is not 1:1, so the electron beam leaves the cathode at an angle closer to perpendicular than it appears. (b) As the number of ions increases, the rate of ionization increases; quite suddenly, the charge in the ion cloud becomes far too large, and self-repulsion expands the ion cloud, resulting in an explosive condition [256].

and ac simulation helps to explain the striking similarity between the observed features of rf and dc craters and starbursts, as described in the above sections.

It would be helpful to predict which emitters, as characterized by their FN properties are prone to breakdown, and at what field level or what total field emission current is necessary for triggering breakdown. A simple analytic estimate [256] shows that breakdown conditions depend on the *product* of the total current and the local gas density. The critical product necessary to initiate breakdown at a field emitter is given by

$$n_\beta I \geq \frac{3\pi\varepsilon_0 E}{\sigma\left[2 + \frac{E_0}{\beta E}\right]} \sqrt{\frac{eEd}{2m_i}}$$

In this equation, E is the electric field, σ is an average cross-section for the ionization of gas by electron impact, $E_0 = 54.6$ GV/m is a parameter from FN emission, d is the distance within which ionization occurs, and m_i is the mass of the ion. For example, to reach the conditions for breakdown at $E = 30$ MV/m, with $\sigma = 3\text{Å}^2$, $d = 4$ µm, $\beta = 250$ and mass 40 ions:

$$nI \geq 1 \times 10^{20} \text{A m}^{-3}$$

If $I = 100$ A, then breakdown occurs for $n > 1 \times 10^{24}/\text{m}^3$.

Thus the analytic estimate of the critical product agrees with computer simulations. But the gas density depends on a number of conditions such as the temperature of the emitting tip and the desorption rate of gas in the vicinity of the emitter, making the desired prediction about which emitters are likely to process very difficult.

Such analytic estimates of the conditions necessary for breakdown are nevertheless helpful. For example, the analytical estimate provides an attractive explanation of "helium processing" of superconducting cavities by the enhancement of gas density at a field-enhancing microprotrusion due to the dielectric polarization of helium.

4.2.5
Model for Helium Processing

Field emission in superconducting cavities can sometimes be reduced by introducing a low density of helium gas in the cavity after the available cw power is found insufficient to reach fields necessary for rf processing. Typically the cavity is filled with about 1 mtorr of helium, as measured at room temperature. A higher pressure would risk rf breakdown of the gas. There are several likely mechanisms for He processing. In cases where field emission decreases after many hours, processing could take place by removal of emitter material via sputtering [240]. In cases where the field emission decreases within a few minutes after introduction of He gas, processing likely takes place by removal of condensed gas. As discussed in [1] condensed gas is responsible for enhanced

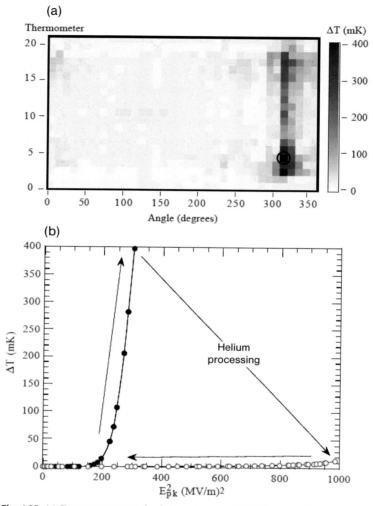

Fig. 4.35 (a) Temperature map for field emission at 17 MV/m in a 1.5 GHz cavity prior to helium processing of the emitter at longitude 310°.
(b) Temperature signal recorded at the circled site [Jens thesis] [243, 258].

field emission [273, 274]. Helium gas can also trigger a spark discharge as in rf processing. In one example, helium processing was attempted for a 1.5 GHz cavity. The temperature maps of Fig. 4.35 show emitter processing upon the introduction of helium gas. After locating the emitter by comparing the observed and predicted temperature maps the cavity was dissected and the processed emitter was analyzed with SEM and EDX (Fig. 4.36).

In a model for He processing, the spatially varying electric field in the vicinity of a field emitter exerts a net force on the dipole moment of the helium atom

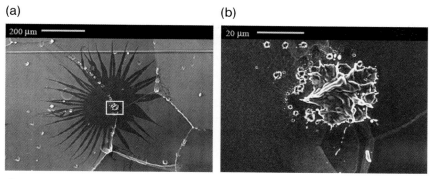

Fig. 4.36 SEM analysis of helium processed site. No foreign materials were detected at the craters by EDX analysis. (a) View of crater and startburst. (b) Enlarged view of crater area.

because the positive charge sees a slightly different field than the negative charge. The electric force on the induced dipole attracts the dipole toward increasing field magnitude. Therefore, the enhanced electric field at a protrusion pulls helium atoms toward the protrusion, creating a local enhancement of gas density given by [256]

$$\frac{n_\beta}{n_0} = \exp\left[\frac{\beta^2 - 1}{2}\frac{aE^2}{kT}\right]$$

With $\beta_{FN} = 100$ and $E = 50$ MV/m, for example, the density enhancement for helium at 2 K is 2×10^4, resulting in sufficient pressure (~2 torr) to trigger local breakdown. With a helium gas pressure of 1 mtorr at room temperature, the number density of helium atoms at 2 K is $4 \times 10^{20}/\text{m}^3$; the enhanced density is $9 \times 10^{24}/\text{m}^3$, bringing the field emission current necessary to initiate breakdown at $E = 50$ MV/m to 100 µA.

4.2.6
Summary for Field Emission

To summarize the extensive topic of field emission and processing, field emitters are predominantly microparticles that can be removed by thorough rinsing with techniques such as HPR. Work continues on better cleaning techniques, such as ethanol rinsing, ultrasonic degreasing, and dry ice cleaning. Emitters left behind or reintroduced during subsequent stages of cavity preparation can be removed if re-HPR is still possible. Otherwise rf processing is the best approach. RF processing is partially successful during the cw, low power testing stage. Further rf processing can be continued using pulsed high rf power, provided thermal or magnetic breakdown is not reached in the processing stage.

There has been major progress in understanding the physics of rf processing through many cavity and dc experiments, supported by simulations. RF processing is essentially voltage breakdown which develops when gas evolved near the emission site becomes ionized by the field emission current. Possible sources for the gas are the melted regions of the microparticle or the adsorbed monolayers on nearby surfaces. A positively charged ion cloud builds above the site which enhances the field and increases the field emission current. Once triggered the process grows in a few rf periods to a microexplosion (spark) which vaporizes the emitting particle and eliminates the field emission, leaving behind a small crater surrounded by a starburst, which is a remnant of the expanding ion cloud.

5
High-Field Q-Slope and Quench Field

5.1
RF Measurements and Temperature Maps

5.1.1
General Description of Q-slopes

The excitation curve (Q versus E_{acc}) of a superconducting bulk niobium cavity in the GHz range generally shows three distinct regions (Fig. 5.1). An increase of Q is often seen at E_{acc} below 5 MV/m, customarily referred to in the field as the low-field Q-slope. It is usually followed by slow Q degradation, up to 20–30 MV/m, called the medium-field Q-slope. The low-field and medium-field Q-slopes were discussed in Chapter 3. At the highest fields, the quality factor starts to decrease rapidly, even in the absence of field emission. This region is called the high-field Q-drop or high-field Q-slope. The onset peak surface magnetic field for the Q-drop falls between 80 and 100 mT (E_{acc} = 20 and 25 MV/m). The signature property of the high-field Q-slope is its improvement with mild baking (100–120 °C for 48 h). Finally, the performance of the cavity is limited by a quench or breakdown of superconductivity.

The plan for this chapter is to first review the many interesting series of cavity results that suggest possible mechanisms involved in the high-field Q-slope, which is generally present for cavities prepared by either of the two popular treatments, buffered chemical polishing (BCP) or by electropolishing (EP). The high-field Q-drop is an area of intense ongoing R&D, so that a review is important to capture the state of our knowledge even though there is no general consensus on a full explanation. After a review of cavity results and a mention of the suggested mechanisms, we turn to a more detailed discussion of various models that are still in play, and evaluate the strengths and weaknesses of these models. Finally, we present surface measurements with a variety of analytical tools, such as XPS, SIMS, and EBSD, to discuss the relevance of surface results on the prevailing models. The complete understanding of the high-field Q-drop remains one of the outstanding issues of rf superconductivity.

In the last section, we present and discuss new information about quench due to defects.

RF Superconductivity: Science, Technology, and Applications. Hasan Padamsee
Copyright © 2009 WILEY-VCH Verlag GmbH & Co. KGaA, Weinheim
ISBN: 978-3-527-40572-5

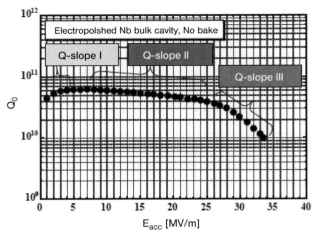

Fig. 5.1 Typical excitation curve of a nonbaked cavity showing low-, medium-, and high-field Q-slope regions [275] (courtesy of KEK).

5.1.2
Magnetic Field Effect

Two kinds of experimental results show that the rf magnetic field is mainly responsible for the high-field Q-drop, rather than the rf electric field. First is a direct measurement. When Q falls at high field, temperature maps of the cavity show heating in large segments of the high magnetic field region. Figure 5.2 shows the Q versus E_{pk} for a single-cell cavity prepared by BCP with a strong Q-drop in the absence of field emission [276]. Temperature maps acquired at field levels in the strong Q-drop regime generally show strong heating in large regions of the magnetic field [277] with significant spatial nonuniformity. The individual thermometer responses (ΔT versus E or B) show power laws with exponents of 10 to 25, as we will discuss in more detail. Similar temperature map results were reported for cavities prepared by EP [278] showing predominant heating in the magnetic field region as well as spatial nonuniformity of the heating.

The strong response of thermometers in the magnetic field regions clearly shows that the high-field Q-drop is a magnetic effect. Figure 5.3(a) shows the response (ΔT versus B) for several individual thermometers with rising field on a log–log plot [163]. For thermometers in the high magnetic field region there is sharp increase in the slope above an onset field (~100 mT), which marks the beginning of the high-field Q-drop. For the same cavity test (Fig. 5.3b), thermometers in the region where the electric field is dominant exhibit no sharp change in the slope, as compared to thermometers in the magnetic field region. The exponent in the power law for temperature rise is 10–25 in the magnetic field region and 2–3 in the electric field region. The low exponent in the electric

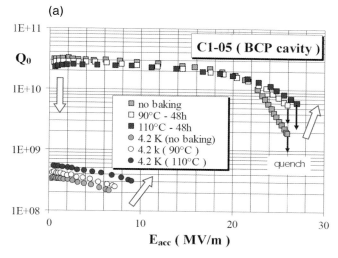

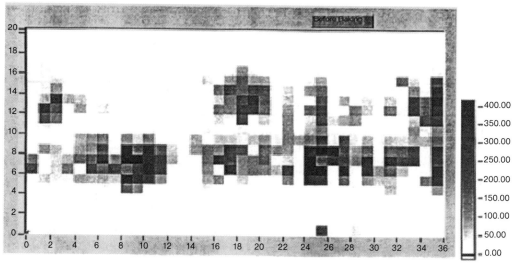

Fig. 5.2 (a) Q versus E_{pk} for a single-cell 1.3 GHz cavity prepared by BCP [276]. There is a slight improvement in the Q-drop after baking (courtesy of Saclay). (b) Temperature map before baking at the highest field (E_{pk}=42 MV/m) shows 400 mK maximum temperature excursions in the high magnetic field region of the cavity due to the high field Q-drop [277]. Note the spatial nonuniformity of the heating. Several but not all regions of the cavity showed a decrease in heating after baking (not shown).

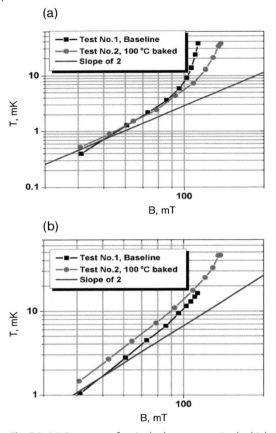

Fig. 5.3 (a) Response of a single thermometer in the high magnetic field region shows a strong Q-slope onset at about 90 mT, which shifts to 110 mT after baking. (b) Response of a single thermometer in the high electric field region shows no strong temperature rise at 90 mT before baking [163].

field region is commensurate with heating due to the thermal feedback effect discussed in Chapter 3 in connection with the medium-field Q-slope.

Another convincing result for the magnetic field dominance comes from a comparison of the high-field Q-drop in TM and TE cavities [279]. If the electric field was the primary cause of the high-field Q-drop, a cavity operating in the TE mode would show no Q-drop, since there is no surface electric field in a TE mode. A single-cell cavity measured in both TE_{011} and TM_{010} modes showed a high-field Q-drop in both modes. The surface preparation used was BCP. Figure 5.4 shows the results for both modes after postpurification when the cavity was able to reach the highest field to exhibit the high-field Q-drop in both modes. The postpurification procedure raised the quench limit by increasing the thermal conductivity, and the accompanying Nb RRR increased from 320 to 725; al-

Heat treatment: 1250 °C, 12 h ramp down to 1000 °C in 20 h. RRR improved from 320 to 725 measured on sample

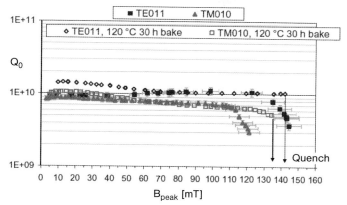

Fig. 5.4 Q versus E curves for the TM (1.5 GHz) and TE (2.8 GHz) modes of the same cavity showing high-field Q-slope in both modes [279] (courtesy of JLab).

lowing high enough fields to exhibit the strong Q-drop. The observed high-field Q-drop in the TE mode can of course be attributed to the surface magnetic field of the TE mode. Baking at 120 °C for 48 h reduced the high-field Q-drop in both modes. Hence both modes showed the signature property of the high-field Q-drop indicating that the mechanism of the high-field Q-drop is likely the same in both TE and TM modes, i.e., it originates from the magnetic field. However, the onset magnetic field of the Q-drop appears higher in the TE mode. A possible reason is a frequency dependence of the onset field [280], which will be discussed further in Section 5.2.3. The TE mode frequency is 2.8 GHz compared to the TM mode frequency of 1.5 GHz.

Based on strong heating in the high magnetic field regions evident from temperature maps, and the presence of the high-field Q-slope in the TE mode, models which attribute the Q-slope to electric field effects, such as the interface tunneling exchange (ITE) model [281], can therefore be discounted. In this model, the oxide layer provides localized states. Electrons tunnel from below the superconducting gap into these states where rf absorption takes place. Because of tunneling the losses increase exponentially with the local electric field. The ITE model will therefore not be discussed further.

5.1.3
Spatial Nonuniformity of Losses

For a cavity operating in the high-field Q-drop regime, there is a large spatial inhomogeneity in the heating, as Fig. 5.5 shows. One may be tempted to conclude that only a few "hot spots" show the high-field Q-drop [153]. However on exam-

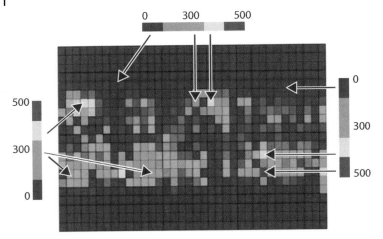

Fig. 5.5 Temperature map at $E_{pk} = 47$ MV/m on a 1.5 GHz cavity prepared by BCP before baking. Note the spatial nonuniformity of the heating as in Fig. 5.1 [282].

ining the individual thermometer responses (log ΔT versus log B), thermometers in the "cooler" regions also show a strong Q-slope with similar onset field to the "hotter" spots, but with a lower power law (Fig. 5.6 a). Every spot in the high magnetic field region displays a high-field Q-slope, but with different steepness. The high-field Q-slope is present everywhere. In the highest magnetic field region, the exponent of B in the power law (for log ΔT versus log B) ranges between 10 and 25 (Fig. 5.6 b), and reflects the changes in B^2. It is also interesting that the onset field of the Q-slope is similar (about 80–90 mT) for the hot and cool spots. In general, the heating at the weld seam is found to be slightly less than at the neighboring equatorial regions. This is partly due to the slight dip in the magnetic field at the exact equator, and partly due to differences in material properties of the weld seam, such as the larger grains. As we saw in Chapter 3, large grains give rise to a phonon peak in the thermal conductivity, and lower surface temperatures.

5.1.4
Comparison of Q-Slopes: Electropolishing and Buffered Chemical Polishing

Over the last few years EP has replaced BCP as the best surface preparation technique to reach the highest accelerating fields for fine grain Nb cavities. There are two major differences between BCP- and EP-prepared cavities. Firstly, EP cavities have a much smoother surface (macroroughness <0.5 µm), as we will discuss in more detail when we compare surface roughness. Secondly, the improvement in the high-field Q-drop after mild baking is far greater for EP cavities. Cavities prepared by both treatments show comparable features in the high-field Q-drops before baking. Figure 5.7 (a) shows several cavities tested

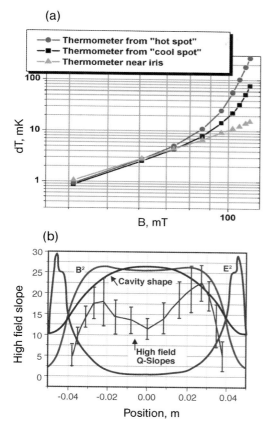

Fig. 5.6 (a) Response of individual thermometers from different regions showing strong heating, weak heating, and no heating in the electric field iris region. (b) High-field Q-slope distribution. The strength of the Q-slope is represented by the exponent in the power law (log ΔT versus log B) for the individual thermometer responses. The bars represent the spreads in the exponents over the 360° azimuth for each location of the cavity profile. These are not error bars. For reference, the figure shows the E^2 and B^2 distributions over the cavity profile [163].

with EP and BCP [283]. Temperature maps for EP cavities (e.g., Fig. 5.7 b) show that the predominant heating causing the Q-slope also stems from the magnetic field region [278], as with BCP cavities. Again the heating shows spatial inhomogeneities, but detailed examination of the individual thermometer response in "hot" and "cold" regions in EP cavities shows a strong rise in losses everywhere past the onset field, as with BCP cavities [284].

A careful comparison of data on a large number of tests shows that the Q-drop onsets and final fields reached for EP cavities are *slightly* (10%) higher than BCP cavities before baking. Figure 5.8 shows the distribution for maximum E_{acc} and E_{acc} at 10 W dissipation for a series of many tests on single-cell

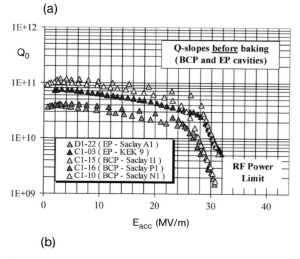
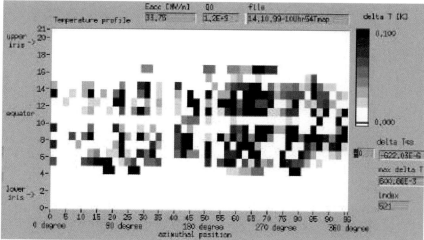

Fig. 5.7 (a) Selection of Q versus E for EP and BCP cavities showing the high-field Q-drop present in both [283] (courtesy of Saclay). (b) Temperature map at $E_{acc}=33.75$ MV/m for a single-cell cavity prepared by EP. Note the dominant heating in the magnetic field region, and the spatial nonhomogeneity, the same as for BCP cavities [278] (courtesy of DESY).

1300 MHz cavities at DESY [285]. The maximum E_{acc} at $Q=10^{10}$ shows that EP cavities are slightly superior than BCP cavities. On the average, EP cavities also reach slightly higher ultimate fields at a power loss of 10 W. This quantity can be viewed as a measure of the steepness of the Q-slope.

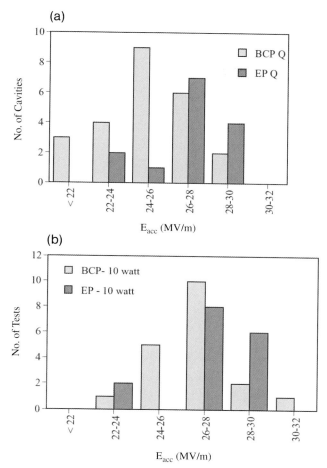

Fig. 5.8 (a) Distribution of maximum fields reached for many BCP and EP cavities, both before bake. (b) Distribution of field values at 10 W dissipation for BCP and EP cavities. Both distributions show slightly higher performance for EP over BCP. (Note: The information from Figs. 2 and 7 from [285] were combined to compare BCP and EP cavities before bake).

Direct comparison of Q versus E curves also shows the slight superiority for EP cavities. Figure 5.9 compares the Q versus E curves of many BCP and EP 1-cell cavities tested at DESY after EP at CERN [286]. Consistent with the histograms of Fig. 5.8 the Q-drop onset of BCP cavities is $E_{acc} \sim 20$–22 MV/m as compared to 20–28 MV/m for EP cavities. Figure 5.10(a) shows the same effect for 9-cell cavities tested at DESY before baking. 9-cell EP cavities also reach a slightly higher field than the BCP cavities [287].

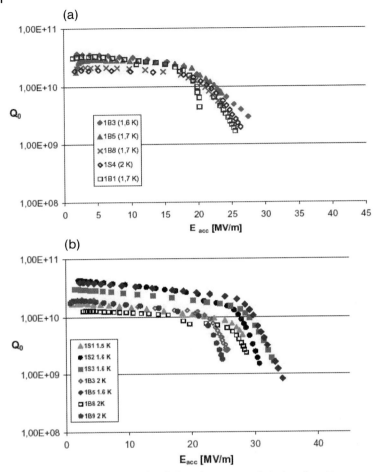

Fig. 5.9 Q versus E curves before baking for a series of single-cell cavities prepared by (a) BCP and (b) EP [286] (courtesy of DESY).

5.1.5
The Mild Baking Benefit

One of the main features of the high-field Q-drop is the surprising beneficial effect of mild baking [276, 288]. Baking temperatures between 80 and 120 °C and baking times of 40–50 h improve the high-field Q-slope for both BCP- and EP-treated cavities, but the effect on EP cavities is far greater [286, 289]. Baking generally eliminates the high-field drop of EP cavities leaving only the medium-field Q-slope. Significantly higher fields are consistently reached for EP-baked cavities, whereas BCP-baked cavities continue to show a Q-drop, or reach a quench limit without an increase in the maximum gradient. Understanding the

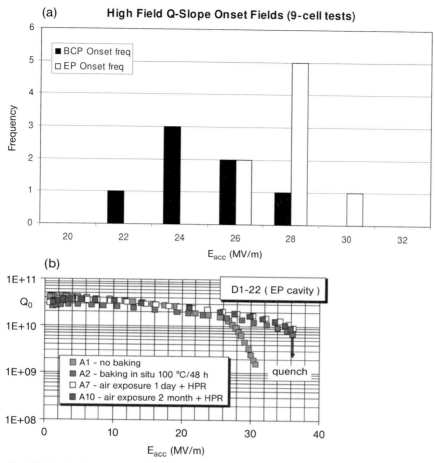

Fig. 5.10 (a) 9-cell test results for BCP and EP without baking [287]. (b) Q versus E curves showing the preservation of the baking benefit on exposure to air and water [290]. This result clearly shows that the baking benefit does *not* come from removal of surface adsorbed water or hydrocarbons (courtesy of Saclay).

baking benefit is the key to understanding the mechanism of the high-field Q-drop. The successful explanation will also need to account for the difference in response to baking of BCP and EP treatments.

Since baking has such a beneficial effect on the Q-drop it is immediately tempting to attribute the Q-drop to the surface contamination layer of adsorbed water and hydrocarbons that normally resides on the surface from the wet treatment, and exposure to (dust-free) air during drying. However, tests at several laboratories show that after the Q-drop is removed by baking, it does not return upon subsequent exposures to clean air or even high pressure rinsing with high purity water. See for example results in Fig. 5.10(b) [290]. Indeed the benefit of

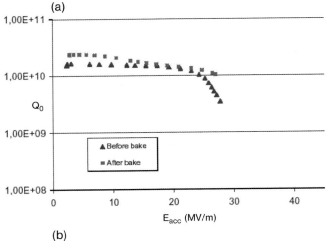
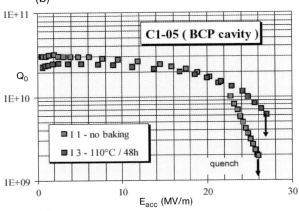
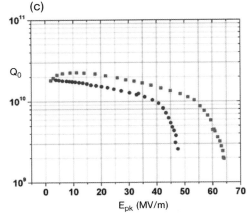

Fig. 5.11 Q versus E curves for BCP-treated cavities before and after baking. (a) From DESY [286]: the magnitude of the high-field Q-drop improvement is masked by an early quench. (b) From Saclay [276]: baking gives a slight improvement but the Q-slope remains. (c) From Cornell [163, 284, 291]: a 1 mm grain size cavity shows substantial Q-drop improvement due to baking, but the Q-drop remains.

baking on the Q-slope remained intact while the cavity sat for four years on the shelf. Therefore, adsorbed water and adsorbed hydrocarbons (from air) are not responsible for the high-field Q-drop.

5.1.5.1 Baking BCP Cavities

There is a wide range of baking benefits with BCP-baked cavities as the three panels of Fig. 5.11 show [276, 284, 286]. We will discuss in Section 5.2 that the large variation possibly stems from a number of sources: degree of surface

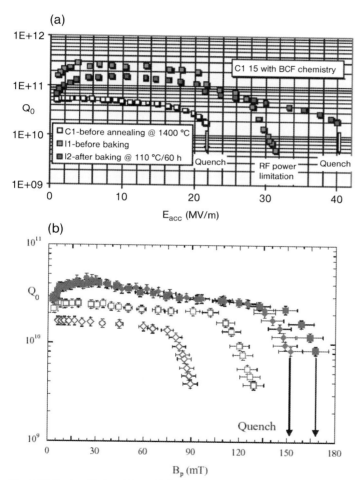

Fig. 5.12 Postpurified cavities with high purity, high RRR Nb, and mm size grains show large improvements in Q-slope due to baking. (a) Example from Saclay [292]. (b) Example for a 2-cell cavity from JLab [143]. The diamond symbols are for the 0-mode, and the square symbols are for the π-mode. Open symbols are for before baking and solid symbols are for after baking.

roughness (5–10 µm), niobium purity (RRR), and grain size (or grain boundary density). As a general rule for baked BCP cavities, a high-field Q-drop remains present after baking, but shows a higher onset field than before baking. Some BCP cavities are limited by quench so that the full Q-slope improvement due to baking is not realized.

Cavities with large grains show Q-drop onsets at higher fields as well as large improvements with baking. There are several examples. A single-cell cavity fabricated from Russian Nb with RRR=500 and grain size of 1 mm shows (Fig. 5.11 c) a large improvement with baking. Figure 5.12 (a) [292] shows the results for a cavity postpurified at 1350 °C with titanium, which doubled the RRR and enlarged the grain size to several mm. The postpurification procedure also increased the quench field. After baking there was a substantial reduction of the high-field Q-drop corresponding to a large increase in the onset field. Similarly, a postpurified 2-cell cavity shows a substantial baking improvement, presumably due to the larger grains (Fig. 5.12 b).

Along with the Q versus E performance curves, the detailed thermometry response (log ΔT versus log H_{pk}) shows that not only does the onset field of the Q-slope increase after baking a BCP-treated cavity but also there is a substantial reduction in the exponent of the power law. The comparison of Fig. 5.13 [163] shows that the exponent for field dependence drops from roughly B^{15-20} to B^{5-10}.

5.1.5.2 Baking EP Cavities

Mild baking yields significantly larger benefits for EP cavities with smoother surfaces (macroroughness <0.5 µm) than for BCP cavities (macroroughness >2 µm). This result has been uniformly established by many laboratories. Figure

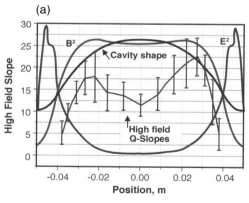
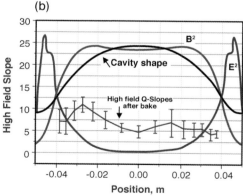

Fig. 5.13 High-field Q-slope distributions (a) before and (b) after baking. The strength of the Q-drop is represented here by the exponent in the power law (log ΔT versus log B) for the individual thermometer responses. The bars represent the spreads in the exponents over the 360° azimuth for each location of the cavity profile. For reference, the figure shows the E^2 and B^2 distributions over the cavity profile [163].

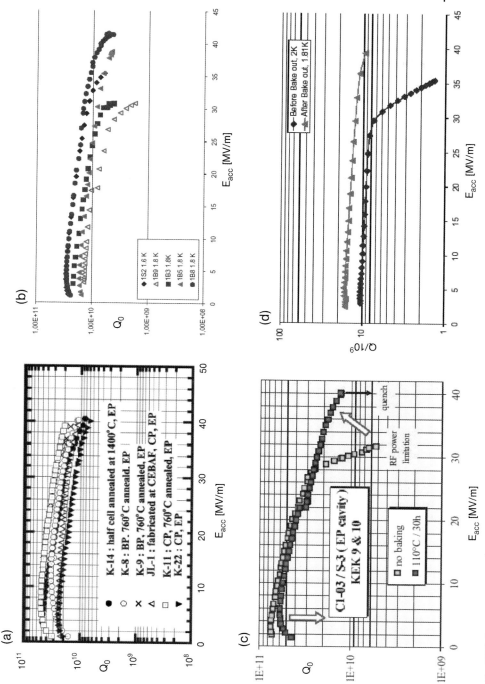

Fig. 5.14 Q versus E curves before and after baking for single-cell cavities prepared by (a) KEK [52], (b) DESY/CERN [286], (c) Saclay [290], and (d) INFN/JLab [293].

(a)

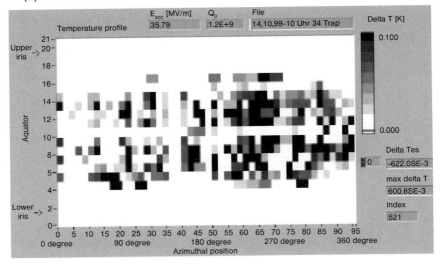

(b)

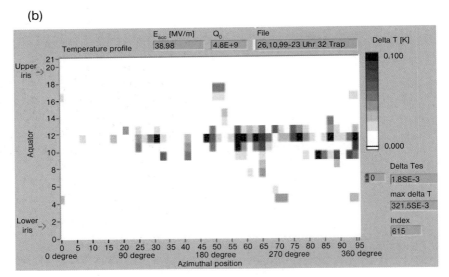

Fig. 5.15 Temperature maps for EP cavities at (a) $E_{acc} = 34$ MV/m, $Q = 1.2 \times 10^9$ before baking and (b) 39.9 MV/m, $Q = 4.2 \times 10^9$ after baking [278] (courtesy of DESY).

5.14 shows results for EP-baked cavities tested by KEK [52], DESY/CERN [286] collaboration, Saclay [290], and INFN/JLab [293]. Single-cell EP-baked cavities consistently reach $E_{acc} = 30$–42 MV/m. After baking, most EP cavities show no high-field Q-drop out to the maximum field. The remaining Q-slope can be viewed as the medium-field Q-slope, discussed in Chapter 3. Temperature maps

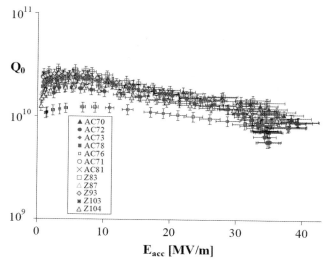

Fig. 5.16 Q versus E curves for many 9-cell cavities prepared by EP followed by baking at 120 °C for 48 h [294] (courtesy of DESY).

before and after baking EP cavities (Fig. 5.15) show substantial reduction in local temperature rise after baking [278]. Temperature maps of baked BCP cavities also show a reduction in heating [284]. In the EP example of Fig. 5.15, the temperature maps after baking are shown at higher fields.

The elimination of the Q-drop in 1-cell cavities has been successfully reproduced with 9-cell structures. Many 9-cell, EP-baked cavities have been prepared and tested to manifest relatively flat Q curves out to 40 MV/m as shown in Fig. 5.16 [294]. This discovery will be of major benefit to accelerator applications demanding the highest possible gradients.

5.2 Candidate Mechanisms for the High-Field Q-Slope

5.2.1 Thermal Feedback

In Chapter 3, we discussed the role of the thermal feedback effect due to BCS and nonlinear BCS resistance on the medium-field Q-slope. These results showed that for high-frequency (>2.5 GHz) cavities, thermal feedback with standard BCS resistance leads to a thermal instability at high fields which could be loosely interpreted as a high-field Q-drop. Upon inclusion of the nonlinear BCS resistance, the medium-field Q-slope becomes strong enough to resemble a high-field Q-slope, even at low frequency (e.g., 1.3 GHz). However, explanations for the high-field Q-drop based on thermal feedback turnout not to be valid. The main reason is that the thermal feedback effect also predicts a clear bath temperature dependence of the Q-slope at high field (Fig. 3.16a). Experiments show, however, that bath-temperature dependence and therefore purely thermal feedback effects can be ruled out as a possible explanation for the high-field Q-drop.

Two separate experiments show the high-field Q-drop remains unaffected by the changes in bath temperature from 1.6 to 2.17 K in superfluid He. Figure 5.17 shows the independence of the high-field Q-slope for cavities prepared by EP

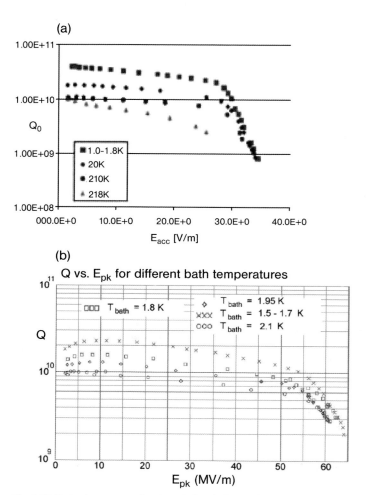

Fig. 5.17 Excitation curves of cavities before bakeout showing the Q-drop at highest fields to be independent of helium bath temperature below the lamda point (courtesy of DESY). (a) Cavities prepared by EP [286]; (b) cavity prepared by BCP [295].

[286] and by BCP [295]. The behavior of the high-field Q-slope appears the same over 20% of the highest field values, independent of the bath temperature in sharp contrast to the behavior expected from the thermal feedback (Figs. 3.8b and 3.16a). Of course, above 2.17 K the role of the thermal feedback comes into play significantly. The transition from superfluid He to normal He substantially increases the medium-field Q-slope due to a severe drop in heat transfer to He–I, as discussed in Chapter 3.

The observed bath temperature *independence* of the high-field Q-drop makes it possible to eliminate the role of the thermal feedback effect with nonlinear BCS resistance as the primary cause. However, it could play a subsidiary role.

5.2.2
Roughness and Q-slope

5.2.2.1 Macroroughness and Microroughness

One of the most remarkable features of the high-field Q-drop is the large difference between BCP and EP-treated cavities *after* baking. The most obvious difference between BCP and EP surfaces is macroroughness on the scale larger than a few microns. Figure 5.18 compares SEM micrographs of surfaces prepared by the two methods [296]. For the EP surface, the average macroroughness is less than 0.5 µm after removal of about 80 µm of material from the raw material stage. During BCP, the etch rate depends on crystal orientation, so that steps form at the boundaries between adjacent grains of different orientation. The average step height increases with material removal and begins to saturate at 5 µm after 100 µm removal (Fig. 5.19) [297]. By contrast, material removal of 80 µm by EP starting from raw material yields a surface roughness of about 0.3 µm [298]. More details of the grain-boundary topography can be seen from 3D AFM images of BCP surfaces in Fig. 5.20. A comparison to the topography

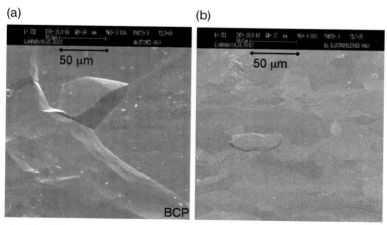

Fig. 5.18 SEM pictures of the surface after 100 µm material removal by (a) BCP and (b) EP [296] (courtesy of DESY and CERN).

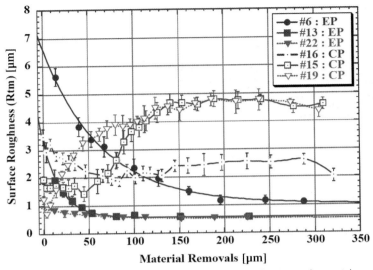

Fig. 5.19 Variation of the niobium surface roughness with increased material removals for both EP and BCP as applied to several samples. In case of EP, the final roughness depends on the initial surface roughness, and exponentially decreases with increased amounts of EP [297] (courtesy of KEK).

of the EP surface is available in the 3D AFM images of Fig. 5.21. Remarkably, the *microroughness* of BCP and EP surfaces within individual grains is very similar as shown in the AFM images (Fig. 5.22) of single grain material treated by both methods [299].

Returning to the topic of macroroughness, further comparisons between BCP and EP have been carried out using profilometry with a stylus (Fig. 5.23) [300, 301]. Figure 5.24 compares the step height distributions after 100 μm BCP and EP [301]. These results show a large distribution in step heights with average values of 5–6 μm for BCP and about 0.5 μm for EP.

The smoothness of EP surfaces is easily destroyed. Starting with raw sheet material, the typical roughness for a BCP surface increases quickly with BCP approaching several μm peak-to-peak after 100 μm material removal [297, 298]. Just 20 μm of BCP etch after EP results in a factor of 2 increase of surface roughness.

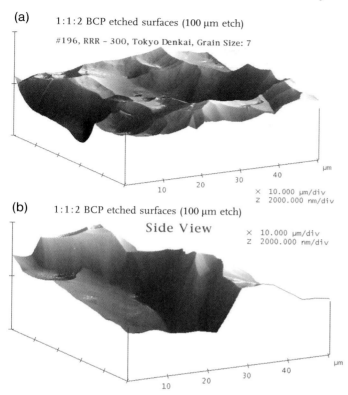

Fig. 5.20 AFM image of a BCP-treated surface showing the sharp edges of grain boundaries. Vertical scale of 2 μm per division: (a) top view; (b) side view [303].

At the electron beam weld, the macroroughness after BCP is even higher, with grain boundary steps as large as 15–20 μm (Fig. 5.25). Moreover, most grains are aligned so that the magnetic field lines at the equator cross grain boundary steps due to the coincidence between the weld beam direction and the magnetic field lines [300, 302]. This orientation turns out to be unfavorable for magnetic field enhancement at the steps.

An early attempt [300] to explain the high-field Q-drop of BCP cavities was based on the field enhancement at the grain boundary steps. We will discuss this model in more detail later, but in a different context. There are several fundamental problems with the roughness model. (a) Single-crystal cavities with no grain boundaries also show a high-field Q-drop, as we will discuss in more detail. (b) EP cavities without the macroroughness associated with grain boundary steps also show a strong high-field Q-drop and (c) the Q-drop decreases with the mild bake (100–120 °C for 48 h). Such a mild bake does not change the macroroughness. A mechanism other than roughness is therefore needed to

(a)

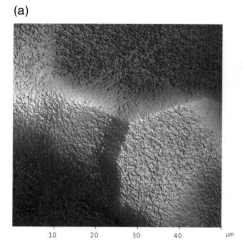

(b)

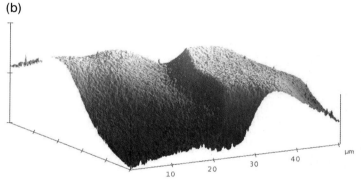

Fig. 5.21 AFM images of an EP surface (a) 50 μm × 50 m; (b) side view, 0.4 μm high by 50 μm × 50 μm [303].

account for the "baking effect." This yet to be determined mechanism must play the dominant role. We will discuss several candidates for the baking effect.

The macroroughness may still play a subsidiary role in the Q-drop for the following reasons: (a) baking has a stronger effect on the high-field Q-drop of the smoother EP cavities; (b) the onset field of the Q-drop in fine grain cavities is about 10% lower than for EP cavities; (c) the onset field of the high-field Q-drop in large grain and single-grain cavities is about 20% higher as we will discuss later.

Once the baking effect has taken hold, experiments on *all baked cavities* suggest a link between macroroughness and the high-field Q-drop of BCP cavities. The onset field of the Q-drop decreases with increasing amount of BCP, which also increases macroroughness (Fig. 5.19). A similar link exists with the quench field for BCP-baked cavities. A series of experiments [53] carried out at KEK

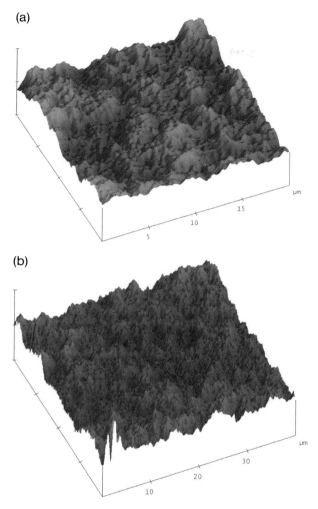

Fig. 5.22 AFM images showing microroughness of single crystal material. (a) BCP 120 μm and (b) EP 120 μm. Vertical scale of 200 nm per division [299] (courtesy of JLab).

(Fig. 5.26) showed that increased amounts of BCP applied to an EP surface reduced the onset field of the Q-drop and reduced the quench field from 37 MV/m (EP) to 28 MV/m with 60 μm BCP, and to 24 MV/m with another 70 μm BCP. To re-emphasize, all results were obtained after mild baking. Earlier we pointed out that BCP cavities respond differently to baking (Fig. 5.11). It is likely that different degrees of surface roughness determine the overall response after baking.

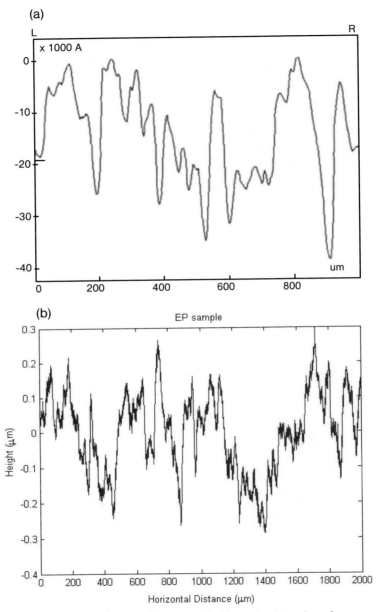

Fig. 5.23 Surface roughness profile measured with a stylus (tip radius of 5 μm and a 60° shank angle. (a) BCP surface, vertical scale of 1 μm/division. (b) EP surface, vertical scale of 0.1 m/division [301].

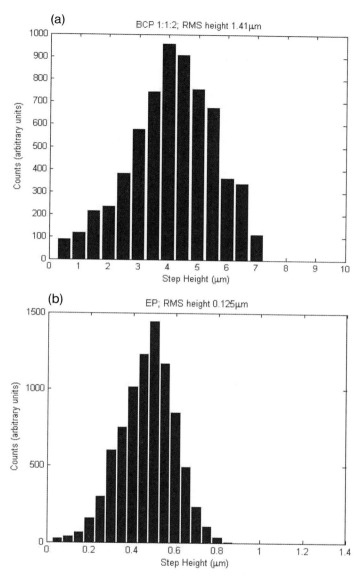

Fig. 5.24 Roughness distribution over the surface (a) BCP and (b) EP [301].

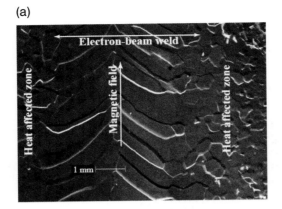

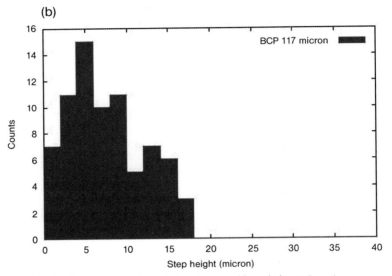

Fig. 5.25 (a) SEM photograph of electron-beam weld (scale bar is 1 mm). (b) Distribution of grain boundary step height measured in the electron-beam weld region with a stylus [302].

The trend in the decreasing onset field was reversed when the surfaces were rendered smooth again with EP plus bake. It was also possible to progressively increase final quench fields by increased EP. However, substantial amounts (100–150 µm) of EP were needed to fully restore performance (and smoothness) after substantial BCP. It is important to re-emphasize that after each BCP and EP treatment the cavity was always baked, so that the Q-drop increases with BCP are due to a cause (such as roughness) that does not respond to baking.

A similar series of experiments were conducted by DESY (Fig. 5.27a) [285]. Once again, BCP lowered the onset field of the Q-drop, and lowered the cavity

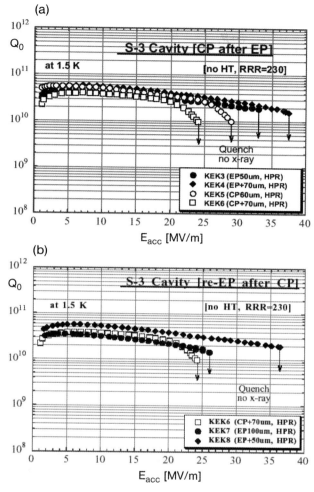

Fig. 5.26 (a) Progressive degradation of Q versus E curves for BCP-baked cavities with increasing depths of BCP, starting with a high performance EP cavity. (b) Progressive recovery of Q versus E curves for EP-baked cavities within increasing amounts of EP, starting with a BCP surface [308] (courtesy of KEK).

quench field. The DESY series ended with an EP-bake to test for recovery. Some of the Q-drop was indeed reversed. However, the amount of EP in this case was not sufficient to completely recover, or to smooth out the roughness introduced by heavy BCP. In another careful series of experiments, (Fig. 5.27b) DESY tracked the performance degradation of 1-cell and 9-cell cavities starting with EP by applying increasing amounts of BCP etch. The maximum field accessible dropped steadily with increasing amounts of BCP larger than 10 μm.

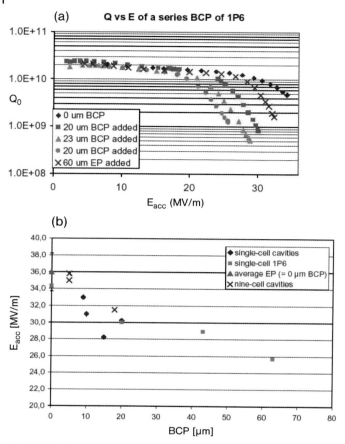

Fig. 5.27 (a) Q versus E curves for progressive amounts of BCP at 2 K. A series of BCP treatments (about 20 μm each) are done on a single-cell cavity. After each BCP, the cavity was baked at 100 °C for 48 h. Each BCP lowers the onset field of the Q-drop. After the third BCP, 60 μm was removed by EP and the cavity baked at 120 °C for 54 h. The cavity performance was partially recovered. (b) Degradation of the maximum field reached with increasing amounts of BCP starting from an EP cavity [285] (courtesy of DESY).

Taken together with surface roughness measurements (Fig. 5.19), these systematic cavity results suggest that increased surface roughness with increasing amounts of BCP does play a role in lowering the onset field of the Q-drop. The converse is true with EP. Several other cases have been reported where BCP cavities show little improvement or even no improvement in Q-slope after 100–120 °C baking (Fig. 5.11).

5.2.2.2 Field Enhancement at Grain Boundary Edges

We now present the detailed model [300], which links macroroughness at grain boundary steps to the high-field Q-drop. The typical onset field is about 80–100 mT, which is far below the superheating critical field for Nb (H_{sh} at 2 K ~200 mT, see Chapter 3). Grain boundary steps provide a geometric field enhancement so that the local magnetic field surpasses the rf critical field to quench the grain boundary edge, and so increase the local power dissipation. Thus comes the contribution of the macroroughness to the high-field Q-drop.

The magnetic field enhancement at a grain boundary microstructure was calculated with the electromagnetic code, SUPERLANS, using a typical step geometry (Fig. 5.28) as measured by profilometry. The enhancement depends on the step height, step angle, and corner radius. A field enhancement factor of 2 was calculated for a 20° slope angle, increasing to 3 for 90°. As the corner radius tends toward zero, the field enhancement factor can become very large. An effective corner radius of 1 μm was assumed based on a simple argument that corners with sharper radii would be normal conducting and the field would penetrate to the normal

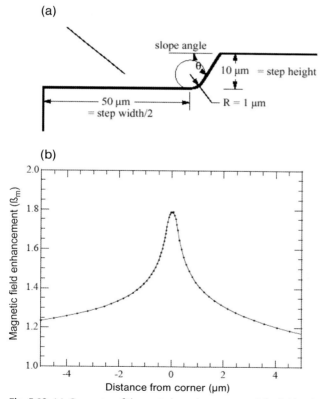

Fig. 5.28 (a) Geometry of the grain boundary step used for field enhancement estimate. (b) Magnetic field enhancement for a 10 μm step with the slope angle of 20° [300].

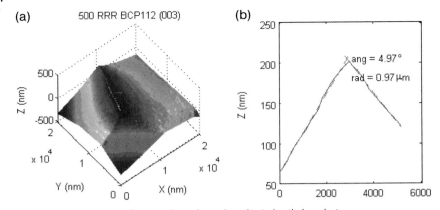

Fig. 5.29 (a) AFM image of a grain boundary edge. (b) A detailed analysis of the geometry of the step reveals the radius of curvature of the edge to be about 1 µm [301].

conducting skin depth ~0.3 µm, resulting in a superconducting corner of radius about 1 µm. Subsequent AFM measurements and analysis [301] showed that corner radii at grain boundaries range between 0.4 µm and 3 µm. Figure 5.29 shows a typical AFM image and the accompanying corner radii analysis.

In conjunction with the field enhancement, thermal simulations with ANSYS determined the temperature profile of the quenched grain boundary, and the nearby region [300]. In particular, the power dissipation and thermal stability of a boundary was analyzed to show that an isolated quenched grain boundary edge does *not* initiate a thermal breakdown of the entire cavity surface. The only effect of the quenched local grain boundary edge is to reduce the Q. As E_{acc} rises, more and more grain boundary edges become normal, and the Q continues to fall. Using a statistical distribution of grain boundaries from profilometry data, the shape of the Q-drop was successfully simulated (Fig. 5.30a). It is important to note that the Q-drop of this BCP-treated cavity did not improve significantly with baking, so the roughness effect was proven dominant in this case, and was successfully simulated. Eventually, the total power dissipated near the most dissipative grain boundary causes cavity breakdown.

The field enhancement model does not provide an explanation for the Q-slope of EP cavities with substantially less surface macroroughness. The grain boundary steps for EP surfaces are less than 0.5 µm (Fig. 5.20). The grain boundary edges of EP surfaces are also rounder than BCP surfaces as shown by the comparison of AFM pictures of Figs. 5.20 and 5.21. Clearly there is another effect (not related to roughness) which dominates the Q-drop and which responds to mild baking.

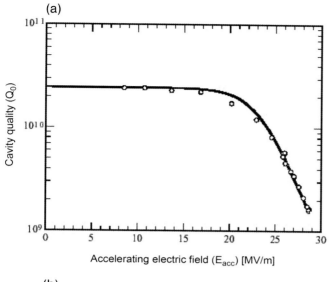

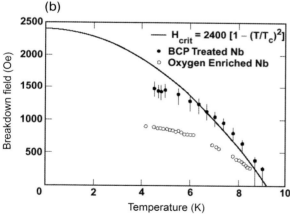

Fig. 5.30 (a) Comparison of high-field Q-drop predicted from grain boundary field enhancement model with experimental data on a 2-cell, 1300 MHz cavity using rf critical field of 200 mT. The Q-drop of this cavity did not improve significantly with baking [300]. (b) Measured rf critical field of Nb cavity enriched with oxygen by anodizing the surface to 80 V (oxide thickness 160 nm) and baking to 200 °C for 2 h [300, 304].

5.2.3
Oxygen Pollution Layer Model

A popular mechanism to explain the baking effect is related to impurities in the penetration depth layer. High concentrations of such impurities can lower the local rf critical field of small regions which would also provide an explanation for the spatial nonhomogeneity of the heating in the Q-slope regime. Small (e.g., 1–10 μm) regions may increase losses or even become normal conducting at the Q-drop onset field without spreading to become a global quench, as in the case of the grain boundary edges. As the field rises, such spots become normal conducting and contribute to the Q-drop. In this model the Q-drop at the highest field does not depend on the bath temperature but on the rising number of such microdefects with increasing magnetic field.

Oxygen is one likely impurity candidate for two reasons. The drop in T_c and thermodynamic critical field H_c of Nb enriched with O was estimated from magnetization data, and is shown in Table 5.1 [305]. The measured rf critical field of O-enriched Nb decreases with increased O content, as shown in Fig. 5.30 [304]. Niobium anodized to 80 V was baked at 200 °C for 2 h and the rf critical field at 4.2 K was found by pulsed high power measurements to drop by 50%. Based on the data of Table 5.1, we can estimate the O content of the rf layer with this treatment to be about 3–4 at.%.

A second reason for O as the pollution candidate to explain the baking effect is that the diffusion constant for O is large enough to account for significant changes in O concentration due to baking at 100–120 °C for 48 h, which corresponds to a diffusion length about 20 nm. By comparison, both N and C diffuse very slowly through bulk Nb, less than 1 nm at 120 °C in 60 h [306] and are therefore unlikely candidates to have a large effect on the healing of the Q-slope due to baking. Faster diffusion of N and C along grain boundaries remains a possibility, however, so that the role of these impurities cannot be completely ruled out. Experiments and arguments that H is not the relevant impurity involved in the baking effect will be discussed later.

The presence of a large concentration of O in the first 10–20 nm of the rf surface layer as the primary mechanism for the high-field Q-drop (especially for EP cavities) is called the "oxygen-pollution model" [143, 173, 280]. The model is illustrated in Fig. 5.31 (a) before baking and Fig. 5.31 (b) after baking. The attractive feature of the oxygen pollution model is the mechanism it provides for the baking effect to reduce the high-field Q-drop. Baking at 100–120 °C dilutes excess oxygen present just below the oxide layer. When O diffuses deeper into the niobium, it reduces the κ_{GL} of the polluted layer, increases the local critical field, and increases H_{c1} toward the value for pure bulk niobium.

Table 5.1 Effect of oxygen concentration on the superconducting properties of niobium.

Alloy (at.% O)	T_c (K)	$H_c(0)$ (Oe)
0.024	9.23 ± 0.01	1910
0.139	9.03 ± 0.02	1854
0.555	8.50 ± 0.08	1717
0.922	8.10 ± 0.05	1613
1.32	6.13 ± 0.10	1528
2.00	7.33 ± 0.12	1399
3.50	6.13 ± 0.13	1102

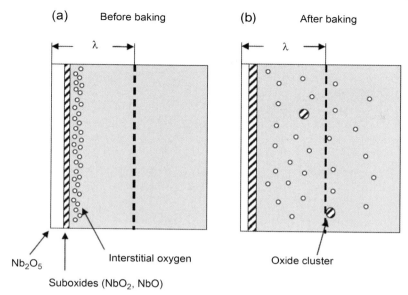

Fig. 5.31 The oxygen pollution model. (a) An enriched layer of O resides below the oxide layer after a fresh surface preparation. (b) Baking dilutes the pollution layer improving the superconducting properties, but lowering the overall mean free path over the penetration depth to account for the drop in the BCS resistance. Baking may also introduce oxide clusters in the rf layer [143, 173, 280] (courtesy of JLab).

The excess oxygen in the pollution layer provides several possible reasons for the high-field Q-drop. Fluxoids enter the rf layer at a reduced H_{c1} and cause losses due to their resistive motion. Another possible explanation is the presence of a large number of micron-size, oxygen-rich regions turning normal conducting at their reduced rf critical fields. In another model [307] there is a large density (a) of superconducting "defects" smaller than the coherence length that turn normal starting at an onset field H_0 that can be related to the low-field Q-slope. After transition, the losses in these normal defects determine the medium-field Q-slope, which depends on the density of defects. When the external rf field reaches the local H_{c1} (depending on the local κ) the high-field Q-slope starts. The virtue of this model is that it is the only one which links the low-field Q-slope, medium-field Q-slope, and high-field Q-drop via three main fit parameters: H_0, and H_{c1}. Several measured Q versus E curves have been successfully fit over the three Q-slope regions.

In Chapter 3, oxygen diffusion also provided a plausible mechanism for the influence baking on the BCS surface resistance. Here the decrease in surface BCS resistance due to reduced mean free path was attributed to the arrival of O from near the surface to the entire penetration depth (~ 40 nm). Hence the pollution model accounts for the drop in BCS resistance due to baking.

There have been a number of interesting experiments with varying baking temperatures and times that tie the baking benefit to diffusion of O [151]. Fig-

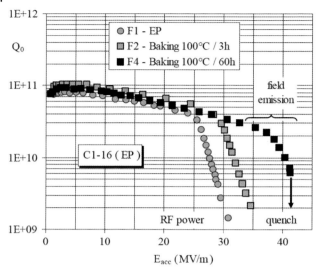

Fig. 5.32 Influence of baking time at 100 °C on the high-field Q-slope of an EP cavity [151] (courtesy of Saclay).

ure 5.32 shows that 100 °C *in situ* baking for 3 h has less effect on the Q-slope of an EP cavity than 60 h baking. Increasing the baking temperature to 145 °C for a short bake of 3 h provides as much benefit to the Q-slope of a BCP cavity as a 110 °C bake for 48 h (Fig. 5.33). Such a comparison would be more convincing for an EP-treated cavity with the generally larger baking benefit. The time and temperature dependences of these results suggest that the healing of the Q-slope is related to O diffusion, since comparable healing is obtained for comparable combinations of temperature and time.

The complete picture of oxygen distribution underneath the oxide layer is still open, and can only be resolved with surface analysis (see Section 5.3). Figure 5.34 shows three different prevailing scenarios. In case (a), the diffusion calculation assumes a finite quantity of O underneath the oxide layer before bake. After 20 h at 120 °C the O concentration in the pollution layer (1–10 nm) falls substantially due to diffusion causing the Q-drop to improve. Over 40 nm (∼penetration depth) the O concentration increases to drop the mean free path and reduce the BCS resistance [173]. In case (b), the O concentration at the surface remains constant (at the solubility limit of O in Nb) during the bake [149] because the surface oxide decomposes upon heating, thus becoming an inexhaustible source of oxygen which can diffuse into the bulk. This scenario would imply that the healing of the Q-slope is due to the *addition* of O into the rf layer, a somewhat counter-intuitive mechanism. The constant O concentration under the oxide is therefore unlikely.

In case (c) [143, 280], also called the "modified pollution model," there is a high interstitial concentration of O present in the pollution layer and it diffuses

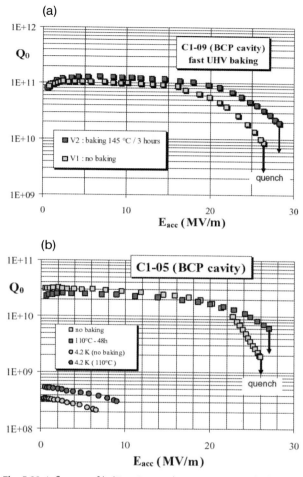

Fig. 5.33 Influence of baking time and temperature on the high-field Q-slope of a BCP cavity. (a) 145 °C, 3 h bake [308] gives similar improvement as (b) 110 °C, 48 h [276] (courtesy of Saclay).

away during baking at 100 °C. In addition, during baking, the oxide contributes more O as it begins to dissociate. The higher the baking temperature and times, the more is the contribution from the dissociation. Since diffusion also increases with temperature and time, the model predicts an optimum baking temperature (near 140 °C) for healing the Q-drop. In support of case (c), XPS analysis of niobium samples in [150, 175, 309] Section 5.3 indicate that baking at progressively higher temperatures does decompose the natural Nb_2O_5 layer to metallic suboxides (NbO, NbO_2) causing an overall reduction of the oxide layer thickness. Presumably O generated from the decomposition diffuses into the bulk.

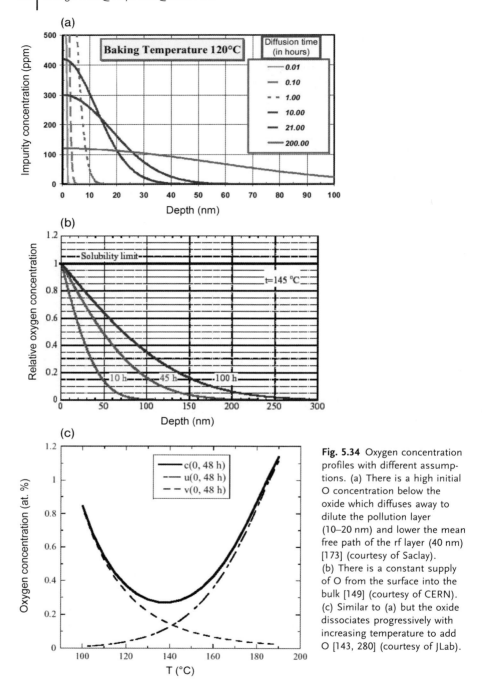

Fig. 5.34 Oxygen concentration profiles with different assumptions. (a) There is a high initial O concentration below the oxide which diffuses away to dilute the pollution layer (10–20 nm) and lower the mean free path of the rf layer (40 nm) [173] (courtesy of Saclay). (b) There is a constant supply of O from the surface into the bulk [149] (courtesy of CERN). (c) Similar to (a) but the oxide dissociates progressively with increasing temperature to add O [143, 280] (courtesy of JLab).

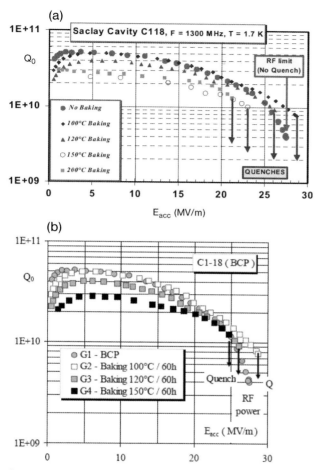

Fig. 5.35 Increase in residual resistance and decrease in quench field due to baking at progressively higher temperatures than 120 °C [151, 173] (courtesy of Saclay).

Several experiments report that baking BCP cavities at temperatures higher than 120 °C cause degradation in several aspects of cavity performance, such as a lower quench field, higher residual resistance, and in some cases a lower onset field with stronger high-field Q-slope. For example, studies in Fig. 5.35 [151, 173] show that after the Q-drop benefit with 100 °C baking, rebaking at 120 °C returns the Q-drop to its nonbaked value, increases residual resistance and lowers the quench field. Increasing baking temperatures to 150 °C and 200 °C further increases the residual resistance and further reduces the quench field. In some cases (Fig. 5.36) there is lower onset field and a stronger Q-drop after 150 °C [284]. These effects are consistent with the modified oxygen pollution model prediction that the Q-slope should return because the oxide breakup pro-

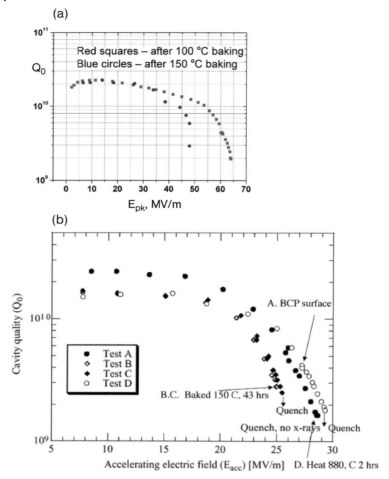

Fig. 5.36 Decrease in onset field of Q-drop and increase of Q-drop due to baking at 150 °C. (a) 1-cell 1.5 GHz cavity [284]. (b) 1.3 GHz, 2-cell cavity [300] shows stronger Q-drop after 150 °C bake, and almost complete recovery of the original Q-drop after 880 °C bake [300].

duces more oxygen to repollute the first 10–20 nm of the surface layer. We will return to the subject of lower quench fields due to higher baking temperatures in Section 5.4.

5.2.3.1 Difficulties with the Pollution Models

If the modified pollution model is correct then any deterioration of the Q-drop due to baking at 150 °C should be reversible with rebaking at 100–120 °C by diffusing away the excess oxygen introduced via the break up of the oxide layer.

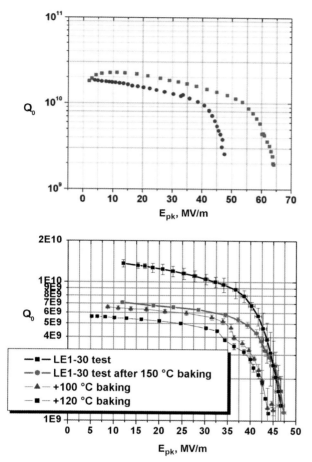

Fig. 5.37 Q versus E curves for a 1-cell 1.5 GHz BCP cavity before and after baking at 100 °C for 48 h. (b) After repreparation with BCP, the same cavity was baked at 150 C for 48 hours. The residual resistance increased and the field value for a $Q=5\times10^9$ also decreased showing that the onset of the high-field Q-drop decreased by about 5 MV/m (13%). Two subsequent attempts to heal the Q-drop by baking at 100 °C for 48 h, followed by another bake at 120 °C for 48 h did not restore the baking benefit, but made the high-field Q-slope even stronger [284].

However the expected recovery was not found for a BCP-treated cavity baked at 150 °C (Fig. 5.37), even after two rebakings at 100 and 120 °C, each for 48 h. In previous tests this cavity showed a significant improvement with 100 °C bake (Fig. 5.11 c). The absence of recovery is troublesome for the modified pollution model. Perhaps a much higher temperature treatment is necessary to remove the added oxygen. A similar degradation of high-field Q-slope due to 150 °C baking could be reversed with heat treatment at 880 °C for 2 h to almost recover the original Q-drop of the 2-cell 1.3 GHz cavity (Fig. 5.36 b) [300].

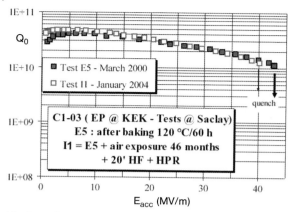

Fig. 5.38 The Q versus E curve of an EP-baked cavity without high-field Q-slope remained without high-field Q-slope after removal of oxide by an HF rinse, and after regrowing a fresh layer of oxide on water rinsing [151] (courtesy of Saclay).

The origin of the O-pollution layer also poses other mysteries. A natural hypothesis is that the oxygen pollution layer forms during the oxidation process. Since oxide grows by voltage-assisted diffusion of O into Nb (Cabrera-Mott process [310]) new layers of oxide grow under the existing oxide, from the Nb–oxide interface. If baking at 100–120 °C dilutes the pollution layer by diffusion of oxygen, one would anticipate that the growth of a brand new oxide layer would form a new pollution layer and restore the Q-drop. But this does not happen.

A variety of tests show that the formation of a new oxide layer does not revive the high-field Q-drop after the reduction or removal of the Q-drop by the baking effect. Therefore, oxide growth *per se* cannot produce the expected pollution layer. In one experiment, after eliminating the Q-slope of an EP cavity by baking, the remnant oxide was removed with HF and a brand new natural oxide layer (typically 5 nm thick) regrown by exposure to water and air [151]. Figure 5.38 shows that the high-field Q-slope *is absent* after growing a new oxide layer. This means that the major changes induced by baking to remove the high-field Q-slope are deeper than the first few nm below the oxide and therefore exclude the oxide and interface as playing a role in the high-field Q-slope. As a reminder, HF rinsing has been shown to affect the low-field Q-slope as discussed in Chapter 3.

In a different experiment (Fig. 5.39), [311] after reducing the Q-drop by 100 °C baking a BCP-treated cavity, the oxide thickness was doubled to 10 nm by anodizing the Nb surface using 5 V. But the strong Q-drop did not return. Similar results were obtained with an EP-baked cavity. Further anodization to 10 V also did not restore the Q-slope previously removed by 100 °C baking. These results were confirmed with a different cavity baked to 120 °C [312]. The Q-slope did not return after 20 V anodizing, but did return after 40 V (Fig. 5.40a). In yet a

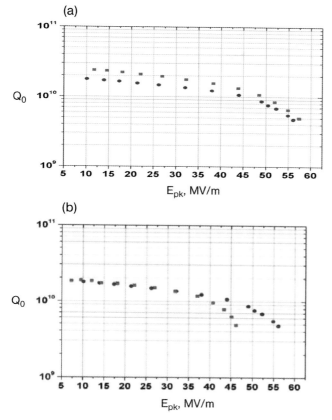

Fig. 5.39 (a) A BCP-baked cavity with healed Q-drop was anodized to 5 V to double the thickness of oxide layer to 10 nm. The high-field Q-drop did not return. (b) After anodizing to 30 V the original high-field Q-slope returned [311].

different approach of thickening the oxide layer (Fig. 5.40b), a baked cavity without high-field Q-drop was baked again in one atmosphere of oxygen at 120 °C for 12 h, followed by another bake at 150 °C for 12 h. Again, the Q-drop did not return [312] despite the attempts to inject oxygen by heating in an oxygen atmosphere.

Upon increasing the thickness of the oxide layer further by anodization to 30 V [311] and to 40 V [312] the high-field Q-drop did finally return (Figs. 5.39 and 5.40). To confirm that the high-field Q-slope had the signature features, baking at 100 °C was shown successful in healing the returned Q-drop for both the 30 V and 40 V anodized cavities. At 30 V anodization, the oxide thickness is 60 nm of which about 20 nm corresponds to the thickness of the niobium consumed in forming the oxide layer. These results imply that the thickness of the baking benefit extends to a depth of about 20 nm below the oxide–metal inter-

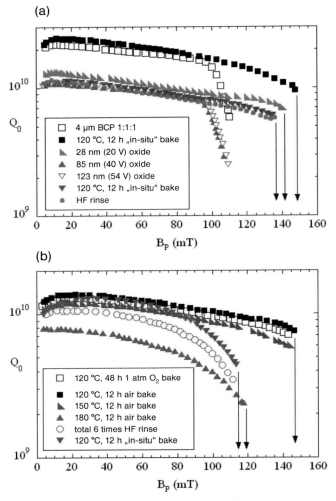

Fig. 5.40 (a) Q versus E curves of a large grain BCP prepared Nb cavity before baking with Q-drop, after baking without Q-drop, after 20 V anodization without Q-drop, after 40 V anodization with Q-drop returned, after 54 V anodization still with strong Q-drop, and after baking to heal the Q-drop once again.

(b) Q versus E curves of the large grain cavity after baking in oxygen at 120°C, 150°C and 180°C. The Q-drop does not return after 120°C or 150°C. The new losses at 180°C do not have the characteristic field dependence of the high field Q-drop and are likely due to creation of suboxides [312] (courtesy of JLab).

face. The 20 nm depth of the baking benefit layer is also consistent with the O diffusion length for 120 °C, 48 h (see Fig. 3.2).

One may still view these results as consistent with the modified pollution layer model if we posit that it takes several cycles of oxidation to restore the pollution layer. If there is a competition between oxidation and interstitial oxygen production during the process of Nb oxidation [313], it is possible that at least 60 nm of oxidation is necessary before the pollution intensity reaches the harmful level. Despite some difficulties, the pollution model remains the most successful explanation to date for the baking effect.

What role does roughness play in the pollution model? For baked cavities, where the impurity diffusion mechanism is no longer active, it is clear from Figs. 5.26 and 5.27 that the rougher the surface, by increased amounts of BCP, the lower the onset of the Q-slope and the lower the quench field. The evidence therefore suggests that *after baking*, the roughness mechanism becomes dominant, leaving BCP cavities with a significant Q-slope, and EP cavities with little or no high-field Q-slope.

In summary, there are two mechanisms responsible for the high-field Q-slope: surface impurities as the main contributor and roughness as a secondary effect.

5.2.3.2 The Nonrole of Hydrogen

Hydrogen is known to enter niobium during polishing and cleaning operations, and is thought to be uniformly distributed throughout the niobium lattice at room temperature [313]. The diffusion coefficient of H is very high so that H diffuses through the bulk at room temperature. However, several surface studies show that there is higher concentration of H just under the oxide at the surface [143]. Hence one may speculate that the enriched H layer is possibly responsible for the high-field Q-slope and that baking at 100–120 °C removes the concentration peak at the surface. However, one of the surface studies also shows that after baking and return to room temperature, the H enrichment at the surface returns is still present, but the Q-drop is gone. This makes H unlikely as the cause of the high-field Q-drop [143].

Another reason that hydrogen dissolved in the Nb is not likely to be responsible is that the high-field Q-drop also occurs in cavities prepared after heat treatment for H removal at 700–800 °C, or by postpurification at 1250–1350 °C. In most cases these cavities were treated with light EP or light BCP after furnace treatment for removal of surface contaminants to avoid field emission. Therefore, one could argue that part of the H may have returned in the chemical process. However, the slow cool-down tests generally show the absence of Q-disease and therefore the absence of significant quantities of H.

5.2.4
Fluxoids

Another impurity-related mechanism for explaining the Q-slope is the penetration of rf magnetic lines (fluxoids) above an onset field [280]. Abrikosov fluxoids generate losses due to resistive motion [314]. The quality factor of the cavity drops rapidly as more fluxoids penetrate with increasing rf field. The typical onset of the high-field Q-drop (80–90 mT) is far below the dc lower critical magnetic field (H_{c1} ~ 1600 Oe at 2 K). However, fluxoids may begin to penetrate at a field much lower than the bulk H_{c1} due to a high surface contamination level which lowers H_{c1} [280]. Interstitial O-rich regions at the metal–oxide interface or near grain boundaries are likely candidates to lower H_{c1}. DC magnetic flux penetration studies using optical polarization [315, 316] showed that grain boundaries are regions for dc magnetic flux penetration when a grain boundary is perpendicular to the rf surface (Fig. 5.41). At other grain boundaries, with large steps or where the grain boundaries were not normal to the surface, there was no preferential flux penetration. The technique uses the strong Faraday effect in Y–Fe garnet to measure the vertical magnetic field component above a sample. The spatial resolution is ~5 µm when the garnet is placed directly on the face of the sample and fields of ~1 mT can be resolved.

If we assume the Q-drop onset at about 90 mT to represent H_{c1} of the niobium surface and use $H_c = 200$ mT from Chapter 3, the corresponding Ginsburg–Landau parameter is 2.4, much higher than the bulk κ_{GL} ~ 1–1.5 measured in dc studies. Measurements of κ_{GL} of niobium for different interstitial oxygen concentrations show that $\kappa_{GL} = 2.4$ corresponds to about 0.56 at.% of oxygen [305]. These results support the oxygen pollution model.

A Q-drop caused by fluxoids may also be consistent with the observed frequency dependence of the onset field: at higher frequencies, the rf field at the surface changes faster than the fluxoid nucleation time. As a result the onset

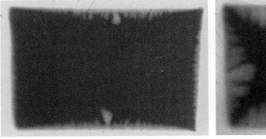
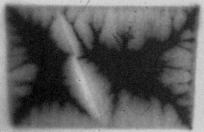

H = 26 mT 　　　　　　　　　　　H = 40 mT

Fig. 5.41 Magneto-optical imaging studies of dc magnetic flux penetration into a niobium sample. Flux penetrates if a grain is perpendicular to the sample surface. The temperature of 7 K was chosen so that the applied field in the cryostat could exceed the upper critical field (H_{c2}) of annealed, pure Nb at 7 K, ~110 mT [315, 316] (courtesy of FSU).

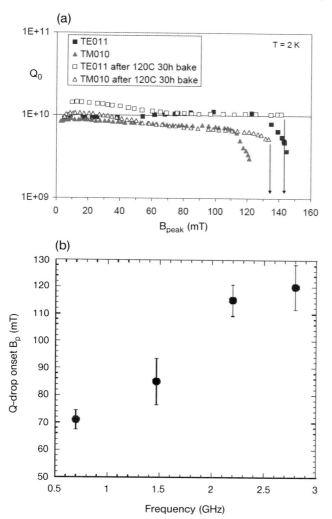

Fig. 5.42 (a) The onset of the high-field Q-slope is higher in the higher frequency TE mode of a TM/TE cavity [280]. (b) Frequency dependence of high-field Q–slope [280] (courtesy of JLab).

field should increase with frequency. This was observed [280] for the Q-slopes in the TE (2.8 GHz) and TM (1.3 GHz) modes of the same cavity (see Fig. 5.42 a). Here the onset field of the Q-slope of higher frequency mode is 30% higher.

Fluxoids in niobium yield losses by Joule heating of their normal-conducting core, reducing the cavity quality factor. The estimated power dissipation due to a single isolated fluxoid in an rf field [317] gives an exponential increase of the surface resistance, once a fluxoid enters the niobium, which qualitatively describes the experimental data. For fluxoids with negligible viscous damping and negligible pinning the field of flux penetration increases linearly with the rf frequency [317]. The Q-drop onset field measured on cavities at different frequencies, between 0.7 and 2.82 GHz, seems to be consistent with this prediction, as shown in Fig. 5.42(b).

Surface roughness may enhance the fluxoid loss mechanism. The existence of a barrier [318] for a smooth surface inhibits flux entry to fields even higher than H_{c1}. The phenomenon of delayed flux penetration was experimentally verified in EP samples of the Type II superconductors Pb-Tl [312]. The experiments also showed that roughening of the surface by scratching and chemical etching destroyed the surface barrier and reduced the penetrating field to H_{c1}. It is conceivable that the surface roughness of BCP cavities reduces the flux entry field to well below the bulk H_{c1}.

In a combined pollution-roughness model one can explain why an EP-baked cavity shows no high-field Q-slope because baking dilutes the pollution layer bringing H_{c1} to its clean value near 180 mT, and the surface barrier for a smooth surface further raises the flux penetration field to H_{sh}. A similar increase to bulk H_{c1} should take place within the individual grains of a BCP-baked cavity. But the roughness of BCP surface eliminates the surface barrier to prevent any further rise from H_{c1} to H_{sh}. Also a typical factor of 2 local field enhancement due to roughness allows fluxoids to enter near the grain boundaries at an average rf surface field of $H_{c1}/2$.

The fluxoid penetration model has its own difficulties. Fluxoids must move in and out of the surface without getting pinned at defects, such as the oxide layer, oxide clusters below the oxide, or grain boundaries. However, the Q versus E curve in the high-field Q-drop regime is generally reversible. This is in contrast to the usually observed fluxoid behavior in dc fields when fluxoids get easily pinned, so that the dc magnetization curve is generally hysteretic (see Fig. 3.27).

5.2.4.1 Role of Grain Boundaries

Apart from high macroroughness at grain boundary steps, there are other potential ill-effects that may arise from grain boundaries. It is well known that the concentration of impurities is higher inside grain boundaries and diffusion of impurities occurs much faster through grain boundaries than the bulk of the grains [290]. Preliminary transport measurements on a bicrystal showed greater normal state resistance and lower superconducting critical current at the grain boundary [315].

We have learned much about the role of grain boundaries from high-field Q-drop studies with single crystal (no grain boundaries) and large grain (very few grain boundaries) Nb cavities. Figure 5.43 shows the high-field Q-slope in a single-crystal 1.3 GHz (TESLA shape) cavity prepared by BCP. The performance shows the signature features of high-field Q-drop and its removal by baking at 120 °C. The remarkable result is the completeness of the baking benefit for a BCP treatment. Polycrystalline cavities treated by BCP improve much less with

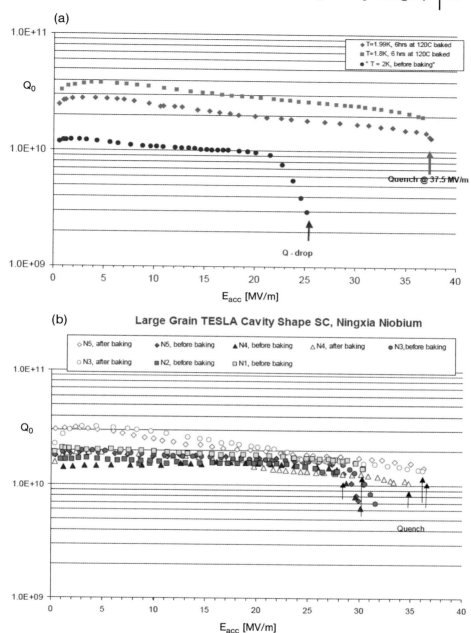

Fig. 5.43 (a) Q versus E curves for single-grain cavities and (b) large grain cavities [320] (courtesy of JLab).

baking (Fig. 5.11) and generally need EP to show comparable improvements in Q-drop. Here again the excellent macro-smoothness of the single grain surface may play a role. The rms roughness of the single crystal surface is less than 10 nm, typical of the microroughness inside single grains. Another remarkable aspect is that only 6 h of baking time was required for the single-crystal cavity. The shorter baking time for single crystal material provides another (yet un-deciphered) clue to the intrinsic nature of the Q-slope.

Cavities from large grain material, i.e., with just a few long grain boundaries, also show a substantial improvement with baking after preparation with BCP treatment as shown in Fig. 5.43 (b). However, EP treatment for large grain cavities generally gives better results, with higher maximum field values, as shown in the composite of Fig. 5.44 [321]. The steps at grain boundaries may be playing the familiar role here. The steps at the grain boundaries of large grain material are rather large as seen in Fig. 5.45, 12–15 µm from crystal orientations 011 to 111 and from 001 to 111, but only about 2 µm from 011 to 001 orientations [321].

A large number of tests (Fig. 5.45 b) show that the onset field of high-field Q-slope in large grain cavities is generally about 20% higher than for polycrystalline cavities, 100–130 mT instead of 80–100 mT [322]. These results support the

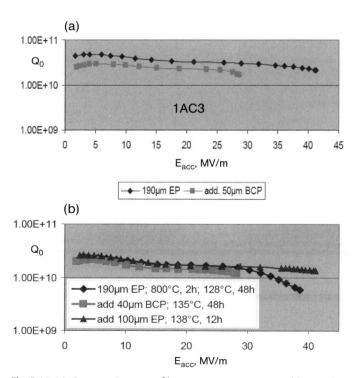

Fig. 5.44 (a) Q versus E curves of large grain cavities prepared by EP, then followed by (b) BCP [321] (courtesy of DESY).

(a)

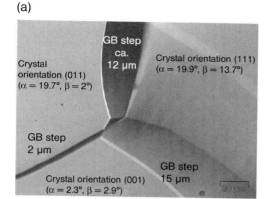

(b)

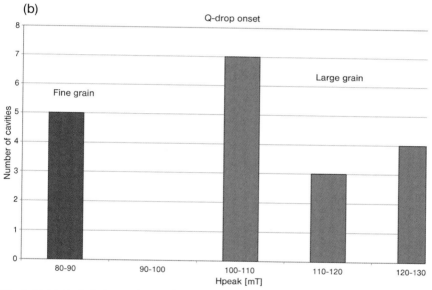

Fig. 5.45 (a) Optical picture showing large steps at grain boundaries of large crystal material [321] (courtesy of DESY) (b) Increase in onset field of the high-field Q-drop of large grain Nb cavities over fine grain cavities [322] (courtesy of JLab).

previous interpretation that grain boundaries play a subsidiary role in the high-field Q-drop. As mentioned before, roughness due to grain boundary steps is a likely contributor, though not the main cause.

Before the advent of large- and single-grain cavities, results with polycrystal Nb cavities already showed hints that the onset field of the Q-drop increases with grain size. Several cavities heated to 1350 °C for postpurification show 10–15% higher Q-slope onset fields [143]. Postpurification increases the grain size from 50 µm to several mm, thereby reducing the number of grain boundaries.

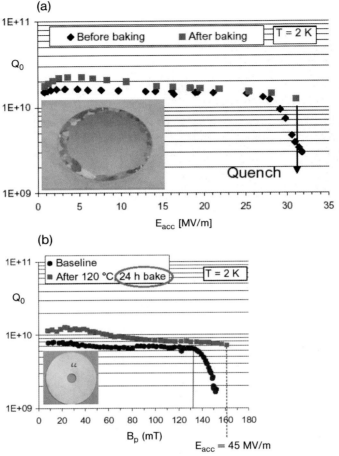

Fig. 5.46 Q versus E curves before and after baking of the first single-crystal cavity at 2.2 GHz [323]. (a) Before postpurification. The onset field of the Q-drop is about 28 MV/m or 120 mT. (b) After postpurification. The onset field increased to above 130 mT and the quench field also increased. The photographs in the inset show the single grain sheet used for cavity fabrication (courtesy of JLab).

The single-grain cavity study revealed another contributing factor to the Q-drop, Nb purity. Postpurification heat treatment raises both the purity of the Nb and the onset field of the Q-drop. For example, Fig. 5.46 [323] shows an increase in the onset field from 100 mT to 130 mT of the Q-slope for the single-grain 2.2 GHz cavity due to postpurification. The final quench field also improves from 115 mT to 160 mT presumably due to purification and accompanying increase of thermal conductivity.

Despite these suggestive results, thermometry studies on large grain cavities with few grain boundaries show that grain boundaries do not provide the domi-

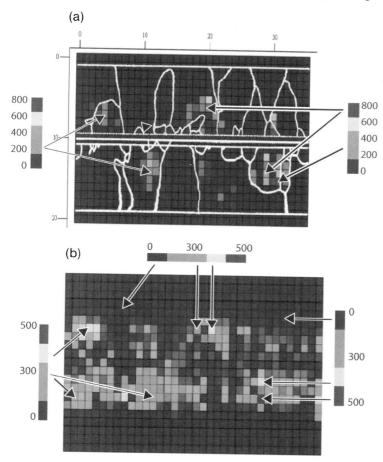

Fig. 5.47 Temperature maps at $E_{pk}=49$ MV/m for heating in the high-field Q-slope regime for (a) a large grain and (b) a small grain cavity [282]. Note the higher ΔT in (a).

nant contribution to rf losses in the high-field Q-drop regime. Figure 5.47 [282] shows that the heating in a large grain cavity does not take place predominantly at grain boundaries. There is strong Q-drop heating in regions without grain boundaries, consistent with the observation of Q-drop in single-grain cavities.

One disappointment is that the highest fields after baking for single grain and large grain Nb cavities are comparable to the best results in polycrystal cavities prepared by EP and baking (Fig. 5.44). Therefore, grain boundaries do not present any limit to the ultimate fields, as long as there are no grain boundary steps. In fact, the best gradients of 53 and 59 MV/m corresponding to surface magnetic fields of 200 to 210 mT were obtained with polycrystalline postpurified material of grain size 1–2 mm (see Fig. 3.32).

5.2.5
Summary of Key Features of High-Field Q-Slope

The high-field Q-drop is present in cavities made from fine grain, large grain, and single grain niobium, prepared by either BCP or by EP. The rf heating takes place in the high magnetic field regions only. Mild baking removes the Q-drop for fine grain cavities prepared by EP, and for large grain cavities prepared by BCP or by EP, with EP results slightly superior. Smoothness plays an important role in the magnitude of the baking benefit. Single-crystal cavities prepared by BCP have a very smooth surface and also benefit greatly by the baking effect. The baking benefit is preserved even after the cavity is exposed to air and water. The baking benefit does not substantially change when the oxide layer is removed and a new one grown. Similarly the baking benefit does not change when the oxide layer is made thicker by baking in air, or by anodizing to about 30 V (60 nm oxide thickness). However, conclusions from overall changes in the Q-drop should be treated with caution since these only contain information about the average behavior over the surface. Detailed information from temperature maps is more telling.

One likely candidate to explain the high-field Q-drop is the presence of many micron size "defects" of reduced rf critical field, which become highly resistive or normal conducting above an onset field defined by the defect with the lowest rf critical field. The regions do not cause a global cavity quench due to their small size. Roughness provides field enhancement and lowers the onset field. Another candidate is regions of reduced H_{c1} where rf flux penetrates and causes losses due to flux flow. An advantage of a smooth surface is that a surface barrier prevents the penetration of magnetic flux into the bulk niobium beyond the lower critical field. It is unclear whether the timescale of the rf is compatible with the nucleation time for fluxoids, and whether the losses associated with fluxoids are large enough to describe the observed increase of the surface resistance.

In both the "numerous defect" or flux entry models, baking improves the superconducting properties of the culprit regions. Excess oxygen content is one likely cause for deterioration of superconducting properties. Diffusion rates of O are consistent with baking temperatures and times. Regions of high strain and crystalline defects like high dislocation density can be another cause, still to be discussed, but here the healing rates are not well known. Besides these likely causes for the primary effect, other effects could make subsidiary contributions, such as field enhancement at grain boundary steps, the higher contaminants inside grain boundaries, and the nonlinear BCS heating at high fields in the clean limit.

5.3
Surface Studies

5.3.1
Nondestructive Methods

5.3.1.1 XPS Studies of the Oxide Structure

We have mentioned a number of possible mechanisms that may contribute to the high-field Q-drop: macroroughness, niobium oxide layer, the metal–oxide interface polluted with interstitial oxygen or oxide clusters, grain boundaries, and microdefects in the crystalline structure. Cavity studies discussed have provided evidence to strengthen or weaken such explanations. In this section we summarize the surface studies that support or deny the role of such mechanisms. Recent reviews of surface studies on Nb can be found in [313, 324, 325].

Many analysis techniques have been used to characterize the niobium surface. We have already seen results of surface roughness measurements by profilometry, atomic force microscopy, and scanning electron microscopy.

X-ray photoelectron spectroscopy (XPS) has been in use for a long time to characterize the chemical composition and bond nature of the oxide, the oxide–metal interface, as well as oxide changes during heating. A mono-energetic x-ray beam excites electrons from the core orbitals of the atoms, and ejects these by the photoelectric effect. Hence their kinetic energy is correlated to their initial binding energy, which is characteristic of the element orbital. Auger spectroscopy is similar in principle, although the excitation source is electrons instead of x-rays. XPS also gives information about chemical bonding, more so than Auger. If the detected element is linked by chemical combination to a more or less electronegative species, its binding energy will be slightly displaced, allowing an inference of the chemical environment. It is therefore possible to distinguish metallic niobium from its oxide(s) and their relative thicknesses, and to also derive information about the presence of any hydroxide.

To determine the thickness of the suboxides, it is necessary to deconvolute the observed XPS spectra on the basis of the spectra for the suboxides. XPS data treatment includes smoothing, energy shift adjustments, background subtraction and curve fitting to distinguish the various components from a peak or group of peaks, and eventually assign those components to individual chemical bonds. But weak signals from NbO, or even interstitials O are difficult to model accurately in the presence of strong Nb_2O_5 and NbO signals, which dominate the spectra. This leads to some variation among studies. The basic conclusions of XPS studies are that there exist small quantities of suboxides present under the natural pentoxide layer. The anticipated oxygen pollution layer underneath the oxide may be connected to the presence of these suboxides at the metal–oxide interface. But its existence has not yet been definitively established by surface studies.

Synchrotron radiation allows varying the incident x-rays energy to provide better energy resolution and depth profiling than conventional XPS with fixed sources. Varying the detection angle also permits the approximate determination of the layer thicknesses [299, 309, 326, 327]. The travel length of photoelectrons belonging to a specific peak is fixed, so that collecting only those exiting at a shallow angle to the surface permits an analysis of the first 2 nm below the surface. The XPS technique generally explores less than 10 nm on the surface, depending on the x-ray energy, the mean free path of the electrons inside the material, and the detection angle. The depth resolution for 1000 eV x-ray energy is about 7 nm. XPS has a sensitivity limit of about 0.1%. The typical microroughness of samples (Fig. 5.22) and roughness variations generally exceed the few-nm escape depth of photoelectrons which gives rise to spot-to-spot and sample-to-sample variations.

The progressive growth of oxides of niobium on a sample was studied by first heating a sample in UHV at 1950 °C and then exposing it to air [328, 329]. After deconvolution, four chemical states were identified: pure Nb, NbO, NbO_2, and Nb_2O_5 in agreement with the Nb–O phase diagram (Fig. 5.48). Angle-resolved XPS showed that the Nb was first covered with a layer of NbO, followed by NbO_2 and Nb_2O_5. On top of these oxides there are adsorbates of carbon with C=O and COH-bonds with 1–2 monolayers of Nb–OH.

These results were confirmed by variable energy XPS [326]. Spectra were collected at photon energies of 352 eV, 520 eV, and 1000 eV, corresponding to sampling depths up to 7 nm. Figure 5.49 shows the results for two highest excitation energies after resolving the spectra to differentiate various oxides present. While the pentoxide is dominant, the peaks for NbO_2, NbO, Nb_2O, and Nb are clearly detectable over a depth corresponding to 1000 eV angle resolved XPS.

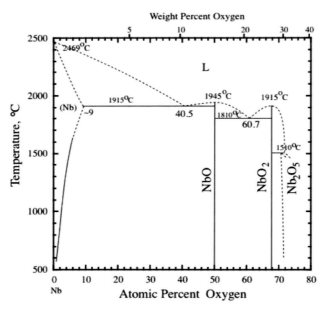

Fig. 5.48 Equilibrium phase diagram of Nb–O system, displaying three different stable oxide phases: NbO, NbO_2, and Nb_2O_5 [330].

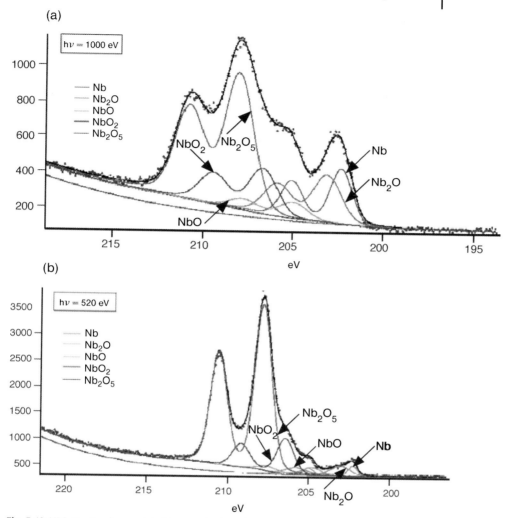

Fig. 5.49 XPS spectra deconvoluted to reveal suboxides.
(a) X-ray energy equal to 1000 eV; (b) 520 eV [326] (courtesy of JLab).

In a controlled study by Palmer [331, 332] the contribution of such surface oxides to the residual resistance was investigated: in these experiments X-band niobium cavities were fired at 1400 °C in UHV, which resulted in the dissolution of the natural surface oxide into the bulk as confirmed by auger electron spectroscopy. The residual resistances of these oxide-free cavities (they were never exposed to atmosphere prior to the test) were 5–10 nΩ – comparable to oxidized surfaces. When oxide layers were carefully regrown under controlled exposure to dry oxygen, the increase in resistance gave a contribution of 1–2 nΩ for the oxide layer.

5.3.1.2 XPS Studies of Baked Samples

XPS studies have been carried out after heating Nb samples in the high vacuum XPS system over a range of temperatures from 100 to 400 °C. Both single crystal and fine grain Nb have been studied.

At 100 °C the spectra showed only minute changes. At higher temperatures the Nb_2O_5 phase progressively dissociated and NbO_2 and NbO developed. Figure 5.50 shows a reconstruction of the decomposition with time for baking at 150 °C [150]. At 180 °C the NbC phase was formed and the final structure of the Nb surface was a thin layer of NbO and NbC stable phases with thickness much less than the initial oxide layer [175]. Upon baking at 250 °C and higher for several minutes, Nb_2O_5 completely disappeared, and only Nb_2O and NbC remained on the surface [175]. The formation of layers of Nb_2O and NbO could be significant for understanding the increase in residual resistance and lowering of quench fields for cavities baked at 150 °C (Figs. 5.35 and 5.36).

To reproduce the optimum cavity baking conditions for curing the high-field Q-slope, XPS studies [299] were carried out on a Nb sample before and after baking to 120 °C for 48 h (Fig. 5.50b). After baking, the pentoxide partially transformed into suboxides, and the total oxide layer became thinner. But after air exposure, the change in the oxide layer observed almost disappeared. Since one of the hallmarks of the baking benefit is survival from an air exposure, these XPS results show that the changes in the oxide created by *in situ* bake are not the underlying cause for the baking improvement.

The high energy XPS studies were also unable to detect any changes with baking in the underlying Nb metal spectra. Such changes would be expected if baking changes the interstitial oxygen content of the oxygen pollution layer by more than a few tenths of a per cent. This result is troublesome for the oxygen pollution model.

A variety of surface studies were conducted to determine the reason for the spatial variation in the steepness of the high-field Q-slope [333]. A fine grain cavity prepared by BCP was studied with thermometry to identify the strong and weak Q-slope regions as seen in Figs. 5.5 and 5.6, and subsequently dissected to separately analyze these regions. High-energy XPS spectra (Fig. 5.51) showed no differences in the oxide composition of the strong regions (called hot) and weak regions (called cold).

In the same studies, XPS spectra did show the presence of N on three of the hottest regions of the fine grain cavity. The chemical shift corresponded to N in the form of a nitrate, so the likely origin is from the nitric acid used in the BCP mixture. After heat treatment at 100 °C for 48 h the N signal was gone. Since the diffusion coefficient of N in bulk Nb is too small to explain disappearance of N [306], the possible mechanism is that N diffused away along grain boundaries upon baking. Of course the N in grain boundary mechanism would not help to explain the high-field Q-drop in general, because for EP cavities no nitric acid is involved in the electrolyte.

Hot spots were also extracted from a large grain cavity prepared by BCP. As mentioned earlier, the hottest spots are not near grain boundaries. These did

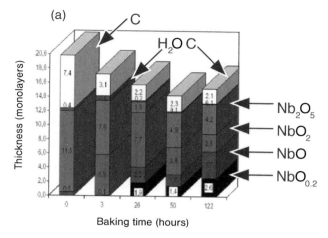
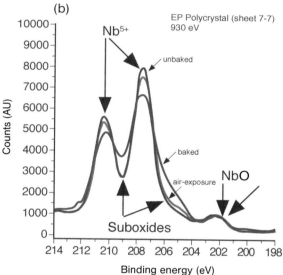

Fig. 5.50 (a) Evolution of the Nb surface with time for a baking temperature of 150°C. [150] (courtesy of INFN) (b) XPS spectra at 930 eV for unbaked, baked (120°C, 48 h) and air-exposed Nb. Changes due to baking disappear on subsequent exposure to air [299] (courtesy of JLab).

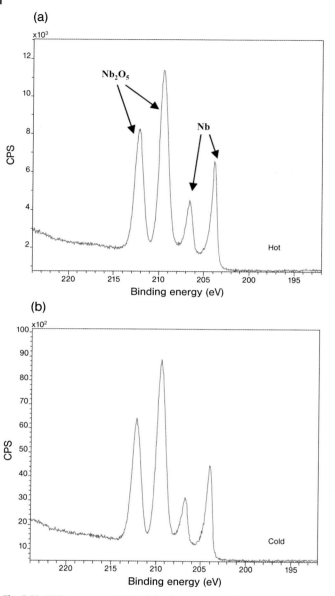

Fig. 5.51 XPS spectra at 2139 eV for the (a) strong (hot) and (b) weak (cold) high-field Q-slope regions of a BCP-treated cavity obtained by extracting the samples from the cavity by dissection [333].

not show any excess N with XPS; therefore N cannot be the dominant explanation for the spatial variation in the Q-drop.

5.3.1.3 Electron Backscatter Diffraction (EBSD)

While crystalline materials diffract the primary electrons, the backscattered electron intensity is slightly reduced along major planes to form a pattern of dark lines. Automated systems are available to index such channeling patterns. From backscattered electrons diffraction patterns the EBSD technique maps crystallographic orientation of the grains on the sample surface to a depth of about 20–100 nm. Since niobium oxide is amorphous and does not contribute to diffraction, EBSD looks directly at niobium underneath the oxide within the magnetic field penetration depth. Figure 5.52 shows an EBSD map of Nb grain orientations.

The spatial nonuniformity of heating in the Q-drop regime is not connected to crystalline orientation of individual grains. EBSD maps of grain orientation distribution in "hot" and "cold" samples showed no major differences. This was true for samples extracted from both a large grain cavity and a small grain cavity, each prepared by BCP.

EBSD can also be used to analyze the crystal defect (vacancies, dislocations) distribution via local crystallographic misorientation maps. There are some hints that crystal defect structure may play a role in the observed variation of heating patterns in temperature maps. Local misorientation distributions for "hot" and "cold" regions were found to be significantly different for a large grain cavity (Fig. 5.53). One mechanism that can introduce such strains and associated vacancies is through oxidation of Nb [161].

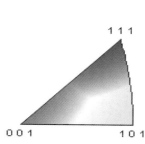

Fig. 5.52 EBSD map showing random grain orientation distribution for fine grain material along with the stereographic triangle which indicates the grain orientation. Black dots appear to be pits [325] (courtesy of W&M).

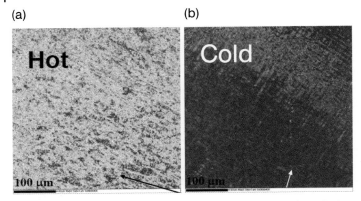

Fig. 5.53 (a) Local misorientation map for a sample dissected from the hot region of a cavity showing high-field Q-slope. (b) Local misorientation map for a cold region of the cavity. Local misorientation is an average misorientation between the pixel and its eight direct neighbors in a 2D map [333].

EBSD local misorientation maps from single grain BCP and EP samples also revealed different microstructures before baking. Mild baking results in the shift of misorientation distribution toward lower misorientation angles in BCP and EP samples, which can be interpreted as a decrease in the intrinsic strain levels or a decrease in vacancy density. Figure 5.54 presents results obtained on single grain EP and BCP samples before and after 110 °C baking for 48 h. There are clear differences between EP and BCP samples as well as differences due to baking [333]. As a cautionary remark, the EBSD technique is a recent application to the problem of the high-field Q-drop so that the results should be considered preliminary.

5.3.2
Destructive Methods

5.3.2.1 Secondary Ion Mass Spectrometry (SIMS)

SIMS and its modification, ToF-SIMS (time-of-flight detector), have been widely used for studies on impurities in materials. A primary ion beam (5–25 keV) erodes the samples at a controlled rate. In the case of ToF-SIMS, a very low-beam intensity is applied and one monolayer is analyzed at the time. Destructive depth profiling allows reconstruction of the distribution of interstitials. SIMS and especially ToF-SIMS is very sensitive with ppb (theoretical) detection capabilities, but there are certain limitations related to surface topography and mixing. Microroughness on the scale of 10 nm blurs the depth profile to the same scale. The ion beam drives some of struck atoms deeper into the solid than their original position (knock-on mixing), also distorting the depth profile. Oxygen is preferentially sputtered from an oxide layer affecting quantitative estimates. Some of the observed species (suboxides) are generated by the sputtering

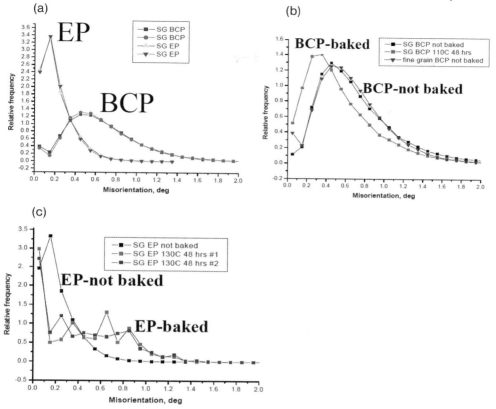

Fig. 5.54 Local misorientation distributions for single grain niobium samples: (a) EP versus BCP; (b) BCP baked/unbaked; (c) EP baked/unbaked [333].

process itself. Ions penetrate the surface, displacing atoms which in turn displace others forming a collision cascade. A few collision trains reach the surface as intact entities and are ejected with near-thermal energy [313, 325].

Figure 5.55 shows ToF-SIMS studies of C, F, O, and H for niobium samples treated by BCP and by EP [313]. A careful examination of the data reveals several interesting features. In the first few nm the natural oxide layer is evident from the drop in Nb and increase in O counts. The quantity of C and H in the oxide layer is depressed, but rises at the surface as expected from the presence of hydrocarbons. F appears to be well mixed in with the O in the oxide layer, which confirms the findings of F and oxide mixing during dc and rf voltage breakdown studies (Chapter 4). No SIMS measurements were carried out in this study with baked samples.

SIMS has also been used to study the secondary ion signals of Nb, and its oxides [312, 334]. Baking in air shows a clear increase in the O profile (Fig. 5.55 b); however, baking in vacuum does not (Fig. 5.56 a). When these results

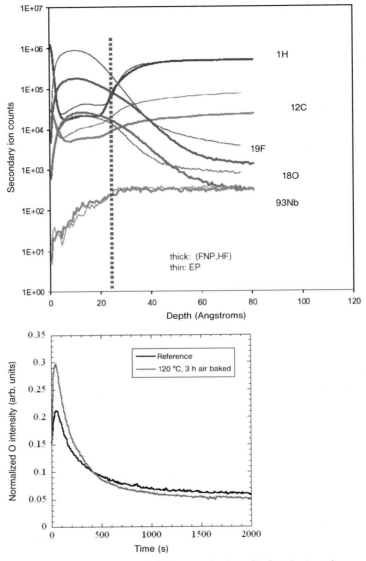

Fig. 5.55 ToF SIMS analysis of Nb showing depth profile for Nb, O, and several impurities [313] (courtesy of Saclay). (b) SIMS shows increase in O due to baking in air [312] (courtesy of JLab).

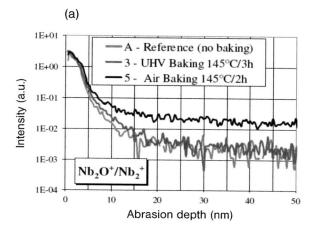

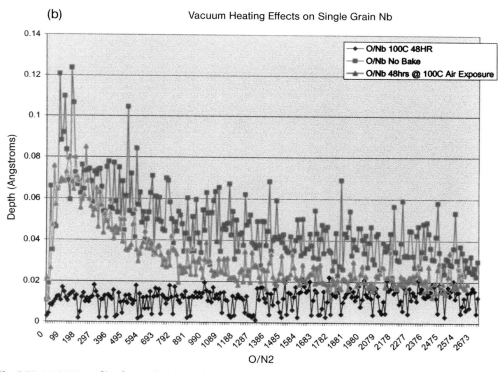

Fig. 5.56 (a) SIMS profiles for O-related signals with samples baked in vacuum and in air. Vacuum baking does not change the profile, but air baking shows O penetrates the rf layer [334] (courtesy of Saclay). (b) O-related depth profile shows a decrease with baking, but the original profile is restored after air exposure [303].

are contrasted with cavity results which show that both baking in vacuum and baking in air remove the Q-drop, the relevance of O to the Q-drop is greatly diminished. Other SIMS results with Ar$^+$ beams [303] do show changes in depth profiles of NbO/Nb (Fig. 5.56 b) due to baking but the original profiles are restored with subsequent air exposure. Since the baking benefit is preserved with exposure to air, these changes may again not be linked to the cavity baking benefit. Overall, the SIMS data do not support O as playing a role in the baking benefit.

New surface analysis techniques coming on line may help to determine whether there is an O-pollution layer present underneath the oxide, and how it changes due to baking. These are transmission electron microscopy (TEM) with focussed ion beam (FIB) sample preparation [312, 335], and atomic probe tomography (APT) [336]. Early results from TEM failed to definitively show any identifiable suboxide regions under the oxide before and after baking. But this work has just started.

APT involves the atom-by-atom dissection of sharply pointed niobium tips, along with their niobium oxide coatings, via the application of a high-pulsed electric field and the measurement of each ion's mass-to-charge state ratio (m/n) with ToF mass spectrometry. The resulting atomic reconstructions, typically containing at least 10^5 atoms and with typical dimensions of 10^5 nm^3 (or less), promise to show the detailed nanoscale chemistry of the niobium oxide coatings and of the underlying high-purity niobium metal. The post analysis removes the H and residual gas atoms from the concentrations to give a better picture of the transition through the oxide layer.

Preliminary results show a nanochemically smooth transition through an oxide layer, more than 10 nm thick, from near-stoichiometric Nb_2O_5 at the surface to near-stoichiometric Nb_2O as the underlying metal is approached (after 10 nm of surface oxide) (Fig. 5.57). Analysis deeper than 5 nm of oxide shows a smooth transition from oxide to metal. The underlying metal, in the near-oxide region, shows a significant amount of interstitially dissolved oxygen (5–10 at.%), as well as a considerable amount of dissolved hydrogen. These results support the presence of an O-pollution layer. Again these studies are also just starting.

5.4
Quench Fields

The prevailing model of thermal breakdown (quench) is that heating originates at submillimeter-size regions of high rf loss, called defects [337–339]. When the temperature of the good superconductor just outside the defect exceeds the superconducting transition temperature, T_c, rf losses increase considerably, as a growing region becomes normal conducting leading to rapid loss of stored energy called "quench." An obvious approach to avoid quench is to prepare the niobium material with great care to keep it free from defects. Performance gains can be expected from scanning the starting niobium sheet for defects, for

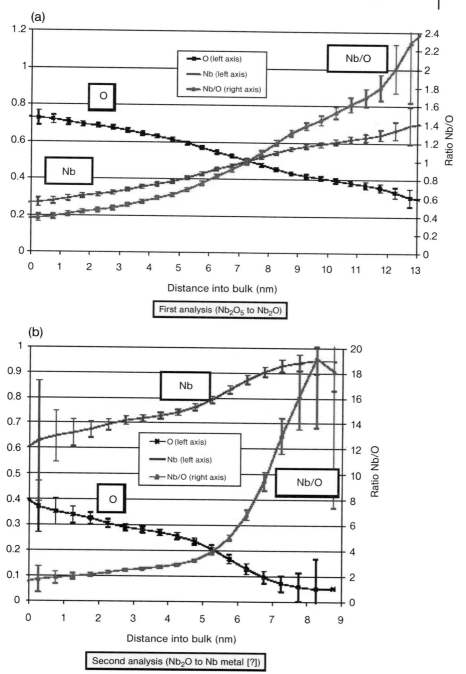

Fig. 5.57 APT results for the O concentration profile.
(a) First 10 nm below the surface; (b) the next 10 nm [336].

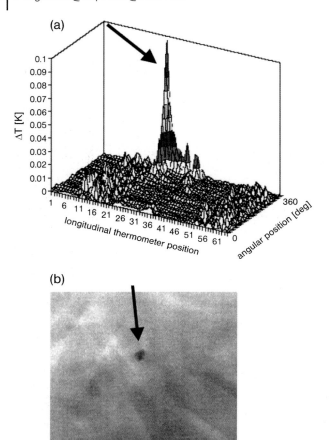

Fig. 5.58 (a) Rare case of a tantalum defect located with thermometry and confirmed by (b) radiography [25] (courtesy of DESY).

example by the eddy-current method, and avoiding suspect sheets which may have inclusions [25]. Chapter 6 on cavity fabrication discusses the advances in material quality control with scanning tools, such as squid-based magnetometers. In one rare find (Fig. 5.58), a Ta defect 0.2 mm in diameter was located by thermometry and subsequently confirmed by x-ray radiography and synchrotron radiation fluorescence analysis [25]. A defect found in a sheet with eddy current scanning was later analyzed to be an inclusion of Cu and Fe [182]. The sheet was rejected for cavity fabrication.

There are many sources of defects. Figure 5.59 shows a variety that has been found at the quench site upon dissecting cavities, and examining the quench region under the SEM. Electron beam welds can have embedded contamination

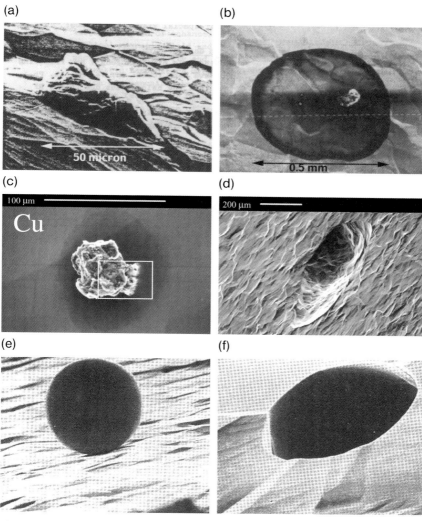

Fig. 5.59 Collection of defects found with SEM after dissection of cavities that quenched. (a) Inclusion with a 50 μm crystal containing S, Ca, Cl, and K. This defect quenched at $E_{acc} = 10.7$ MV/m [339]. (b) A chemical or drying stain 440 μm in diameter. The small crystal on the right side contains K, Cl, and P. This defect quenched at $E_{acc} =$ 3.4 MV/m [339]. (c) A copper particle-like defect found in a 1.5 GHz cavity [243]. The base of the particle was melted in the rf field. (d) A pit with sharp edges found in a 1.5 GHz cavity at a quench field of 93mT [243] (e) A Nb ball found in a 3 GHz cavity [339]. (f) A weld hole found in a 3 GHz cavity [339]. Note the sharp edges.

at the seam before welding [25]. Welds sometimes have voids with sharp edges that enhance the magnetic field. If done poorly, welds can produce spatter that deposit as Nb balls (Fig. 5.59e). Residues can persist from chemical treatment (Fig. 5.59b). Although quality control at every step of fabrication and processing has made progress toward defect-free surfaces, it is generally impossible to ensure that there will be no defects, especially in large area cavities, or when dealing with hundreds of cavities, or perhaps nearly a thousand cavities for an x-ray free electron laser or nearly fifteen thousand cavities for a future linear collider.

One form of insurance against thermal breakdown is to raise the thermal conductivity of niobium by raising the RRR. With high thermal conductivity metal, any large (100 µm diameter) defect can tolerate more power before driving the neighboring superconductor into the normal state. Normal material quality control and treatment procedures should avoid such large defects. However with 10 000 cavities there is always a small chance of encountering a few.

A simple thermal analysis of thermal breakdown [1] shows that the quench field H_{max} depends on defect size (a), defect resistance (R_n), and Nb thermal conductivity (κ):

$$H_{max} = \sqrt{\frac{4\kappa(T_c - T_b)}{aR_n}}$$

Here T_c (9.2 K) is the critical temperature and T_b the bath temperature. The thermal conductivity of Nb has two components, one due to phonons and the other due to electrons which have not yet condensed into Cooper pairs [166, 340]. Below T_c the thermal conductivity falls exponentially as electrons condense into Cooper pairs, which do not carry heat. The phonon contribution is relatively small due to large electron–phonon scattering, so that the electron component dominates from T_c to about 3 K. The magnitude of the electron component of the thermal conductivity is inversely proportional to the residual resistivity of Nb due to impurity scattering of the electrons. Hence the thermal conductivity increases linearly with the RRR, which primarily depends on interstitial impurities such as oxygen, nitrogen, carbon, and hydrogen. Below 3 K the phonon contribution starts to become important as most of the electrons freeze out into Cooper pairs. Phonon scattering due to lattice defects and grain boundaries determines the size of the phonon peak (Fig. 3.6). Large grain Nb can have a substantial phonon peak. However, a large phonon peak does not significantly improve the quench field because the temperature of the superconductor in the vicinity of the defect rises close to T_c (H), where the electron component dominates.

The simple thermal model ignores many subtle features of heat transport, such as the temperature dependence of the thermal conductivity and the Kapitza resistance. Numerical thermal model simulations have been carried out to include such effects [338, 339]. Figure 5.60 shows thermal model results for the breakdown field of normal conducting defects [167]. For a 100 µm normal conducting defect the breakdown field is about 10 MV/m for RRR=40 Nb. The

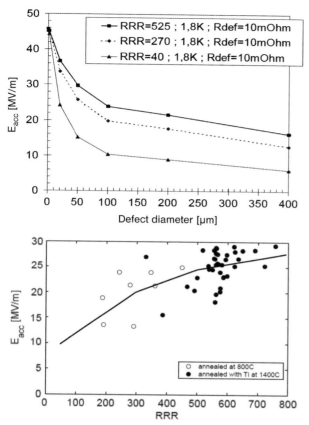

Fig. 5.60. (a) Thermal model calculation results for breakdown field versus defect size for a few RRR choices. The defect is assumed to be normal conducting Nb [167] (courtesy of DESY). (b) Spread in quench fields for BCP-treated 9-cell cavities (unbaked) tested at DESY [25]. There is a tendency for quench fields to saturate above 25 MV/m due to the high field Q-drop (courtesy of DESY). Line is for 100 μm defect.

RRR must be raised to 300 by using higher purity Nb in order to reach $E_{acc} = 20$ MV/m. With the higher thermal conductivity, the large defect will tolerate more rf dissipation before driving the neighboring superconductor into the normal state. Raising the RRR to 500 will allow $E_{acc} = 25$ MV/m. Raising the RRR to 500 will allow $E_{acc} = 25$ MV/m.

To compare theory with data, Fig. 5.60 (b) shows a large set of data on 9-cell cavities prepared by BCP [25]. The large spread in quench fields at any given RRR is indicative of the spread in defect size (or resistance). There is also a saturation of quench field around 30 MV/m, independent of RRR, even though the rf critical field is above 40 MV/m. This is due to the role of the high-field Q-drop above 25 MV/m prevalent for BCP-prepared cavities. Excess heating due to a strong high-field Q-drop raises the rf surface temperature and lowers the quench field. The simple thermal model for quench field does not take into account the surface temperature rise due to the high-field Q-drop, and therefore does not predict saturation at 30 MV/m.

As defects get smaller than 100 µm due to better material, inspection, and treatment procedures, performance near the rf critical field and the impact of larger RRR gets smaller. For gradients of 35 MV/m and higher, defect diameters need to become smaller than 20 µm! For such defects, doubling the RRR from 270 to 525 improves the gradient by less than 10%.

As a pleasant surprise, the introduction of EP and baking to address the high-field Q-drop also raises quench fields, as first discovered with 1-cell cavities [53], and shown in Fig. 5.61 (a). This conclusion is supported by the distribution of 9-cell results of Fig. 5.61 (b) [341]. Note here that EP-baked cavities also show a large spread in quench fields, as do cavities prepared by BCP.

One possible interpretation is that the EP treatment results in smaller defects, which would suggest that most defects come from the final preparation rather than from the material. Field enhancement at grain boundary steps for BCP cavities provides another possible explanation, if the edge of a defect coincides with a grain boundary edge.

Figures 5.26 and 5.27 for BCP-baked cavities show that increased BCP not only increases the high-field Q-drop but also decreases the quench field. Correspondingly, profilometry studies show increasing roughness with increasing BCP (Fig. 5.19). We can expect field enhancement factors as high as 2–3 for extensive BCP treatment (Fig. 5.28).

The recrystallized grains along the equator electron-beam weld are even more likely to cause breakdown in BCP-prepared cavities because of the greater roughness of the large-grain weld region where the step heights at grain boundaries is 30 µm or more, much greater than the 2–5 µm steps on the cavity surface. Moreover, the grain boundaries in the weld region are not randomly orientated, but nearly perpendicular to the magnetic field in the TM mode, which yields the largest geometric field enhancement. Grains have elongated dimensions, comparable to the half width of the weld (Fig. 5.25). In the heat-affected region, grains have smaller dimensions compared to the weld region and take random orientations. Indeed BCP cavities tested with thermometry at Saclay and KEK [53] show that quenches almost always occur at an equator weld seam or its vicinity. The quench location moves to other spots in the equator region after additional BCP, but the quench field does not improve much. However, after electropolishing the cavity for about 50 µm or more the breakdown field increases, the Q-slope decreases and the breakdown location shifts to a random region in the cavity, rather than at the e-beam weld. EP is well known to reduce the dimensions of surface irregularities and sharp edges are rounded. See for example Fig. 5.62 for the smoothening of the grain boundary step due to EP [302].

In summary, there is a large scatter in quench fields due to the large variety of defects that may be encountered. Results from BCP-treated cavities built from 300 RRR Nb suggest that defects are about 100 µm in diameter. The thermal model shows that improving the RRR to 600 by postpurification should in-

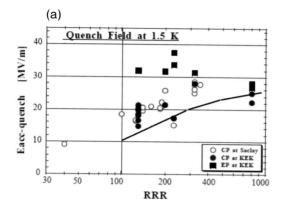

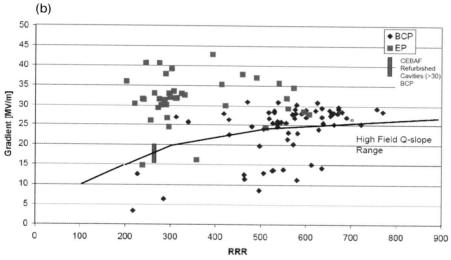

Fig. 5.61. Distribution of quench fields versus RRR for cavities prepared by BCP and by EP. (a) Single-cell cavities tested at KEK and Saclay [53]. The lines show the thermal model prediction for 100 μm diameter, normal conducting defects. (b) 9-cell cavities tested at DESY [341]. The large scatter reflects the spread in defect size and resistance typically encountered. Data on more than 30 CEBAF refurbished 5-cell, 1.5 GHz cavities are included [342]. In order to make a proper comparison to DESY results, the accelerating field for the CEBAF re-furbished cavities (250RRR) is re-normalized to $H_{pk}/E_{acc} =$ 42 mT/mV/m (instead of 47) for CEBAF cavity shape. The BCP cavities are more frequently limited by high-field Q-slope so the quench fields become limited at 30 MV/m [343]. The line shows the thermal model prediction for 100 μm diameter normal conducting defects.

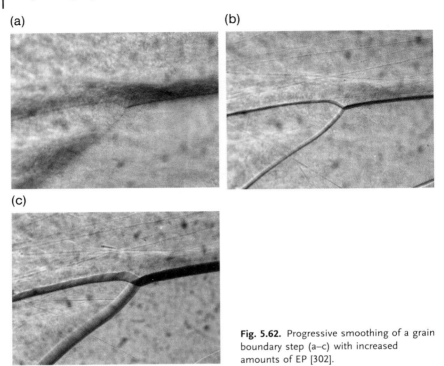

Fig. 5.62. Progressive smoothing of a grain boundary step (a–c) with increased amounts of EP [302].

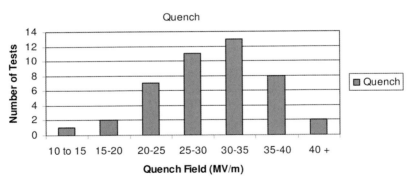

Fig. 5.63. Spread in quench fields for a large number of 9-cell cavities all prepared by EP [343].

crease the quench field to above 30 MV/m, but the high-field Q-drop saturates the quench field to around 30 MV/m. Electropolishing improves the situation dramatically, either by reducing the defect size or by providing a smooth surface without regions of strong field enhancements, such as at grain boundary edges, especially on the equator weld. The spread in the quench field for EP-prepared cavities is still rather large as shown by results of Fig. 5.63.

**Part II
Technology**

6
Cavity Fabrication Advances

6.1
Overview

The main cavity fabrication procedure, briefly reviewed here, has remained essentially the same as described in [1]. We will point out some advances and potential developments that are likely to take place in the coming years. We restrict our discussion to methods which have been successfully used to fabricate cavities with rf test results, and leave out many interesting new ideas that have not yet matured to the point of high quality cavity results.

Niobium sheets can be ordered from several suppliers to meet well-defined cavity specifications. The sheets arriving at the cavity vendor (or controlling laboratory) are inspected for flatness, uniform grain size, near-complete recrystallization, RRR value, and good surface quality, such as absence of scratches. Since the many fabrication stages from ingot to sheet can embed defects, each sheet is scanned with eddy current scanning, or more sensitive squid scanning. Scanning developments are reviewed in Section 6.2.2. The better side of the sheet is selected for the rf surface. Half-cell cups are stamped, spun or hydroformed, checked with the coordinate measuring machine (CMM) for correct shape, then final trim machined for weld preparations. Soaking the cups in dilute hot sulfuric acid for a few hours is a simple method to remove any embedded iron particles (e.g., from the rolling mill). The cups are checked by a rust test for any remaining iron inclusions by soaking in water for 12 h. Cavity parts are given a light etch (20 μm) before electron beam welding (EBW), rinsed, dried in a clean room, and kept clean in sealed nylon bags before welding. Improved welding parameters developed for a 2.8-mm-thick niobium equator and iris welds are discussed in Section 6.3. To avoid RRR degradation, the vacuum in the electron beam welder should be better than 2×10^{-5} torr. Sufficient time should be allowed between welds for parts to cool down to avoid the likelihood of holes with the standard welding parameters. Automation of sequential welding is desirable for large-scale cavity production required in future projects. All welds are inspected for complete, smooth underbead, flat on the inside, and no weld spatter (which is rare for the correct parameters). In general, welds carried out with the correct parameters do not limit performance as shown by the absence of temperature map signals on single-cell R&D cavities. However, weld defects have been found on occasion, especially during the development stages for new cavity vendors. Above 25 MV/m, BCP cavities tend to quench at equator e-beam welds as discussed in Chapter 5. Dimensional checks are carried out on the structure all through the fabrication stages and after completion.

The sheet-metal forming, machining, and welding method for cavity fabrication is time-consuming and labor intensive. Therefore, development continues on several alternate cavity fabrication methods for large-scale cost reduction, but none of these methods have reached the stage of maturity to supplant the standard fabrication methods. The high cost of the end-groups containing couplers reduces the impact of cost savings introduced by seamless cavity fabrication methods. Single cells from 500 MHz to 6 GHz have been successfully spun from Nb sheets or tubes. 9-Cell, 1.3 GHz cavities have been spun from a single sheet and from a tube, but there are thickness variations which need to be controlled in the future. Using numerically controlled spinning machines and spinning from two diametrically opposite points simultaneously promises improved reliability. The approach of spinning seamless multicell cavities has advanced by spinning from tubes instead of sheets. Hydroforming single-cells and 3-cells has also made progress. The number of cells is presently limited by the capacity of the hyrdoforming machine, but the general multicell techniques have been developed and proven. Spinning and hydroforming single-cell cavities have been successful from niobium–copper composite sheet material where the composite is fabricated by a variety of methods including explosion bonding, hot rolling, and hot isostatic pressing (HIP). Seamless 500 MHz single-cell cavities have been spun from composite Nb–Cu sheets. We will briefly discuss these seamless forming techniques and results in Section 6.5.

Thin-film deposition by sputtering continues to suffer from the problem of Q-slope which starts at low fields. Research continues to identify the possible causes as does work on alternate methods of coating. Several different approaches are under trial to increase the energy of Nb atoms arriving at the substrate. These include dc bias sputtering, bias magnetron sputtering, vacuum arc deposition, and ECR deposition. We will briefly discuss these techniques at the end of this chapter.

6.2
Niobium Material

6.2.1
Specifications

The key specifications of sheet niobium are RRR > 300, with Ta content less than 500 wppm, yield strength greater than 50 MPa (N/mm^2), tensile strength > 100 MPa and percent elongation greater than 30%, uniform grain size ASTM 6 (50 µm), thickness variations ±0.1 mm, and planarity tolerance < 0.5 mm.

Table 6.1 Technical specifications to niobium sheets for XFEL cavities [a].

Concentration of impurities in ppm (weight)				Mechanical properties	
Ta	≤500	H	≤2	RRR	≥300
W	≤70	N	≤10	Grain size	≈ 50 μm
Ti	≤50	O	≤10	Yield strength, $\sigma_{0.2}$	$50 < \sigma_{0.2} < 100$ N/mm² (MPa)
Fe	≤30	C	≤10	Tensile strength	>100 N/mm² (MPa)
Mo	≤50			Elongation at break	30%
Ni	≤30			Vickers hardness HV 10	≤60

a) No texture: The difference in mechanical properties (Rm, Rp0.2, AL30) orthogonal and parallel to main rolling direction <20% (cross rolling).

Table 6.1 gives the specifications asked for by DESY [182, 344] to fabricate 1.3 GHz, 9-cell cavities for XFEL, which will be the next large-scale fabrication project (1000 cavities).

Generally, the yield strength of Nb decreases with increasing grain size. Yield strength also increases dramatically at low temperatures with typical increase of a factor of 5 between room temperature and 77 K [345]. Depending on the RRR, the yield strength of fine grain material drops after the 800 °C heat treatment required for H removal. This effect prompted some labs to use 600 °C, 10-h heat treatment for H removal instead. Of course, substantial yield strength drop occurs after postpurification treatment at 1300 °C.

Reproducibility of mechanical properties from sheet to sheet still presents issues related to microyielding [346], and spring-back. Half-cells formed from different sheets or prepared in different batches finish with different resonant frequencies, grain size distribution, and grain texture.

6.2.1.1 Availability of High RRR Nb

Nb material has been produced to the required specifications by various companies, ATI Wah Chang (USA) [347], W.C. Heraeus (Germany) [348], and Tokyo Denkai (Japan) [349], and used for cavity fabrication to yield high performance prototype and production cavities. Other companies interested in niobium production for cavities are Plansee (Germany) [350], Ningxia (China) [351], Cabot (USA) [352], CBMM (Brazil) [353], H.C. Starck (Germany) [354], and Giredmet (Russia) [355]. Qualification of further niobium vendors is underway at DESY [356] and various cavity laboratories around the world.

Typical pure Nb production capacity is 10 tons/year for a 500-kW electron beam melting furnace [357]. With 40 000 tons annual production for a range of broad applications such as high strength alloys, there is no shortage of raw Nb material. But the availability of high purity, high-RRR sheet is limited by the number of high-purity melting facilities. The ILC needs of 500 tons over a few years could be supplied by existing facilities.

Fig. 6.1 Electron beam melting furnace at Wah Chang in Albany, Oregon [358].

Fig. 6.2 Finished electron-beam melted ingot (diameter 15 inch) from Wah Chang [358]. Inset: Typical grain structure of standard ingot [182].

6.2.1.2 Sheet Production from Ingot

Figure 6.1 shows an electron beam melting furnace at Wah Chang in the USA [358]. High RRR is obtained by electron beam melting the ingot several times. The final RRR is generally limited by the residual vapor pressure in the chamber [359]. Tokyo Denkai melts six times to achieve a narrow spread in RRR between 295 and 370 [357]. As a result of the multiple melts necessary, electron beam melting purification becomes the largest component of the material cost. There is potential for reducing this part of the cost by improving the vacuum in the furnace [359]. As mentioned in Chapter 5, except for the large (>100 μm) defect, the benefits of RRR > 300 diminish rapidly as performance approaches critical field levels due to reduction of defect size through good quality control on materials, and EP.

Starting from a large ingot (Fig. 6.2), sheets are produced after several stages of forging, grinding, rolling, cleaning, and annealing, as depicted in Figs. 6.3 to

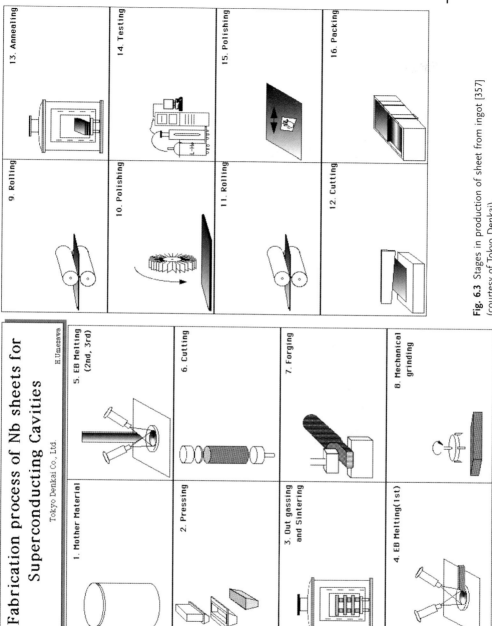

Fig. 6.3 Stages in production of sheet from ingot [357] (courtesy of Tokyo Denkai).

Fig. 6.4 200 ton press forge [360] (courtesy of H. C. Starck).

Fig. 6.5 Cold rolling mill at Wah Chang [358].

6.5. Final annealing between 750 and 800 °C is essential to recrystallize the material as shown in Fig. 6.6. The annealing time and temperature must be chosen carefully taking into consideration the RRR so that recrystallization will be nearly complete, but the temperature must not be too high so as to allow fine and uniform grain size (50 µm) for best forming properties.

6.2.2
Eddy Current and SQUID Scanning

An important step in the characterization of niobium sheets is eddy current scanning, a nondestructive technique, first developed by DESY [25, 361]. The basic principle (Fig. 6.7a) is to detect the alteration of the eddy currents with a

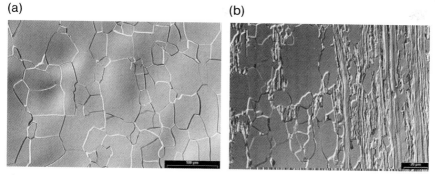

Fig. 6.6 (a) Grain structure for fully recrystallized sheet; scale bar is 100 μm. (b) SEM micrograph of Nb sheet with incomplete recrystallization; scale bar is 20 μm [182] (courtesy of DESY).

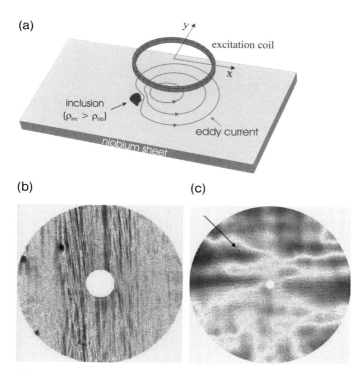

Fig. 6.7 (a) Principle of eddy current scanning. (b) Rolling marks and a defect are visible in an eddy current scan [182]. (c) An eddy current scanning defect analyzed using Synchrotron Radiation Phosphorescence Analysis to have Cu and Fe [182] (courtesy of DESY).

double-coil sensing probe to identify inclusions and defects embedded under the surface. One key element for high-performance scanning is the size and stability of the gap between sensing head and sample. In general, the nearer the head is to the sample, the stronger and sharper is the signal generated by a defect. Several methods are used to keep the gap stable in the presence of sheet unflatness: pneumatic probe levitation and compressed air blown through a spring mounted probe. With a sheet size of about 300×300 mm^2 and a line width of 1 mm, a scan of one sheet lasts about 15 min.

In one commercial version, designed to scan disks, the sample is fixed to a rotating table, attached to a linear slider, while the sensing head is fixed. The sample rotates several times allowing for data acquisition, before the relative position of the head changes by sliding the table to repeat data acquisition over a different circumference. Multiple frequencies can be used for simultaneous investigation over several depths.

Pits, inclusions, scratches, and roller marks are the typical defects found, but many of these can also be detected by simple optical inspection. Figure 6.7 (c) shows an inclusion later identified to contain Cu and Fe using synchrotron radiation phosphorescence analysis [182]. Marks left by the rolling stages, usually not visible with the naked eye, are not a cause for sheet rejection. Such defects are not always eliminated by the subsequent processing (etching and annealing) of the material.

The conventional eddy current system has limited sensitivity (~0.1 mm depth). SQUID detectors for measuring the eddy current's secondary magnetic field improve sensitivity and provide an excellent signal/noise ratio [362, 363]. The low-T_c SQUID is situated within a 1.5-L fiber glass He cryostat fixed along the z-axis (Fig. 6.8). The coil diameter is selectable between 1 and 3 mm. In order to maximize the resolution, the magnetic field of the excitation coil can be minimized at the sensor location by a fine adjustable compensation current. The SQUID is used with a flux-compensating amplifier. The amplifier output is then processed by a lock-in amplifier. The sensitivity of the method is remarkable. Even with an operating frequency of 90 kHz, when the skin depth of eddy

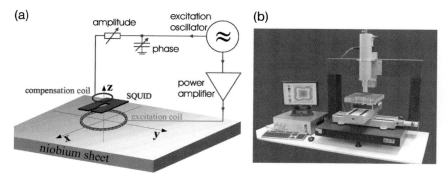

Fig. 6.8 (a) Principle of the SQUID scanning technique.
(b) SQUID scanning apparatus at DESY [362, 363] (courtesy of DESY).

6.2 Niobium Material

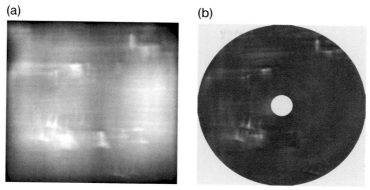

Fig. 6.9 Comparison of (a) SQUID and (b) eddy current scanning of the same Nb plate [362, 363] (courtesy of DESY).

currents in Nb is 0.16 mm, scratches on the back of the sheets (thickness: 2.8 mm) can be easily detected! By contrast, for standard eddy current scanning at 170 kHz (skin depth 0.12 mm) none of the scratches at the back of the sheet are visible. Figure 6.9 compares the results of higher resolution SQUID scanning with the more conventional eddy current scanning.

The entire SQUID-based system works in a nonshielded environment. The scanner is on an xyz table with 300 mm×300 mm travel area. The Nb sheets are fixed by a vacuum sample holder in order to keep them as flat as possible. The distance between the sensor system and the surface is measured, and the signal is corrected for the difference occurring due to a lift-off between the sample and excitation coil. Because of SQUID's low noise, measurement speeds of up to 100 mm/s are possible, and with distances of 5 mm between the scan lines. As a result, a 30×30 cm^2 sized sheet of Nb can be scanned within less than 5 min.

6.2.3
Large-Crystal and Single-Crystal Niobium

Starting at JLab [323, 346] and followed by many laboratories, niobium cavities have been produced from large grain ingot material and from single-crystal Nb, with starting sheets cut directly from the ingot by either wire EDM or diamond saw cutting. Several Nb producers can now provide large grain sheets: CBMM, Ningxia, W.C. Heraeus, and Wah Chang. Figure 6.10 shows a large grain ingot with a large central, single crystal, and cuts with a diamond saw.

6.2.3.1 Material Properties
The availability of large grain and single grain Nb is an exciting new development, with the potential of simplifying the sheet production sequence and consequently the material cost. Initial experience with large grain niobium indi-

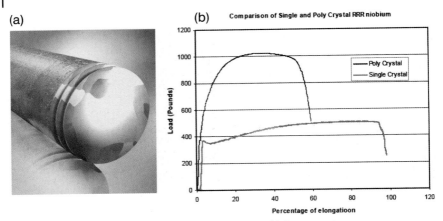

Fig. 6.10 (a) Large grain ingot from W.C. Heraeus [364]. (b) Stress–strain curve comparing the elongation of single crystal to fine grain material [323, 346, 368] (courtesy of JLab).

cates that very smooth surfaces can be obtained using the BCP etch process only. This feature opens the possibility of replacing the complex EP procedure with simple chemical etching for gradients less than 30 MV/m. Figure 5.22 shows AFM pictures of the comparable microroughness of single-grain material after BCP and EP treatments.

The process of producing sheet from ingot material has expenses associated with many stages of forging, grinding, rolling, cleaning, and annealing (Fig. 6.3) with loss of material in some of these steps. There is also a danger of introducing defects at any of these stages. One cost analysis [365] shows that the dominant cost of high purity niobium production comes from the multiple electron beam melting operations, with about 15% of the cost from the forging, rolling, and annealing steps. On the other hand, the cost of slicing ingots has not yet been fully determined. Wire EDM which provides a very smooth starting finish is very slow, taking 1 day per sheet. Diamond saw cutting is faster and has proven successful [364]. But the overall cost advantage of large grain material is still unclear.

Figure 6.11 shows a few example sheets sliced directly from large grain ingots. One has a large single crystal in the center and was used to make 2.2 GHz single-crystal cavities discussed in Chapter 5.

Mechanical properties of the large grain and single-grain material provide some superior and some inferior features compared to fine grained material. The yield strength for single-crystal material is lower (Fig. 6.10b) than for as-received fine grain material. But microyielding in fine grain complicates such a direct comparison. A single-crystal sample from CBMM ingot showed an extremely large elongation of nearly 100% – a factor of 2 larger than obtained with fine grain polycrystalline material as indicated in Fig. 6.10. This discovery at JLab [346] led to the first interest in exploring large grain and single grain niobium for cavities. However,

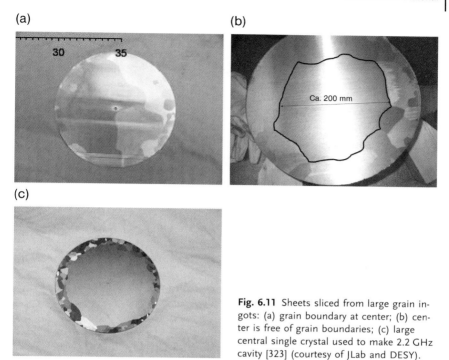

Fig. 6.11 Sheets sliced from large grain ingots: (a) grain boundary at center; (b) center is free of grain boundaries; (c) large central single crystal used to make 2.2 GHz cavity [323] (courtesy of JLab and DESY).

the percentage elongation depends significantly on the load direction as shown in Fig. 6.12 [366]. Moreover, the elongation for large grain material in two-dimensional deformation determined from the bulge test is rather low (<15%). The final rupture takes place close to a grain boundary.

6.2.3.2 Fabrication from Large Grain

Half-cells from large and single grain can be formed from sheet by spinning or by deep drawing. The single crystal forms very well. But several difficulties show up during the fabrication of cavities from large grain materials (Figs. 6.13 and 6.14). Large grain sheets deep draw with ragged edges at the equator. This type of "earing" is an indication of the dependence of the mechanical properties on crystal orientation. A displacement (0.5 mm) often occurs between adjacent grains with large steps at the equator region. Sometimes the material thins or rips at the irises, if the grains "meet" in these areas. Therefore, it is recommended to start with sheets without any grain boundaries in the center. There is some spring-back after the deep drawing, making the half cells "oval." The same happens after trim machining for EBW due to internal stresses built up.

Assembly for EBW is sometimes more difficult than with fine grain material. However, there is no problem with single crystals. The variation of the half-cell

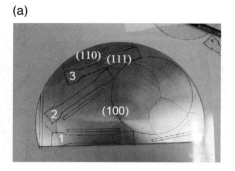

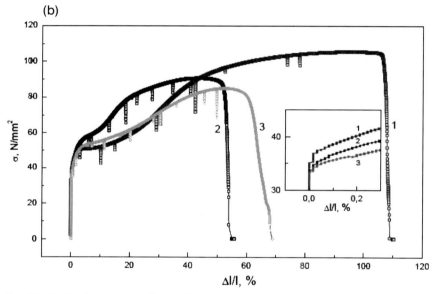

Fig. 6.12 (a) Single-crystal samples cut from three orientations of the large grain disk. One disk-shaped sample with grain boundaries was cut for the bulge test. (b) Stress–strain curves for different orientations [366] (courtesy of DESY).

shape is slightly more pronounced than cells formed from polycrystal material. This makes equator welds a bit more complicated. Assembly takes more time and needs special tooling [367]. The dependence of mechanical and etching properties from the crystal orientation need to be better understood.

The maximum gradient possible with large or single-grain Nb cavities is no higher than for fine grain cavities. The spread in gradients for large grain cavities is also comparable to fine grain cavities. Many single-cell 1.3 GHz cavities have been fabricated from large grain niobium and tested by several laboratories. Most cavities have been treated by BCP and a few by EP. The mean surface magnetic field reached after about 50 tests on single cells at JLab after BCP

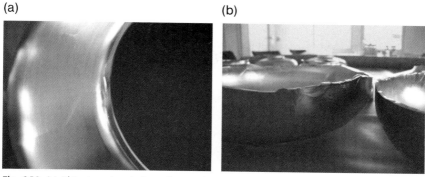

Fig. 6.13 (a) Thinning at grain boundary in the iris region; (b) strong earing and grain steps at the equator region [367] (courtesy of ACCEL).

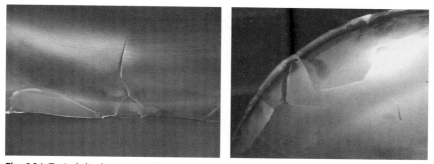

Fig. 6.14 Typical displacements (0.5 mm) found between neighboring grains [367] (courtesy of ACCEL).

is about 140 mT, 10% lower than the mean of similar number of tests on single cells with fine-grained material prepared by EP (Fig. 6.15). Better EP treatments discussed in the next chapter show far narrower spread for single cells prepared by fine grain material. Most of the tests in Fig. 6.15 were limited by quench between 120 and 160 mT, which casts doubt on the initial high expectation that starting from ingot material is likely to reduce the occurrence of defects that quench. While many fine-grain, single-cell, and 9-cell cavities have reached H_{pk} between 150 and 170 mT, only a few large grain single cells have exceeded 150 mT, and some of these were treated by EP. In one series of tests [182], large grain single-cell cavities were treated by EP, followed by BCP, followed by EP (Fig. 6.16). The results show higher gradients for EP treatment, and decrease in gradient for BCP. This is most likely due to the grain boundary steps, as discussed in Chapter 5

Several large grain 9-cell cavities have also been fabricated and tested at DESY (Fig. 6.17) and JLab. Test results at DESY after BCP show maximum gradient limited to about 30 MV/m [182, 371, 372] and can be improved to 40 MV/m with EP (Fig. 6.16).

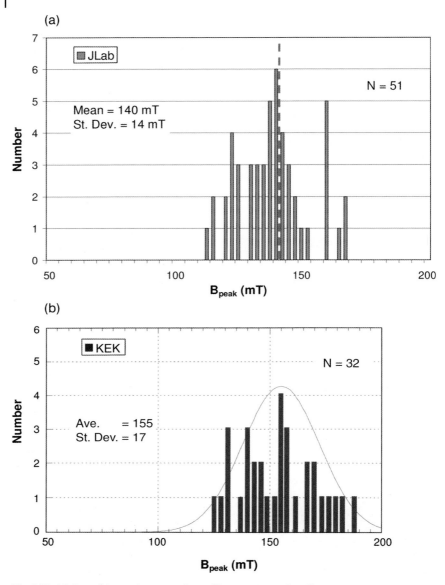

Fig. 6.15 (a) Spread in maximum gradient of large grain single-cell cavities prepared by BCP only [320, 368] (courtesy of JLab) (b) Spread in maximum gradient of fine grain single-cell cavities prepared by EP. Improved EP treatments give a much narrower spread [369, 370] (courtesy of KEK).

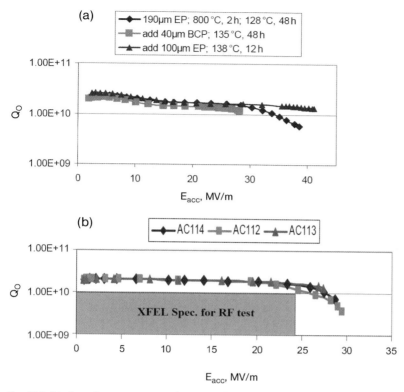

Fig. 6.16 (a) Q vs. E curves measured at DESY for three large grain 1-cell cavities after EP, BCP and EP. Best gradients are obtained after EP. All tests carried out after mild bake [182]. (b) Q versus E curves measured at DESY for three large grain 9-cell cavities after 100 μm BCP, 800 °C, 20 μm BCP, and HPR [371, 372] (courtesy of DESY).

Fig. 6.17 A large grain 9-cell cavity fabricated at DESY [182].

6.2.3.3
Single-Crystal Nb

The largest single-crystal sheet (Fig. 6.11c) obtained is about 18 cm in diameter instead of 25 cm required for a 1.3-GHz half-cell. Smaller crystals form near the outer perimeter presumably due to cooling at the copper crucible, which holds the ingot during the melt. Perturbations such as thermal gradients, vibrations, and fluid flow cause nucleation off the crucible wall.

Single-crystal cavities at 2.2 GHz were first successfully fabricated at JLab using the central single grain of a large grain sheet (Fig. 6.18) [366]. The best result after postpurification with titanium at 1300 °C, BCP, and *in situ* bake (120 °C) is a surface magnetic field of 160 mT. Before postpurification, the cavity quenched at lower gradients.

Another approach to obtain large enough material to make single-crystal 1.3-GHz cavities is to roll a small diameter thick sheet from the central single crystal to the required thickness and diameter with intermediate annealing steps [372]. If the thickness reduction is less than a factor of 2, it is possible to maintain the single-crystal structure, and to heal the dislocations by annealing at 800 °C. At DESY, two half-cells made by this method were electron beam welded together. By careful orientation of the halves before welding, it was possible to obtain a single-crystal cavity without a grain boundary in the weld seam (Fig. 6.18). The cavity was tested at JLab. The first set of tests after fabrication with 40 μm and 60 μm etching by BCP showed field emission which could be processed. A test after 80 μm BCP was stopped by a quench at 126 mT. After several successive BCP treatments, the cavity reached 150 mT (37.5 MV/m). The total amount of material removed was about 110 μm. After another 20 μm etch, the cavity quenched at a lower field of 134 mT.

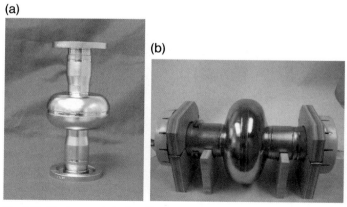

Fig. 6.18 Single-crystal single-cell cavities: (a) 2.2 GHz at JLab [323]; (b) 1.3 GHz at DESY [372].

6.3
Advances in Welding

With the expansion of superconducting cavity application to accelerators, EBW capability for cavities has increased worldwide. Figure 6.19 shows the beam welder at JLab preparing for a multicell cavity. A good vacuum in the electron beam weld chamber is essential to maintain RRR > 300. Figure 6.20 shows the degradation with chamber pressure. Extensive measurements [182] have shown that the welding does not deteriorate the RRR if the vacuum in the welding chamber is better than 2×10^{-5} mbar. Other studies have yielded more relaxed vacuum specifications. Samples welded at 4×10^{-5} mbar indicate no drastic change of RRR across the weld region [373].

A smooth weld underbead is essential for high field performance. This can be achieved with defocused beam welding or by using a raster with a rhombic or circular pattern as described in [1]. The circular deflection pattern is more favorable for thicker welds. The resulting defocused welds have a wide molten puddle which produces a smooth underbead due to high surface tension. However, there are thickness limitations for defocused welds. While it is possible to weld thicker materials, the beam current must be increased close to the regime of blowing a hole. Tolerances become tight. Small variations in material thickness or fit-up between mating parts lead to blowing a hole. Undesirable heavy underbeads appear.

One solution for welding thicker material is to reduce the weld thickness to 2 mm or less to achieve the desired inside smooth finish. A combination of inside and outside welds can be used to reduce the chance of blowing a hole but this requires multiple setups. Multiple welds also make shrinkage and distortion prediction difficult.

Another approach to weld thicker material and still maintain a flat underbead is to add a small undercut to the underside of the weld (Fig. 6.20b) [374]. The properly sized undercut is filled in by the normal molten bead drop-through,

Fig. 6.19 Electron beam welder showing vacuum chamber, multicell cavity, and electron gun [182] (courtesy of JLab).

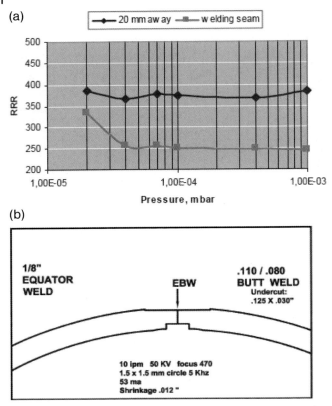

Fig. 6.20 (a) Degradation of RRR with vacuum level in the electron-beam welder [182] (courtesy of DESY). (b) Weld geometry for smooth defocused weld on thick material [374].

and leaves a near-flat surface. By this technique full penetration welds with no protruding underbead can be obtained with nearly 3-mm-thick niobium. The method also gives more reproducible shrinkage, minimal distortion, and wider margin for the beam current needed to produce a good weld. Still it is important to maintain tight tolerances and precise weld setups. The edges of the undercut should be rounded in case the weld does not fully consume the step area. All full penetration welds should be inspected for smooth underbead and no spatter. Figure 6.21 shows a special optical inspection apparatus used at DESY.

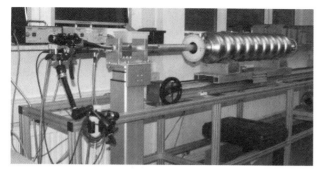

Fig. 6.21 Optical inspection apparatus for nine-cell [182] (courtesy of DESY).

6.4
Low-β Resonators

Most low-β resonators are made from bulk niobium with high RRR (150–300). The resonator parts are formed from Nb sheets, or machined from rods and plates. Parts are joined together by EBW in high vacuum (10^{-5} torr). Even rather complicated shapes can be obtained with this technology. Sometimes the final structures include a He vessel made of normal grade niobium or stainless steel. Figure 6.22(a) shows the components of a 115 MHz QWR for $\beta = 0.15$

For a more specific account of the fabrication sequences, we discuss a 3-spoke resonator shown in Fig. 6.22 [376]. Except for the beam ports, which were machined from bar stock, all niobium elements were formed from 1/8-inch sheet. All niobium, except for the support ribs, was selected with RRR >250. The cavity spherical end walls and the spoke elements were hydroformed. The spokes were formed in halves and seam welded together. Transition rings were formed and welded to the ends of the spokes to provide a transition to the cylindrical

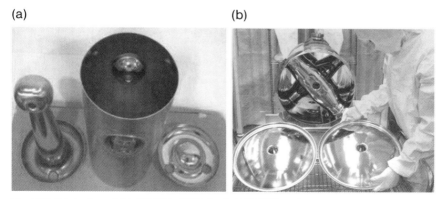

Fig. 6.22 Fabrication of (a) 115 MHz QWR for $\beta = 0.15$ [375] and (b) a three-spoke niobium resonator [67] (courtesy of ANL).

housing with a blend radius of 1/2 inch. All electron beam welds were performed at pressures below 10^{-5} torr, and the joined parts were cooled in vacuum below 50 °C before venting the welding chamber. Tuning and preliminary surface processing were performed when the three major subassemblies of the cavities were complete. The subassemblies are the two spherical end walls, complete with beam ports and support ribs, and the cylindrical body of the cavity with coupling ports and all three spokes. Prior to welding the three sections together, each was heavily electropolished, removing 150–200 µm of material to eliminate any damage caused by forming and machining.

The niobium cavity shell is contained in an integral stainless-steel helium jacket. Two 2-inch diameter coupling ports and a helium port can be seen at the top of the cavity. Also visible are several of the niobium ribs welded to the exterior of the niobium shell for mechanical stiffening. The jacket is joined to the niobium shell at the beam ports and at the coupling ports using a copper braze joint. After EBW the three niobium sections together, the helium jacket was clam-shelled into place and welded together. The niobium–stainless braze transitions were then welded to the niobium shell to complete the assembly.

6.5
Seamless Cavities

6.5.1
Hydroforming

Since the equator-to-iris diameter ratio of typical elliptical cavities is about 3, any seamless forming technique presents significant challenges [377]. Hydroforming a cavity starts with a tube of a diameter intermediate between iris and equator. The forming takes place in two stages: reduction of the tube diameter (necking) to form the iris area and hydraulic expansion (hydroforming) for the equator area. Figures 6.23 and 6.24 show the necking principle and the apparatus used. Figures 6.25 and 6.26 show the hydroforming principle and the appa-

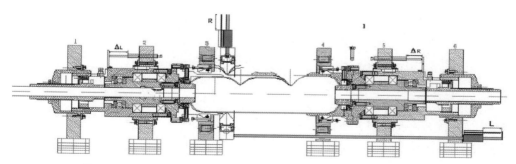

Fig. 6.23. Principle of "necking" tube diameter reduction in the iris area [378] (courtesy of DESY).

Fig. 6.24 Apparatus used at KEK for the necking stage [378, 379].

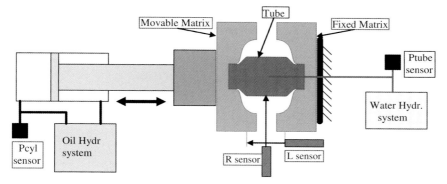

Fig. 6.25 Principle of hydroforming [182] (courtesy of DESY).

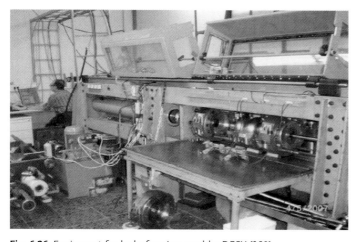

Fig. 6.26 Equipment for hydroforming used by DESY [182].

ratus used. A larger than iris initial tube diameter results in moderate work hardening, easing the second stage of hydroforming. On the other hand, a large starting tube diameter will increase the roughness at the iris area during the first stage of diameter reduction. Optimization of hardness and roughness yields a tube diameter between 130 and 150 mm for 1.3-GHz cavities. Material properties are carefully checked for selection of the best available properties. During hydroforming, the main elongation occurs in the circumferential direction. Therefore, a conventional tensile test on samples in the orientation parallel to final cavity circumference is necessary for final material qualification. Here the two-dimensional hydraulic bulge test is very helpful where a disk sample is hydraulically expanded into a spherical form because it checks the anisotropy of plastic properties.

The best necking results were achieved using a specially profiled ring moving in radial and axial directions. During hydraulic expansion, with the internal pressure applied to the tube, a simultaneous axial displacement is necessary to form the tube into an external mold of the cavity shape. The hydraulic expansion relies on the use of the correct relationship between applied internal pressure and axial displacement, assuming that the plastic limit of the material is not exceeded. A problem of excessive thinning on the end iris has been solved by proper choice of the forming die shapes. Tensile tests show that elongation can be increased nearly 30% by applying periodic stress fluctuations (pulse regime), instead of a monotonic increase of stress. In addition, it is necessary to control the strain rate so that the deformation is slow. By keeping the strain rate below 10^{-3} s^{-1}, a 10% higher strain can be obtained.

After forming, the arithmetic mean roughness, R_a, inside the cavity is normally several μm larger compared to deep drawn cells. In many cases, the surface roughness was reduced to acceptable values after standard BCP or EP treatment. However, the combination of hydroforming and centrifugal barrel polishing (CBP) before EP or BCP results in an overall better smoothness [377].

Several single-cell cavities have been manufactured without intermediate annealing starting from spun or deep drawn tubes with an inner diameter of 130 or 150 mm. The highest achieved accelerating gradients are comparable for both seamless and welded versions as shown in Fig. 6.27 [377]. The accelerating field is 43 MV/m with a high Q-value of $>1.5\times10^{10}$. Several two-cell and 3-cell Nb hydroformed cavities successfully followed (Fig. 6.28). A 9-cell hydroformed cavity has been completed by EBW three 3-cell cavities along with stiffening rings (Fig. 6.29).

6.5.1.1 Niobium Clad Copper Hydroformed Cavities

The laminated Nb/Cu approach offers the advantages of the bulk Nb performance since the Nb layer is 0.5–1 mm thick, combined with the high thermal conductivity and stiffness of the copper backing. The reduced quantity of Nb promises significant cost savings [377]. Stiffening against Lorentz forces and microphonics is possible by increasing the thickness of a Cu layer. The Nb layer

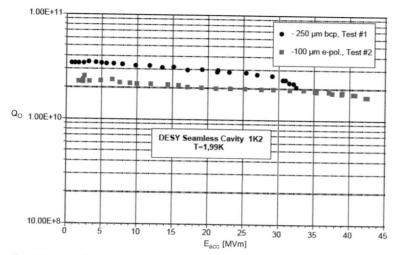

Fig. 6.27 One of the best results of hydroformed bulk Nb single-cell cavities [377] (courtesy of DESY).

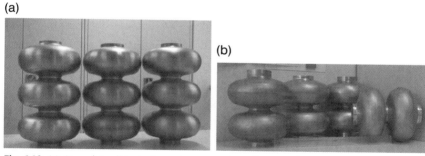

Fig. 6.28 (a) Several 3-cell hydroformed cavity sections for a 9-cell cavity [378]. (b) Several Nb-Cu hydroformed cavities [377] (courtesy of DESY).

Fig. 6.29 A 9-cell hydroformed cavity by welding together three 3-cell sections [378] (courtesy of DESY).

has the same microstructure and properties of bulk Nb. Therefore, surface treatment such as BCP and EP can be the same as for bulk Nb cavities, but heat treatment for H removal can be a problem at 600 °C where copper softens and may diffuse significantly into the bulk niobium. Cu–Zr alloy has been studied as an alternative.

Hydroforming has been successfully used to produce seamless high gradient single-cell and 2-cell cavities at DESY and KEK [377, 379]. Nb/Cu-laminated material with 3 mm of copper and a 1 mm Nb layer was first formed into a tube. Explosion bonding, back extrusion, and hot rolling techniques were successfully used to produce the starting composite sheets and tubes [380, 381]. Many areas are still under further development including the Nb–Cu bonding technique. The best method of making the composite is not yet clear among several available choices: explosion bonding (DESY), hot rolling (KEK-Nippon Steel Co.), back extrusion (DESY). In explosion bonding, an explosive charge drives a high-velocity angular impact of two metal surfaces at very high speeds creating huge contact pressure. For example, a back-extruded seamless Nb tube (4 mm wall thickness) has been explosively bonded with an oxygen-free Cu tube (wall thickness about 12 mm) and then flow formed into a Nb–Cu tube of 4 mm wall total thickness (1 mm Nb plus 3 mm Cu).

In hot bonding, a composite of three concentric tubes or three plates Cu/Nb/Cu is prepared with a vacuum of 10^{-5} to 10^{-6} mbar between the layers. The composite is welded. The combined procedure is called "canning." Thus, the Nb layer is protected against contamination at high temperatures by a Cu shield. Subsequently, hot rolling or hot extrusion produces a copper clad niobium tube or sheet with reduced wall thickness. The extra layer of copper is removed and the composite is ready for hydroforming or spinning cavities [380].

Being highly plastic, the Cu plays the lead role during the forming process. Although the niobium layer is harder and less plastic, it is much thinner so that Nb follows the Cu during forming, as long as the bond is strong. However, reduction of the tube diameter in the iris area (during the necking stage) has a tendency to form cracks. Intermediate annealing of the copper layer is an option. Copper can be fully recrystallized by annealing at 560 °C for 2 h with the grain size about 30 μm, and acceptable for hydroforming elongation.

Attaching end flanges or end groups containing coupler ports requires welding to the thin niobium layer. The copper backing must be completely removed and cleaned at the weld joints before EBW, which presents some risks of copper contamination resulting in leaky joints. The Nb tube of thickness down to 0.7 mm has been successfully welded to 2 mm thick niobium tubing. An alternative approach would be to attach the end groups fabricated from bulk Nb using superconducting joints, under development at JLab [382].

Examples of Nb/Cu single-cell cavities produced at DESY by hydroforming from an explosively bonded tube are shown in Fig. 6.30 [383]. The highest accelerating gradient achieved on seamless Nb/Cu clad single-cell cavities is comparable to that reached on bulk Nb cavities. The preparation and testing was carried out at JLab, 180 μm BCP, annealing at 800 °C, baking at 140 °C for 30 h.

(a)

Some prototypes of Nb/Cu laminate cavities fabricated at DESY. Singer SRF-05

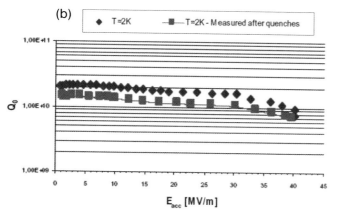

Fig. 6.30 (a) Single-cell Nb–Cu cavities fabricated by hydroforming [377]. (b) Best test result on a Nb–Cu hydroformed cavity [383] (courtesy of DESY).

One of these cavities achieved 40 MV/m (Fig. 6.30). A similar result was achieved at KEK from a hot rolled bonded tube [377]. Another Nb/Cu single-cell cavity produced at DESY from sandwiched hot roll bonded tube (provided by Nippon Steel Co.) was prepared and rf tested at KEK, also with excellent performance results [380].

6.5.2
Spinning

Spinning cavities from a sheet has been discussed in [1]. Several single-cell cavities have been spun starting from disks and from seamless tubes [384]. After spinning, cavities must be tumbled or mechanically ground at least 100 μm to remove surface fissures before chemical treatment. Q-values over 10^{10} and accelerating fields over 40 MV/m have been reproduced with spun single-cell cavities.

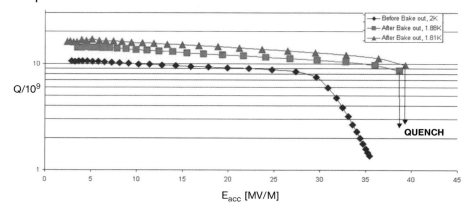

Fig. 6.31 The best result of a single-cell spun cavity [182, 378] (courtesy of INFN and DESY).

The best result is shown in Fig. 6.31 which provides the proof-of-principle soundness for the method.

Spinning multicell cavities from a single sheet presents many challenges. After the forming operation of each cell, the material remaining to be formed becomes cone-shaped, progressively smaller in diameter, and thinner with increasing number of cells. Several collapsible mandrels must be used as the multicells evolve. The spinning parameters for each cell must be separately adjusted.

When spinning from a tube, each cell is formed exactly in the same way. Still there is some variation in required force as the tube material hardens progressively through the multicell formation. A 9-cell Nb cavity was spun from a circular Nb blank 550 mm in diameter, 12-mm thick by first flow-turning a seamless tube (Fig. 6.32). The entire cavity spinning process took 30 h, but a signifi-

Fig. 6.32. The first 9-cell seamless cavity spun from a tube first made by flow turning [384] (courtesy of INFN).

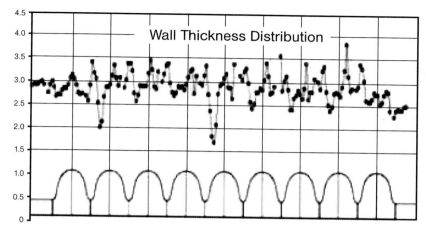

Fig. 6.33 Thickness distribution of a 9-cell seamless cavity by spinning [384] (courtesy of INFN).

cant part of the fabrication time was for mandrel and set-up changes. The thickness distribution is plotted in Fig. 6.33. The average final thickness is 2.8 mm. Some of the thickness variations can be reduced by automation as well as by the use of multiple rollers that can push the material forward and backward to reduce set-up changes. Unfortunately, the first 9-cell cavity was damaged during CBP and could not be tested.

6.5.2.1
Tube Forming

Tubes of the right size are difficult to obtain from niobium suppliers, due to significant investment needed for tooling, but tubes of any size can now be made from a disk by deep drawing or flow forming. Tubes can be formed from sheet by flow turning and deep drawing, which can also be carried out in reversal. Figure 6.34 shows the sequence of steps. The primary operation in reversal deep drawing is similar to standard deep drawing. In the redrawing steps, the punch pushes the tube bottom, and the inside and outside surfaces are reversed.

Tubes of 1.4 m long can normally be formed without any intermediate annealing from either disks of 8 mm thickness and 780 mm diameter, or disks of 12 mm thickness and 550 mm diameter. Seamless Nb tubes of 208 mm in diameter and up to 700 mm in height were successfully deep drawn from 3 mm disks of 800 mm diameter. In the case of direct deep drawing, four steps were needed, and no intermediate annealing was required.

Bimetallic tubes can also be successfully formed both by flow turning and by deep drawing. For deep drawing, the original thickness of the slab is maintained in the tube. By flow turning instead, the original thickness is reduced, but the thickness ratio between the two metals is maintained.

Fig. 6.34 Flow forming of a tube from a sheet in several stages [384] (courtesy of INFN).

6.6
Nb–Cu-Sputtered Cavities

6.6.1
High-β Cavities

As discussed in [1], thin films of niobium on copper offer several advantages over bulk Nb cavities, but the phenomenon of Q-slope which sets in at low fields remains the main obstacle to more wide-spread application. The advantages are higher thermal stability from the high thermal conductivity of the oxygen-free, copper substrate, and of course the reduced material cost. There is also a fortuitous order of magnitude lower sensitivity of the Nb layer's resistance to external dc magnetic fields, reducing the magnetic shielding requirements for cryomodules which house the Nb–Cu cavities. The simple explanation of the insensitivity of Nb on Cu films to dc magnetic field penetration is that the Nb films have a low electron mean free path and high H_{c2}. The typical RRR of the films is 10–20. The Nb/Cu cavities are also not affected by Q-disease, since no bulk chemistry is necessary, which eliminates the path to H contamination [385].

The best films are obtained by cylindrical magnetron sputtering on to a copper cavity substrate. Great care has to be taken to produce a smooth and clean surface without pores and inclusions. Electropolishing and chemical polishing techniques have been developed to produce copper surfaces of roughness less than 20 nm. This is particularly important since the overlaying Nb grain size is of the order of 100 nm. Crystals are predominantly in the 110 direction resulting in an oriented film. It is not clear whether this particular grain orientation is the best for rf properties. More than 300 cavities (350 MHz) for LEP II and sixteen 400 MHz cavities for LHC have been successfully applied [257]. Based on this success, Nb–Cu technology was adopted for the 16 LHC cavities at 400 MHz. Figure 6.35 shows the best performance of a 400 MHz LHC cavity

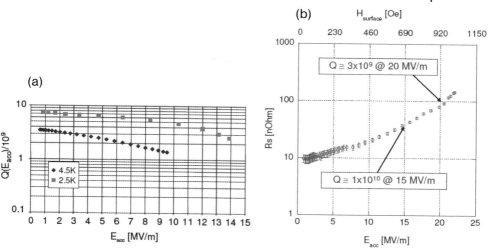

Fig. 6.35 (a) Typical performance of industry-produced LHC 400 MHz cavities. Specification values are $Q=2\times10^9$ at $E_{acc}=$ 5.5 MV/m and 4.5 K [386]. (b) Best results for 1500 MHz Nb sputtered on Cu cavities [385] (courtesy of CERN).

[386]. Nb–Cu cavities produced by magnetron sputtering have also been used in several new accelerator facilities, such as SOLEIL [387–389].

In an effort to evaluate Nb–Cu technology at higher frequencies (>1 GHz) desired for future large-scale applications such as ILC and XFEL, CERN continued development with 1.5-GHz single-cell cavities [390]. Figure 6.35 (b) shows the best results at 1500 MHz [390, 391]. The Q falls from 10^{10} at 15 MV/m to 3×10^9 at 20 MV/m. Since these results are still substantially below the best results for bulk Nb cavities, sputtered Nb films are not yet competitive for high-frequency applications. For relatively low gradient application, a CERN–Saclay collaboration developed a Nb–Cu cavity at 1500 MHz for third harmonic bunch lengthening [393]. These cavities have been installed and operated successfully in ELETTRA and SLS (see Chapter 11) [393].

6.6.2
Medium- and Low-β Cavities

When the magnetron sputtering technique was applied to medium-β elliptical cavities, the Q-slopes were found to increase with decreasing β geometry (Fig. 6.36). This has been attributed to the average and peak incidence angle of the niobium atoms arriving at the substrate during film growth, the more grazing the arrival angle, the stronger the Q-slope. We return to this topic in the next section on Nb–Cu R&D.

Several sputtered Nb–Cu 160 MHz $\beta=0.13$ QWRs (Fig. 6.36) have been successfully fabricated and operated in ALPI at INFN Legnaro [395]. The base reso-

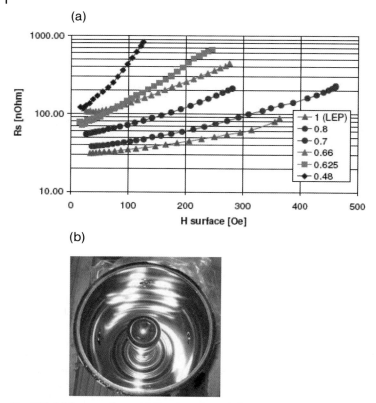

Fig. 6.36 (a) Best results at 4.2 K with medium-β elliptical cavities, compared to LEP $\beta=1$ cavities. The Q-slope increases as the geometry of the cavities changes with lower β values [385, 394] (courtesy of CERN). (b) Nb–Cu quarter-wave low-β resonator for the ALPI project [395] (courtesy of INFN).

nator was made of thick OFHC copper, with rounded shape optimized for sputtering, and no penetrations in the high current regions. Generally, the base structure should not have any braze joints where adhesion becomes poor. However, many cavities for ALPI had been constructed for Pb plating a long time ago and had braze joints. DC bias sputtering rather than magnetron sputtering was chosen for the complex QWR shape to deposit a thin layer of Nb. The slower deposition rate of the dc diode configuration was compensated by the use of bias, which promotes impurity release during the film growth, reduces film contamination, and improves film quality.

The cathode is a Nb cylinder ending with a sharp edge, which creates a high surface electrical field, to promote the impingement of Ar ions. The distance between the end of the cathode and the shorting plate is critical. The cavity body is negatively polarized for proper bias; cylindrical stainless steel nets, coaxial to the cathode, provide the ground plane. The nets have to be substituted after a few sputtering cycles because the deposited film can fragment and powder can deposit on the rf surface. The Cu structure surface is prepared by chemical and electropolishing, with HPR between cleaning stages. After evacuating to 10^{-6} mbar, the cavity is baked at 600 °C for a couple of days using infrared lamps. At the end of the process, the vacuum reaches the high 10^{-9} mbar. A lower quality vacuum has been correlated to a decrease of resonator performance.

The sputtering process is performed in many short steps to keep the temperature below the brazing temperature for the parts of the Cu base structure. Generally, a high substrate temperature is favorable for obtaining a clean film. The power sustained by the discharge is 5 kW. It is crucial to avoid sparks in the plasma to avoid defects which spoil the film. The entire sputtering cycle takes a full week. The optimum film thickness is 2–4 µm.

Initial prototypes exceeded 7 MV/m at 7 W dissipated power. Encouraged by these initial results, 44 Pb–Cu resonators of ALPI medium-β section (160 MHz, $\beta=0.11$) were stripped to the copper base, and the Pb coatings replaced by Nb coatings. A Nb–Cu cryostat of several higher β-type cavities was also installed in ALPI. The mechanical stability of the Cu base structure made the cavities less prone to mechanical resonances and insensitive to deformation due to He bath pressure drifts, simplifying the resonator rf control system. The high thermal stability allows convenient rf processing through field emission and multipacting.. However, sputtered Nb–Cu cavities do not always have comparable performance to bulk Nb cavities, due to the pronounced Q-slope.

The results on properly designed and built substrates show Qs of 1–2×10^9 and reach maximum accelerating fields exceeding 11 MV/m (Fig. 6.37), corresponding to an impressive peak electric surface field exceeding 50 MV/m, and to a peak magnetic field of about 110 MT [395].

Another laboratory pursuing low-β Nb–Cu resonators is the Institute of Heavy Ion Physics in Peking which reached 5 MV/m.

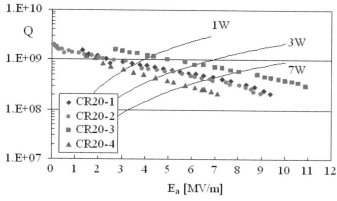

Fig. 6.37 Q versus E curves for some of the best performing Nb–Cu QWR [395] (courtesy of INFN).

6.6.3
Research for Better Film Quality

With the incentive for future large-scale applications, basic research efforts have advanced the understanding of the ubiquitous Q-slope for thin-film cavities. The experience with medium-β elliptical cavities (Fig. 6.36a) suggests that films sputtered at low grazing angle have lower RRR, and higher microroughness, which also increases the granularity of the film. Shadowing generates an increase in film roughness due to lower atom flux and reduced surface mobility in shadowed areas. These effects lead to poorly connected grains.

The fine (100 nm) grain size is probably not the dominant problem for the large Q-slope. Larger grain films have been grown on oxide-free surfaces (obtained by reverse sputtering), and these follow the grain structure of the underlying substrate with an average grain size equal to the copper grain size (~µm). However, the Q-slope was still found for the large grain films [390].

Sputtered films have low RRR (10–20) due to trapped impurities such as argon and due to electron scattering from microstrains. The associated low mean free path means reduced H_{c1} further depressed by demagnetization due to surface roughness. Thus Abrkosov fluxons can nucleate at rather low fields causing a Q-slope as one model suggests [397]. RRR of large grain Nb films as high as 44 have been obtained when sputtered with Kr instead of Ar [390]. Nevertheless the Q-slope was still present.

A careful study has been carried out to analyze the effect of the deposition angle on film morphology and superconducting properties using a planar magnetron source [398]. Both RRR and transition temperature were found to decrease monotonically as the arrival of Nb atoms approached a grazing angle. The highest RRR was 17 for $T_c = 9.4$ K when the atomic flux was at $90°$. The RRR dropped to 5 and T_c dropped to 9.0 at a $15°$ flux angle. X-ray diffraction showed normal flux deposited films were more preferentially oriented in the 110 direction. But at grazing angles, the films became amorphous. The average surface roughness increased from 2 nm for normal incidence to 10 nm for grazing incidence.

Several efforts are underway to improve the quality of films by increasing their smoothness either by ion bombardment, or by ionizing the arriving Nb atoms. Adding bias to the classical magnetron configuration to promote ion bombardment during film growth has not been successful so far to improve film properties. An evolution (HPPMS) of the magnetron sputtering approach pursued at CERN [385] is to impose ~100 µs, 1 kV voltage pulses which deposit a huge power density onto the target (of the order of a few kW/cm^2 compared to a few W/cm^2 of the standard dc process) producing a highly dense plasma in which Nb atoms are partially ionized. These can be attracted to the substrate with a suitable bias allowing deposition with a normal angle of incidence everywhere, and thus suppress self-shadowing.

As an alternative to thin films, INFN is pursuing thicker cathodic arc coatings, where an electric arc is established over the cathode's surface triggered by

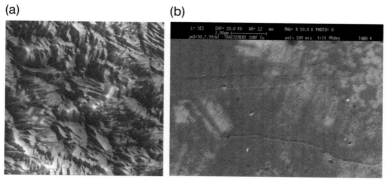

Fig. 6.38 (a) AFM picture of the cathodic arc deposited sample surface [401] (2×2 μm area). The surface roughness is about 400 nm as compared to EP bulk Nb (200 nm) and BCP bulk (2 μm) [400]. (b) FESEM image of a smooth ECR-deposited film. The scale bar is 2 μm.

a high voltage or laser pulse [399]. An adequate power supply sustains the arc to produce a high density plasma plume in which all Nb atoms are ionized and attracted to the substrate at the desired energy with a bias. There is no need for a discharge gas which eliminates trapped impurities typical for sputtering. Because of the explosive nature of the process a stream of microparticles also arrives at the substrate. Magnetic filtering steers the ion flux to the substrate and filters out the droplets. Nb coatings have been deposited on flat samples and 1.3 GHz single-cell cavity surfaces. Microwave characterization shows promising results providing incentive to continue [400]. Polycrystalline films with μm size grains have been obtained with very smooth surfaces and free from microstrains. Figure 6.38 shows the surface microstructure with AFM.

An alternate approach under investigation is electron cyclotron resonance (ECR) deposition [401]. An electron-beam evaporation source creates a large flux of Nb atoms which are subsequently ionized by rf via the ECR process. The ions are then steered to the suitably biased substrate by magnetic guidance. Compared to arc deposition there are no macroparticles. The coating rate is low and demands an extremely clean vacuum environment. Only samples have been produced.

7
Cavity Treatment Advances

7.1
Overview of Cavity Treatment Procedures

We start with a brief overview of the treatment steps for a cavity after fabrication is complete and before the cold test. Later sections will discuss each step in more detail.

A completed cavity goes through many steps before it is ready for installation into a cryomodule. At first, the inside surface is inspected optically, especially the weld quality. Mechanical measurements ensure straightness and correct dimensions. The field profile is checked and adjusted. The usual goal is 98% field flatness.

If the welds have imperfections, tumbling or centrifugal barrel polishing (CBP) can be used for smoothening. Although it is used only at a few laboratories, this procedure provides a fairly uniform surface by removing imperfections such as roughness at welds, pits, and mild scratches remaining from the starting sheet material. The rate of material removal is highly dependent on the tumbling medium and rotation speeds. A light step of buffered chemical polishing (BCP), usually about 50 µm, removes the tumbling abrasive embedded in the surface. The danger of H pick-up during subsequent chemistry is greater after mechanical abrasion.

The surface damage layer due to sheet rolling and cell stamping (or hydroforming) has been established to be 100–200 µm deep [402] so that a bulk chemistry to remove this amount of material is necessary. In this step, the vapor from welding deposited on the inner surface (25 µm) is also removed to arrive at a clean surface. BCP is used for the heavy etch for many applications, but EP is needed to provide a smooth surface for those applications demanding surface fields above 80 mT, corresponding roughly to $E_{acc} > 20$ MV/m, as discussed in Chapter 5.

BCP has been covered extensively in [1]. Here we will cover EP in detail. Both BCP and EP procedures carry a risk of H absorption so that furnace treatment is necessary to ensure a hydrogen-free cavity (see Chapter 3 for a discussion of the H–Q disease). Hydrogen contamination risk for BCP is greatly reduced by keeping the acid temperature below 15 °C. The danger for H contamination

RF Superconductivity: Science, Technology, and Applications. Hasan Padamsee
Copyright © 2009 WILEY-VCH Verlag GmbH & Co. KGaA, Weinheim
ISBN: 978-3-527-40572-5

with heavy EP is especially great since copious quantities of H are evolved at the cathode during EP, and should be guided out using special measures described later.

The safe annealing parameters to remove H contamination are 600 °C for 10 h to prevent a serious drop in yield strength due to grain growth for 300 RRR material. However, many labs are using 800 °C for 2 h. One concern about RRR degradation due to a heavy etch and 800 °C anneal has been put to rest. A control study [373] showed that the bulk RRR (400) of Nb samples did not change after several preparation steps: 100 min BCP (1:1:2) etch, a high-temperature heat treatment (800 °C, 5 h) in vacuum ($<10^{-6}$ torr), followed by another 20 min BCP polish.

Due to nonuniform material removal during chemistry and possible deformation during heat treatment, the cell-to-cell field profile needs to be measured and adjusted by mechanical tuning. The usual goal is 98% field flatness.

The final chemical treatment is a light etch (about 20–30 µm material removal) either by BCP or EP depending on the target field level. If done properly, there is little risk of H recontamination. After thoroughly rinsing the acid residues with high-purity water, the cavity is transported to a Class 10–Class 100 clean room where the inside surface is given a high-pressure (100 bar) rinse (HPR) with high-purity water jets for many hours. The main goal is to scrub the chemical residues and particulate contaminants which may cause field emission or thermal breakdown. Among the recent developments for more effective removal of sulfur residues from EP are ultrasonic degreasing with soap and water and ethanol rinsing before all the HPR steps.

After HPR, the cavity dries in the clean room for 1 or 2 days before it is assembled with components for testing. Heating lamps are used on occasion to speed up the drying process. To reach the highest fields, an electropolished cavity needs to be baked at 120 °C for 48 h to remove the high-field Q-drop (see Chapter 5). This bake is normally carried out with the inside of the cavity in a good vacuum (about 10^{-8} torr). Developments are underway for more efficient final baking procedures, such as shorter times and slightly higher temperatures (e.g., 145 °C, 3 h). The baking also improves the BCS Q due to lower mean free path (Chapter 3).

Another heat treatment procedure is sometimes helpful to raise the quench field of a cavity is to raise the RRR by a postpurification heat treatment at 1300 °C with titanium. This procedure was covered in detail in [1]. If used, another 80–100 µm of material needs to be etched from the inside to remove the titanium layer, and at least 20 µm from the outside surface of cavity to remove the excess titanium film for best helium bath cooling [404]. Since higher fields can generally be reached by EP, there is less motivation for postpurification, especially since the high-temperature treatment drops the niobium yield strength substantially, making the cavities vulnerable to plastic deformation and frequency detuning. Yield strength can decrease from 70 MPa to 40 MPa or even lower [405]. Postpurified cavities with RRR in the range of 600–800 are also more susceptible to the H Q-disease as expected from the depletion of interstitial impurities which serve as H-trapping centers [406].

7.2 Inspection

Meeting dimensional tolerances is important for a cavity to properly interface with various ancillary components such as the helium vessel, tuner, input coupler(s), to properly match up with other cavities in a string assembly, as well as to meet the fixed points of the cryomodule that will house the cavity string. The mechanical axis of the cavity should not deviate from the beam axis beyond specified tolerance [407]. In addition to standard CMM measurements, Fig. 7.1 shows special dimensional control apparatus developed at DESY for a 9-cell, 1.3 GHz cavity and Fig. 7.2 the optical inspection apparatus developed at KEK for the inside surface and especially the welds [408].

Fig. 7.1 Special apparatus for measuring the cavity eccentricity [182] (courtesy of DESY).

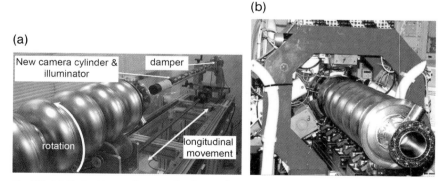

Fig. 7.2 (a) Optical inspection gear for the inside cavity surface [408] (courtesy of KEK). (b) Tuning machine at DESY [409].

7.3
Barrel Polishing

CBP (also called tumbling) has been applied to niobium cavities mostly at KEK [370, 410] and to some extent at DESY [411]. CBP effectively removes irregularities like scratches and especially any roughness at electron beam weld seams (Fig. 7.4). CBP is also needed for removing the fissures in the initial surface preparation of hydroformed and spun cavities, as discussed in Chapter 6.

At KEK, a cavity is partially filled with liquid soap and plastic chips embedded with ceramic abrasive powder (Fig. 7.3). The cavity is rotated in a horizontal orientation at 100 rpm so that the chips press onto the cavity inside surface. The removal rate, which depends on the abrasive selection and rotation speeds, can range from 10 μm/day to 25 μm/h. The largest removal rate at the equator section is about 250% of the average material removal. The finished roughness is 1–2 μm. Figure 7.4 compares the finish of the equator weld region before and

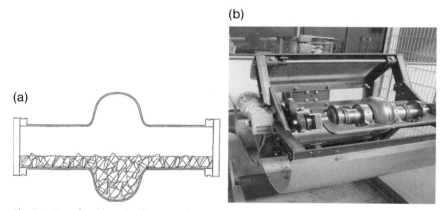

Fig. 7.3 Centrifugal barrel polishing: (a) schematic and (courtesy of KEK); (b) single-cell setup [411] (courtesy of DESY).

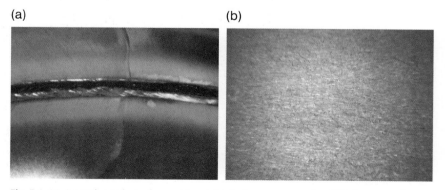

Fig. 7.4 (a) A rough inside surface weld and (b) its improvement after CBP (courtesy of KEK).

after CBP. When CBP is applied before chemical treatment, a thinner Nb layer needs to be later removed by EP or BCP. KEK experience shows that the removal of 60 µm instead of 150–200 µm is sufficient. Some of the abrasive compound gets buried in the polished surface as with buffing with polishing wheels. The particles penetrate into Nb to a depth of about 50 µm but completely disappear after removing a layer of 60 µm.

After CBP the electron beam welding seam becomes completely invisible (Fig. 7.4). CBP poses an increased risk of H contamination [410]. The continuous mechanical abrasion of the natural protective oxide layer leads to H pick-up from water in the polishing medium. The hydrogen content can be reduced by using a fluorocarbon based liquid (FC-77, C8F18, and C8F16O) [410]. However, the niobium surface remains highly sensitized during barrel polishing. Minute surface defects introduced by mechanical grinding remain likely paths for hydrogen absorption during subsequent chemical etching. Consequently, large amounts of hydrogen can be absorbed by a tumbled surface.

At DESY, CBP takes place in two stages [411]. During preliminarily grinding, the cavity is filled with chips with plastic binding and abrasive compound. CBP is carried out in the vertical position and orientation reversed half-way through the process in order to remove the material more uniformly. The rotation speed of the cavity around the machine axis is 120–140 rpm, and of the cavity around its own axis is 360–400 rpm. A layer of 140–150 µm can be removed by in 5–6 h. This is sufficient to completely smooth the hydroforming defects, welding areas, protrusions, scratches, and other surface defects. Higher rotating speed increases the centrifugal chip pressure on the cavity wall which results in an enhancement of the removal rate. In the final polishing stage, the rpm of the cavity around the machine axis is reduced to 70–100 rpm, and the speed of the cavity around its own axis to 100–120 rpm. Duration of the fine polishing treatment is 1–2 h.

The highest efficiency is achieved with chips of pyramidal shape and plastic binding (type RKS–10P). Chips with ceramic binding are heavier and more stable than those with plastic binding. But ceramic chips have a higher friction against a viscous niobium surface at high velocities. The abrasion rate is lower than plastic chips. Ceramic chips can be applied only for the fine grinding operation.

7.4 Electropolishing

7.4.1
A Short History of EP

Electropolishig niobium cavities started in the 1970s by Siemens Company and became a standard cavity treatment process for cavity preparation at many laboratories until the simpler BCP process replaced it in the 1980s. The Siemens EP process continued in use for low-β split-ring and quarter-wave resonators at

Argonne National Lab. KEK also remained one of the few laboratories that continued EP for high-β cavities but improved the Siemens process for application to large cavities, such as the TRISTAN 5-cell, 508 MHz cavities. KEK made the process continuous, as compared to the Siemens intermittent process which required short periods (5–10 min) of EP, followed by similar short periods of electrolyte stirring.

Although the BCP is described in detail in [1], a summary here is appropriate. BCP consists of two alternating processes: dissolution of the Nb_2O_5 layer by HF and reoxidation of the niobium by a strongly oxidizing acid, such as nitric acid (HNO_3). To reduce the etching speed, a buffer substance is added, for example phosphoric acid H_3PO_4, and the mixture is cooled below 15 °C to ensure low pick-up of H. The standard procedure with a removal rate of about 1 μm/min is called BCP with an acid mixture containing 1 part HF (40%), 1 part HNO_3 (65%), and 2 parts H_3PO_4 (85%) in volume.

BCP gives rise to a major problem for gradients higher than 20 MV/m: the development of steps at grain boundaries due to different etch rates of niobium grains with different orientations. The steps are typically a few μm, and increase with grain size, approaching 10–15 μm inside the large grain electron beam weld seams. Surface roughness is responsible for premature quench and possibly plays a role in aggravating the high-field Q-drop (see Chapter 5). By now, there exists compelling evidence that BCP limits attainable accelerating fields in 1.3 GHz cavities to about 20 MV/m for high Q, and 30 to 32 MV/m at Q of 10^9.

With the success of EP and baking to remove the high-field Q-drop, EP returned worldwide to play a major role in the drive for higher gradients. With EP, it is possible to achieve an average surface macroroughness below 0.5 μm. As discussed in Chapter 5, EP and surface smoothness alone do not solve the high-field Q-slope. EP surface preparation must be followed by a mild bake at 100–130 °C for 48 h. The reasons for the high-field Q-drop and the benefits of mild baking are still under active investigation, as discussed extensively in Chapter 5. With large (> many cm) grain material it is still possible to use BCP alone to obtain Q-slope free cavities out to 35 MV/m.

7.4.2
Basics of Electropolishing

We will discuss practical EP setups for single cells and accelerating structures in subsequent sections. Here the focus is on principles.

Figure 7.5 shows a basic setup for EP for an open cavity half-cell [412, 413]. The niobium cavity is the anode (+) in an electrolytic cell and the cathode (−) is made from pure aluminum (1100 series). The electrolyte is a mixture of hydrofluoric and sulfuric acid in a volume ratio of 1:9, using typical commercial strengths HF (40%) and H_2SO_4 (98%). The associated molar concentrations are 2.29 mol/L of HF and 16.08 mol/L of H_2SO_4. As current flows through the electrolytic cell, the niobium surface absorbs electrons and oxygen to convert to niobium pentoxide which subsequently dissolves in the HF present in the electro-

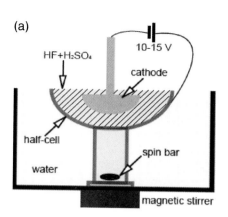

Fig. 7.5 Electropolishing geometry for an open half-cell. (a) Schematic: Acid should cover the full area that needs polishing. A remote magnetic stirrer drives a teflon-coated magnetic spin bar placed inside the half-cell and provides continuous slow acid agitation. (b) The setup: To contain the acid inside the half-cell the top equator surface is sealed against a teflon ring as is the bottom iris surface. The sealed half-cell assembly is immersed in bath water with temperature regulation [413].

lyte. HF converts Nb_2O_5 into soluble NbF_5 or H_2NbOF_5 to re-expose fresh niobium metal for further oxidation and dissolution resulting in continuous niobium etching. Hydrogen evolves at the (negative) cathode and rises to the electrolyte surface to form a thick foam of bubbles. Neutral H atoms and the stream of H_2 gas can also be entrained in the electrolyte to reach the anode, so that the cathode should not be positioned too close to niobium anode. The half-cell is cooled on the outside by immersion in a temperature-controlled water bath.

Under optimum conditions of voltage, temperature and stirring, the current shows oscillations (0.5–5 Hz at 12 V and 30 °C). Figure 7.6 shows the current oscillations in the original Siemens method [414]. The current oscillates due to the growth of a resistive viscous layer (decreasing current) during niobium oxidation and the reduction of the viscous layer due to diffusion of the soluble niobium fluoride compounds. The amplitude of the current oscillations damps out in about 20 min. At this stage, the voltage is turned off and the electrolyte stirred for about 5 min to dissolve the dark-colored insulating layer which develops on the niobium surface and to revive current oscillations on voltage turn-on.

In the Siemens intermittent polishing method, there is no stirring during current flow. The polishing action can be made continuous by gentle stirring during the current flow. A magnetic spin bar provides sufficient agitation for uniform material removal. Under optimum conditions of voltage, temperature and stirring, the current also shows oscillations (0.5–5 Hz at 12 V and 30 °C). Figure 7.6 compares oscillations in the continuous method [415–417] with those of the

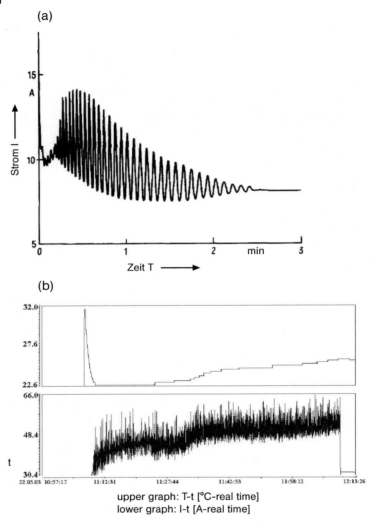

Fig. 7.6 Comparison of current oscillations for a single-cell cavity electropolished by (a) the intermittent Siemens method [414], and (b) the continuous method. A constant voltage (18 V) was applied and the process duration was limited to 60 min, respectively a temperature limit of the acid of 35 °C [415] (courtesy of DESY).

original Siemens method. However, current oscillations are not an absolute must to get a smooth surface. At DESY, a single-cell cavity electropolished without oscillations showed a bright surface and the cavity reached $E_{acc}=37$ MV/m [415].

The alternating steps of polishing and stirring make application to large-scale cavities difficult. KEK [418, 419] developed the continuous process. Figure 7.7 shows the schematic [286] and the setup [420] for horizontal EP of a complete

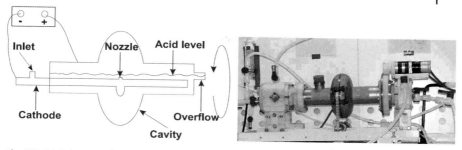

Fig. 7.7 (a) Schematic for continuous EP developed by KEK [286] (courtesy of DESY). (b) Single cell cavity EP set-up at Saclay [420] (courtesy of Saclay).

single cell. Rotation of the cavity polishes the entire surface, while continuous flow of the electrolyte between the cavity and an acid reservoir provides the acid circulation to prevent the formation of the insulating layer. A circulation system establishes electrolyte flow between the cavity and a large volume acid barrel with heat exchanger to regulate the acid temperature.

There is a significant danger of H contamination with the intermittent EP method where the metal surface is in contact with electrolyte for long periods without being polarized and thus without a protective oxide layer. In the continuous method, the continuous presence of the positive voltage protects the anode from picking up H^+. However, if the niobium surface has significant mechanical damage after grinding or CBP, H released from the cathode will enter the Nb cavity. A perforated teflon coaxial tube (or porous teflon cloth) surrounds the hollow aluminum cathode to inhibit the evolving hydrogen gas from mixing with the electrolyte and reaching the niobium surface. Dry nitrogen gas floods the space above the electrolyte to prevent water vapor from reacting with the hygroscopic H_2SO_4.

Great care must be exercised to achieve reproducible operating conditions so that it is important to continuously monitor a large number of key parameters. Table 7.1 shows a range of voltage, current density, temperature, and flow rate over which continuous EP gives a smooth surface. KEK explored a range of parameter variations for highest brightness and minimum macroroughness [418].

Table 7.1 Range of working parameter sets for horizontal EP.

Parameter	Range
Current density	10–50 mA/cm^2
Voltage	8–16 V
Electrolyte temperature	25–35 °C
HF concentration	60–90 cm^3/L
Rotation speed	1–2 rpm
Acid flow rate	6–12 L/min

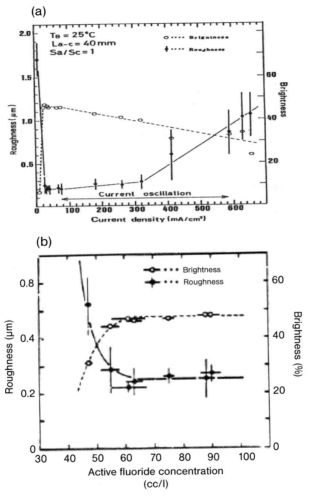

Fig. 7.8 Brightness and roughness dependence on EP parameters: (a) current density and (b) fluoride concentration [418] (courtesy of KEK).

For example, Fig. 7.8 shows that the best current density is between 30 and 100 mA/cm² and the HF concentration should not drop below 60 cm³/L.

One of the key parameters characterizing the EP process is the current (I)–voltage (V) characteristic curve. Figure 7.9 shows the ideal and typical the I–V characteristics for EP [421]. Depending on the voltage applied, it is possible to obtain pitting, polishing, or gas evolution. For voltages less than V_b, the surface preserves its mechanically worked appearance and shows some signs of pitting. A solid brown film forms on the anode surface for voltages between V_a and V_b but disappears at higher current density. This region is not recommended for EP. Just above V_b, there occur fluctuations in both voltage and current and a

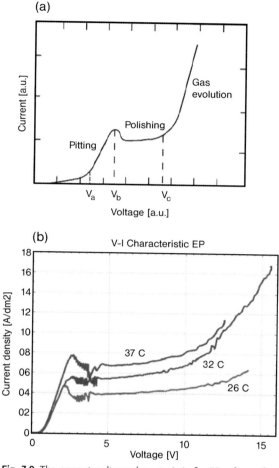

Fig. 7.9 The current–voltage characteristic for EP, often referred to as the polarization curve. (a) Ideal [421] (courtesy of INFN); (b) measured at different temperatures for a simple half-cell setup [298] (courtesy of FNAL).

simultaneous drop in current density. Between V_b and V_c, a current plateau (about 50–100 mA/cm^2) appears, usually attributed to diffusion-limited phenomena. Over the plateau, the current density remains constant even though the voltage increases. A polishing effect is observed between V_b and V_c, but the best results are obtained near point V_c. The first bubbles of gaseous oxygen appear on the anode at V_c. At higher voltages, evolution of oxygen accompanies the dissolution of metal, and pitting may occur due to oxygen bubbles trapped on the anode surface.

Under the polishing condition, a thin layer of electrolyte assumes a bluish color in the vicinity of the anode. The current density is inversely proportional to viscosity of the layer. Moderate agitation of the electrolyte reduces the thickness of the viscous layer so the current density increases, and the voltage drops. Vigorous agitation reduces the thickness to a few tenths of a millimeter and voltage drops even more. But intense stirring gives a rougher surface finish rather than polishing. At about 60 mm/s acid flow rate the surface roughness was measured to be 0.15–0.25 µm, but the roughness increased to ~1.55 µm at 160 mm/s [418].

Since the plateau region in the I–V characteristics gives the best polishing conditions, it is important to consider the effect of key process parameters on the plateau. These are electrolyte temperature, acid concentration, viscosity, and stirring. Figure 7.9 shows the effect of temperature on the I–V curve. With higher flow rate, the onset of the voltage for the plateau region gets higher and the voltage range for the plateau gets narrower. In addition, the onset voltage for the gas evolution regime gets lower. Work is in progress to develop techniques to automatically find the optimum EP conditions, for example by the use of flux gate magnetometry to keep close track of the current density [422].

There is not yet one complete theory of EP. An attractive but naive explanation is that electric field is higher at edges and projections than inside wells and craters so that material removal takes place preferentially at protrusions. More generally accepted EP theories are based on the formation of passive films at the anode. In one model [423], the anodic film has higher viscosity and correspondingly higher electrical resistivity than the bulk of the electrolyte. Above protrusions, the film is thinner than in the valleys resulting in a higher current density and more rapid dissolution. An alternate proposal [424] is that surface leveling occurs due to the diffusion of anodic products from the anode through the film, driven by differences in the concentration gradients of metal ions.

The nominal HF content for the 9:1 mixture is ~100 cm^3/L (Fig. 7.8). Maintaining the relative composition of HF and H_2SO_4 is important for best polishing, and also to avoid synthesis of solid sulfur. HF is consumed by the Nb dissolution process, while some HF evaporates during the process. There is also some early HF loss during mixing and pouring. The conditions for best surface roughness and brightness correspond to an HF content of >60 cm^3/L [416]. A simple method to track the HF content is to monitor the voltage needed to obtain a current density of 50 mA/cm^2. When the HF content becomes too low, more than 14 V is required. Another indicator of electrolyte aging is the Nb content in solution. After the electrolyte gains 5–7 g/L Nb, the surface starts to degrade in brightness [416, 425, 426]. But this is also partially due to a decrease of HF content. Several methods are under investigation for improved electrolyte quality control. These are [415]: titration, inductively coupled plasma optical emission spectroscopy, ionic-chromatography, nuclear magnetic resonance, Fourier transformation-infrared spectroscopy-attenuated total reflection. In addition, there are ion selective electrode method [428], capillary electrophoresis method [429], and online reference cells. The best monitoring method is not yet established.

7.4.3
Horizontal Electropolishing

Most laboratories position the cavity in a horizontal orientation for EP. Figure 7.10 shows a 9-cell EP setup [415], and Fig. 7.11 a generic acid flow diagram [430] with pumps, reservoir, and heat exchanger. One advantage of the horizontal orientation is that H gas produced at the cathode can be efficiently swept out of the volume of the cavity with nitrogen gas flow, resulting in minimal exposure to the niobium surface in contact with HF, and therefore without the protective oxide layer. This minimizes the danger of H absorption into the bulk niobium. About 2 L/min of H gas will be evolved at 300 A [431]. A drawback of the horizontal arrangement is that since only half the surface is immersed in electrolyte, the cavity must be rotated to polish the entire surface. Leak tight rotary sleeves at the flanged ends are essential to contain the acid mixture.

For EP 9-cell structures, DESY installed two heat exchangers in the acid barrel with nominal 20 kW heat exchange capacity. The total volume of acid stored in the barrels is 150 L (cavity volume is 30 L), and the acid circulation rate is 10–12 L/min. A membrane pump drives the acid mixture through the cooled barrel, and through a 1 μm pore filter before it reaches the inlet of the hollow cathode. From here, the electrolyte fills the center of the cells through openings in the hollow cathode. The acid returns to the storage tank via an overflow. The acid level is controlled to keep 53% of the cavity surface immersed.

All components of the EP system exposed to acid are made typically from teflon (PTFE) which is inert against acid. Other viable fluoroplastics are polyperfluoro alkoxyethylene (PFA) and polyvinylidene fluoride (PVDF). The EP system is placed in a vented area where the exhaust gases are pumped out through a neutralization system to avoid environmental hazards. The 9-cell resonator is connected to a handling frame which allows lifting the cavity into the vertical position for acid drainage after EP. The current leads are made from copper. The electrical connections to the cavity are sliding contacts made from a copper carbon alloy.

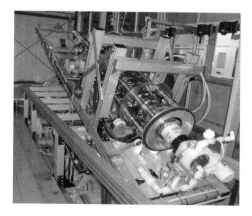

Fig. 7.10 EP setup at DESY [415].

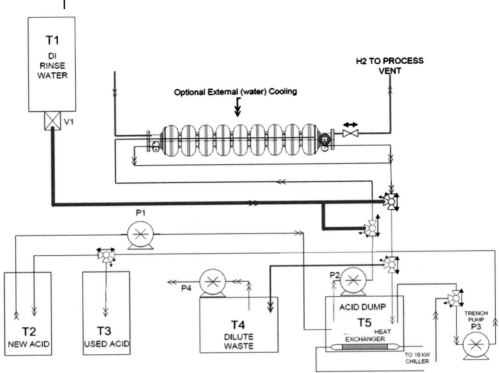

Fig. 7.11 Acid flow schematic for horizontal EP [430] (courtesy of ANL).

After establishing the equilibrium level for the electrolyte, the cavity rotation is switched on and a fluid leak check is done. The current–voltage characteristic is measured and the voltage set for current oscillations. As mentioned, oscillations are a good guide to optimum conditions, but not absolutely necessary to get the best results [418]. Polishing results are good for an oscillation amplitude of 10–15% around the mean value. The typical voltage is 17 V and current 300 A ± 20 A for a 9-cell cavity. The temperature in the storage barrel stabilizes to 30–35 °C in about 1 h. Typical start temperatures are 20 to 23 °C. The typical current of 300 A is stabilized within about 1 h, and the set point temperature (30–35 °C) is reached. The temperature of the acid must be maintained constant to 1 °C to keep the current stable. Temperatures above 40 °C must be avoided as they result in etch pits on the surface. The total time for one polishing sequence is about 4 h to remove about 100 µm. Multiple polishing sequences must be carried out if more material removal is necessary. Twenty-five sensors are installed to ensure safe and reproducible operation.

The average material removal is determined by the integrated charge, ultrasonic thickness measurement, or by weighing the cavity accurately before and after EP. The ideal removal rate is 0.67 µm/min for 50 mA/cm^2 current density [431]. The actual removal rate is about 0.4 µm/min since only a fraction of the cavity area is in contact with the acid. Due to the proximity of the cathode to the irises, the etch rate at the equator is about 60% the average removal rate. When the desired amount of niobium material has been removed, the current is switched off. The rotation is stopped and the cavity is put into the vertical position to drain the acid mixture. After rinsing the cavity and electrode with pure water, the cathode is removed. Another low-pressure water rinsing follows. A 2-h rinse with ultrapure water completes the EP process before the cavity is removed from the EP apparatus. After cleaning the cavity exterior with ultrasound, the wet cavity is transported to the ultrapure high-pressure water rinsing facility to remove remaining chemical residues from the rf surface. In one of the latest innovations to remove chemical residues, in particular S particles, the cavity is ultrasonically rinsed in a dilute surfactant (Micro-90) for several hours followed by thorough rinsing with water [432]. Another approach is to rinse the cavity for 1 h in ethanol, which is known to dissolve S [433].

7.4.4
Vertical Geometry Electropolishing

Vertical EP of niobium cavities has been developed at Cornell University and applied successfully to single-cell, 1.3–1.5 GHz niobium cavities [413, 434]. The method has been adapted to process 9-cell cavities. The schematic layout for 9-cell cavities is illustrated in Fig. 7.12. Acid circulation is achieved by stirring the electrolyte with a series of paddles located within the cavity cell. Agitation at the equator is significantly enhanced due to the extension arms (Fig. 7.13). Table 7.2 lists the main parameters. 5.8 kW are generated within the cavity–electrolyte system. Cooling is accomplished by flowing a thin sheet of water over the exterior surface of the cavity. Flowing water covers the entire cavity surface including higher mode and fundamental coupler ports. Each cell receives a flow of "fresh" coolant. 30 °C water preheats both the cavity and electrolyte during acid

Table 7.2 Parameters of vertical EP.

Cathode	Aluminum >99.5%
Stir tube	PVDF
Paddles	PVDF
Seals	Viton
End groups	PTFE, HDPE
Electrolyte	24 L
Maximum use	9 g/L dissolved Nb
Current	400 A
Voltage	14.5 V
Temperature	30 to 35 °C
Stir-tube transparency	>50%
Stir frequency	1 Hz
EP rate at equator	0.5 µm/min
EP rate iris/equator	<1.5 µm/min

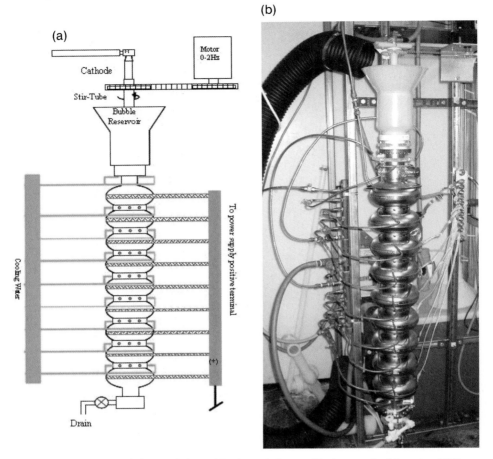

Fig. 7.12 Vertical electropolishing: (a) schematic layout; (b) photograph of the setup [434].

filling. This minimizes the time (typically <10 min) to reach stable operation at 50 mA/cm^2. Periodic current oscillations are seen after thermal equilibrium has been reached. The temperature is allowed to rise to 32 °C. As the HF concentration diminishes, the current decreases from the starting 400 A. EP takes place continuously over 50 min after which the electrolyte is discarded since it saturates with more than 7 g/L of dissolved niobium. When the electrolyte is used for longer periods of time, there are significant sulfur deposits in the system (Section 7.4.5). Hydrogen bubbles evolving at the cathode exit vertically along the stir-tube through the same series of holes (Fig. 7.13) intended for the electrolyte to reach the cathode. There is no teflon barrier mesh used as for the horizontal system, since a layer of bubbles accumulating between the barrier and cathode increases the electrical resistance. No signs of grooves or bubble tracking have been seen in a finished cavity, which testifies to the efficiency of stir-

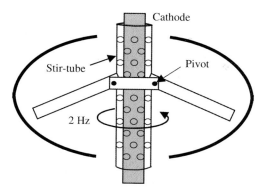

Fig. 7.13 Cathode and stir-tube assembly for each cell.

ring paddles sweeping out any bubbles entrained in the electrolyte. Bubbles exiting the cavity through the top beam tube collect in a funnel reservoir at the top. The KEK formula for electrolyte which includes a trace quantity of nitric acid [435] generates approximately half the volume of hydrogen bubbles as for the standard EP electrolyte. Gravity affects the diffusion of the viscous layer. As a result, less material is removed from the lower half than the upper half of each cell. This necessitates flipping the cavity and EP an equal amount in the opposite orientation for uniform material removal.

This method offers potential advantages over the standard rotating EP in the horizontal orientation. Since the cavity and cathode are both fixed, so are the electrical connections, which eliminate the complexity of the rotating system and the sliding electrical contacts. The acid mixture is confined to the cavity volume, so the overall plumbing is simpler, and there is no need for a storage barrel and heat exchangers for cooling. However, the 25 L acid volume must be changed more frequently before HF exhaustion. Since etching takes place over the entire cavity surface, the polishing rate is about twice as fast.

7.4.5
Contamination Issues

Early practitioners of EP recognized some of the key contamination issues. XPS studies [436] showed the presence of S and F. The presence of S is also evident from the persistent odor of H_2S on electropolished surfaces. During the preparation of cavities for the TRISTAN phase at KEK, sulfur contamination of the bath was encountered when the amount of niobium in solution exceeded 4 mg/L [418]. Figure 7.14 shows a S particle captured on a Nb monitor sample placed in the EP electrolyte [437]. Niobium oxide particles have also been found. Such particle deposits may become field emitters. S accumulation in the heat exchanger piping from previous treatments can also move to the cavity during acid circulation. Hence the EP system needs to be thoroughly cleaned at regular intervals.

Maintaining the relative composition of HF and H_2SO_4 is important to avoid excessive synthesis of solid sulfur. Therefore, the aging of the EP bath should

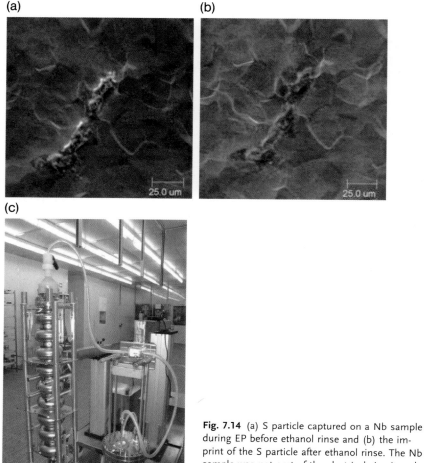

Fig. 7.14 (a) S particle captured on a Nb sample during EP before ethanol rinse and (b) the imprint of the S particle after ethanol rinse. The Nb sample was not part of the electrical circuit and so did not etch [437]. (c) Ethanol rinsing apparatus at DESY [409, 433].

be monitored for HF concentration to determine the maximum EP time for a particular setup. During EP, some of the HF evaporates and some is consumed by the Nb dissolution process. Enhancing the HF proportion helps to reduce the sulfur generation [425]. However, the proper relative concentration is also important to minimize the corrosion of the aluminum cathode [426]. Both sulfuric and hydrofluoric acids corrode pure aluminum cathode slowly. Being slightly soluble in water, the fluoride salts are less harmful. Alternate cathode materials, such as copper and platinum, are not suitable as these tend to leave metallic particles.

Several rinsing techniques have been found helpful against S (Fig. 7.14). These include rinsing with ethanol [433], ultrasonic rinsing in surfactant (soap and water) [432], and ultrasonic rinsing in hydrogen peroxide. Ethanol is known to slowly dissolve S particles, but leaves a residual imprint (Fig. 7.14) [437]. The solubility of sulfur in ethanol at 20 °C is about 1% by weight. Surfactant rinsing reduces the adhesion of S to the Nb surface. The peroxide rinse promotes full oxidation of reaction compounds on the surface to soluble products. A final treatment called oxipolishing removes the oxide layer and embedded contaminants, growing a fresh new oxide layer. Oxipolishing consists of several steps of anodizing in an ammonium-hydroxide solution and stripping the pentoxide in hydrofluoric acid under ultrasonic agitation [438]. Another option under investigation is final EP (3 µm) with fresh acid sealed in the cavity volume and closed off from the S-contaminated acid recirculation system. This has proven to be helpful in giving field emission free results [369].

7.5
Heat Treatment and Hydrogen Degassing

EP generates copious quantities of hydrogen at the cathode which can penetrate into the niobium if the setup does not effectively remove it. The danger increases when H bubbles come into direct contact with the Nb surface exposed to HF, due to the absence of the protective oxide layer. With more than 100 atomic ppm H dissolved, there is a danger of niobium hydride precipitation during cool-down, if the cavity remains around 100 K for several hours (see Chapter 3), especially for high RRR cavities. Some experiments show [435] that H absorption during EP can be reduced by adding a few parts per thousand of oxidizer (such as nitric acid) to the EP electrolyte. Hydrogen concentration can be measured by hydrogen gas chromatography upon melting the niobium.

A furnace treatment at 800 °C for 2 h or 600 °C for 10 h reduces the hydrogen concentration to a few atomic ppm in the bulk and ≤1 at.% in the surface layer. The lower temperature annealing avoids the danger of grain growth (Fig. 7.15) and yield strength drop, especially for high-RRR material (Fig. 7.15) with RRR ≥300 (see Section 3.4.1). To prevent contamination of the niobium and the UHV oven during the annealing process, the cavities exterior is normally

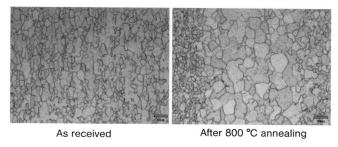

Microstructure is Heterogenous (Banded)

Crystal grains change with the 800 °C heat treatment for dehydrogenization and stress release

As received   After 800 °C annealing

Fig. 7.15 Grain growth with 800 °C annealing. The scale bar is 100 µm [442].

cleaned by a 20–40 µm BCP treatment before furnace installation (Fig. 7.15). It is generally believed that 800 °C heat treatment also helps cavity performance by relieving stresses and annealing dislocations introduced during rolling the sheet material and during stamping (or hydroforming). DC magnetization curves show reduced hysteretic behavior after such annealing (Fig. 3.27). Nevertheless there are many cases of good rf results without 800 °C annealing [439, 440]. For example a 9-cell cavity reached $E_{acc} = 26$ MV/m limited by the high field Q-slope after BCP 150 µm (Fig. 7.16a). Similarly four single-cell 1.3 GHz cavities reached $E_{acc} > 25$ MV/m with BCP and without annealing [441].

7.6
Final Cleaning

7.6.1
High-Pressure Rinsing

Microparticle contamination has been identified to be the leading cause of field emission. This stresses the importance of cleanliness in all final treatment and assembly procedures. Rinsing with high-pressure ultrapure water (HPR) is the most effective tool to remove microparticles and therefore reduces field emission [443–445]. HPR is also effective in reducing field emission which cannot be processed during an rf test. Figure 7.16b shows the reduction of field emission with HPR after an rf test limited by strong field emission.

HPR must be carried out in a Class 10–100 clean room to prevent recontamination with dust [446–448]. For best cleaning, it is important to avoid drying between final water rinse after chemistry and the start of the first HPR. Particles adhere more strongly to the surface as the water evaporates. Before entering the clean room, the outside of the cavity must be degreased with soap and high-purity water using an ultrasonic bath for about 30 min, followed by high-pressure rinsing at 100 bar. In some cases, especially after postpurification, the outside of the cavity goes through a light BCP etch (20 µm) to improve He bath cooling during rf operation.

Figure 7.17 shows an HPR system for a 9-cell cavity [440]. Typically HPR systems work with a water flow rate between 5 and 20 L/min and a pressure between 80 and 150 bar, which allows removal of particles larger than a few micrometer [449, 450]. The final particle filter (pore size ≤0.2 µm) is placed as close to the nozzle as possible with no moving parts (i.e., valves) or dead ends between the filter and nozzle. It is important to monitor the erosion of the nozzles (Fig. 7.18a) under the SEM, especially if the shower head is made from stainless steel. The preferred nozzle material is sapphire. For one system (DESY), the water exits through eight nozzles in the head at the top of the rinsing cane, which moves up and down inside the cavity while the cavity rotates. During a 2-h rinse up to 2000 L of water are sprayed onto the surface. At DESY,

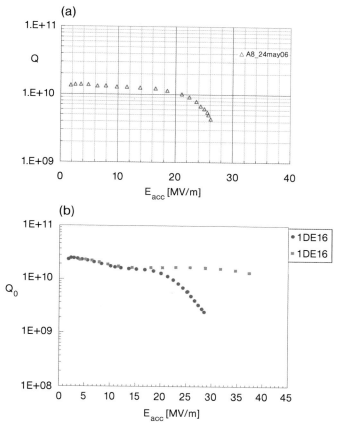

Fig. 7.16 (a) Q vs. E curves for a cavity prepared by BCP and without 600 °C or 800 °C annealing [434]. (b) Q vs. E curves for a single-cell cavity showing strong-field emission which was subsequently removed by additional HPR. For the first test, the cavity was prepared by EP, bake, and HPR but showed strong field emission with the maximum accelerating gradient of 28.1 MV/m. After another cycle of only HPR (no additional chemistry) the field emission was gone (no X-rays) and the maximum gradient increased to 37 MV/m [178] (courtesy of DESY).

9-cell cavities are rinsed once after the BCP or EP treatment, and additionally up to five times after the assembly of the flanges.

Water quality deserves close attention [447]. The standard practice to monitor the pure water supply systems is to measure resistivity, particle count, bacteria, total organic content (TOC), and residues. Particle counters are proven tools for liquid quality control and monitor particles in the sub-µm range.

One study monitored the HPR cleaning effect by measuring the particles rinsed out of the cavity using a particle counter [265]. During the first HPR after BCP or EP, a large number of particles were rinsed out of the cavity. Many materials used in the cavity preparation and associated systems were found to

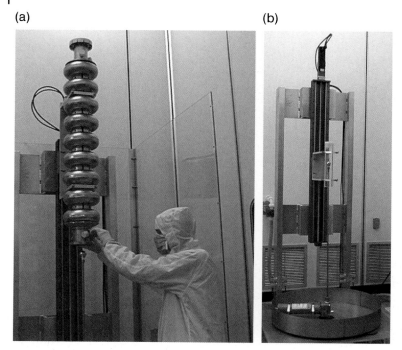

Fig. 7.17 (a) The HPR system for a 9-cell cavity [440]. (b) Layout of the HPR system showing rotating filter, rinsing and cane and linear bearing for vertical motion.

be present such as, viton, copper, and steel. In some cases, particles >100 μm were found. These would be dangerous for thermal breakdown if left in the cavity. In subsequent rinses, the particle number decreased. Figure 7.18(b) shows a correlation between a number of particles rinsed out of the cavity during the final rinse to the onset of field emission in the first power run of the following vertical cavity test.

A systematic study has been carried out on forces generated at the cavity surface due to HPR with 10 L/min flow at 100 bar [451]. The high-pressure water is filtered by a 40 nm–100 bar filter and then sprayed by a head with six nozzles. Each nozzle is 0.55 mm in diameter with a sapphire diaphragm. A load cell placed at proper distances from the nozzle characterizes the forces exerted by the water jet. At 100 mm from 100 bar jet the HPR force is 3.5 N. The impact spot size is 0.5 mm at 60 mm from the head axis. The power density is 250 W/mm^2 at 35 mm from the axis. Wedges installed on the load cell measure the angular dependence of the water force profile, which is found to follow the expected sinusoidal dependence. Such force is sufficient to shear off burrs from scratches as shown in Fig. 7.19 [452].

HPR is generally accompanied by the formation of ozone, as well as electrostatic charging of the cavity. If the HPR water jet is parked at a fixed spot for a

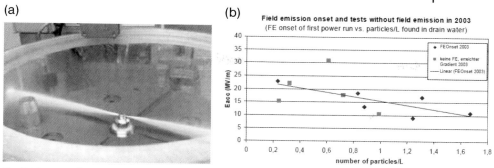

Fig. 7.18 (a) High-pressure water jet stream emerging from a nozzle [265]. (b) Correlation between field emission onset field and particle density count from rinsing water [265] (courtesy of DESY).

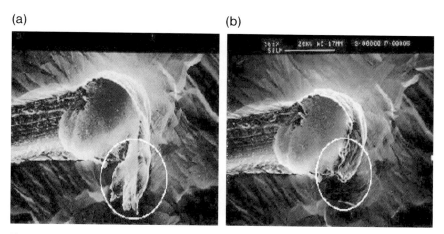

Fig. 7.19 SEM pictures of a scratch on a Nb surface (a) before and (b) after HPR. The scale bar is 50 μm [452] (courtesy of Saclay).

few seconds colored rings appear on the Nb surface. These arise from the rapid growth of Nb_2O_5 [451, 453]. The growth can become severe enough to cause field emission and even quench [454].

7.6.2
Alternate Final Cleaning Methods

Among the alternatives to HPR still under development are dry-ice cleaning [252] and megasonic rinsing [455].

Relaxation of a liquid CO_2 jet emerging from a nozzle (Fig. 7.20) at a pressure of 50 bar results in a snow–gas mixture with nearly 45% snow at 194 K. In addition, a supersonic jet of N_2 (pressure 12–18 bar) surrounds the CO_2 jet to

Fig. 7.20 Snow cleaning with a CO_2 jet [252] (courtesy of DESY).

accelerate and focus the jet. The nitrogen also helps prevent condensation of humidity onto the cooled cavity surface. The mechanical cleaning is based on several effects: shock freezing the contaminations, strong impact of the snow crystals, and a 500 times volume expansion at sublimation. Contaminations get brittle and start to flake off from the surface. As the snow particles hit the surface and melt at the point of impact, a chemical cleaning effect also takes place. IR heaters are installed around the cavity to prevent moisture condensation. Since the entire process is dry and particles are entrained in the gas flow out of the cavity, snow cleaning may be possible in the future for cavities in the horizontal orientation as installed inside the final cryomodule.

The principle of megasonic cleaning is similar to ultrasonic, but with frequencies around 1 MHz. The cleaning effect is based on high power pressure waves inside the cleaning solution, and less on cavitation. Particles down to 0.1 µm can be removed from surfaces.

Future application to cavities needs the development of an oscillator inside the cavity to realize a high transmission of megasonic power. The transportation of particles out of the cavity requires a high liquid flow-rate, which is suitable for an open cavity, as for HPR.

7.7
Dust-Free Assembly

After HPR, drying in the Class 10–100 Clean Room follows for about 12 h. In some cases, a second HPR takes place after assembly of some of the necessary flanges, and the field monitor probe. Several final assemblies follow: attaching flanges, installing input couplers, and assembling cavity strings. Great care must be exercised to avoid recontamination during the subsequent cavity handling, component assembly, installation, and operation of the accelerator modules [449]. Handling and assembly time of an open cavity should be kept as short as possible. Movement of personnel should be slow and monitored by particle counters to avoid kicking up dust, especially from the floor. Clean room staff must be well trained and highly motivated to keep the discipline required for the high level of cleanliness necessary.

All components, including fastening hardware (screws and nuts), are pre-cleaned in an ultrasonic bath and rinsed with pure water. After drying in the clean room the residual particle contamination is controlled before installation of components to a cavity. All components are blown carefully by ionized and particle filtered (0.02 µm) nitrogen gas. Satisfactory attenuation of the number of particles blown off is monitored by an air particle counter. During assembly, all component flanges are attached to cavity flanges with just two screws to make a particle-tight seal at each connection and to keep the creation of dust at a minimum. After sealing the last flange all the remaining screws are installed and tightened. For power coupler assembly, the cavity is filled with filtered argon at 100 mbar over pressure. This allows clean gas flow out of the cavity flange during opening a flange connection and will prevent particle movements toward the cavity. After disassembly of the blind flange, the power coupler is guided toward the cavity and fixed with two screws to close the connection particle-tight. Finally the remaining screws are set and tightened.

The basics of dust-free clean room technology can be found in dedicated textbooks [456, 457]. The requirements of design, construction, commissioning, and maintenance of high quality clean room installations is found in industry standards such as ASTM, ISO, JIS, VDI, and others. A thorough discussion of aspects relevant to cavity assembly is given in [446]. Proven tools for quality control are particle counters for air and liquids down to the sub-µm range. For laminar air flow conditions, the standard value of air velocity should be set to 0.5 m/s with a tolerance of ±20%. Shock waves produced by fast opening and closing valves can dislodge particles; therefore, valves have to be cleaned and activated slowly. The air flow conditions of the clean room, the influence of movement (doors, personal) can be visualized using a fog generator, so special attention can be paid to avoid turbulent air flow at the cavity and components couplers [447]. Evacuation of clean components should be slow enough to avoid turbulent flow and movement of any residual dust.

8
Input Couplers

8.1
Overview of Requirements and Design Principles

An input coupler is a device that efficiently transfers rf power from the generator (source) to a beam-loaded cavity by providing a good impedance match between the two, as depicted in Fig. 8.1 (a).

The coupler must operate over a wide range of load impedance, which varies from a matched impedance at full beam loading to full reflection when there is no beam. For a superconducting cavity, the input coupler is normally inserted at the beam pipe just outside the end cell of the accelerating structure (Fig. 8.1 b) rather than inside the cell to avoid field enhancements that may lower the quench field, or field perturbations that may initiate multipacting (MP) in the cell. As an auxiliary device, the coupler design must fulfill numerous requirements and functions. It must preserve the cleanliness of the superconducting cavity, provide a vacuum barrier between the cavity and the feeder waveguide, allow some mechanical flexibility for alignment and thermal contraction during cool down, permit variable coupling strength (external Q) in desired cases for different operating modes with beam, and serve as the thermal transition from room temperature to cryogenic temperature with minimal static and dynamic thermal losses. In addition, the coupler must be equipped with diagnostic elements to allow safe operation. These requirements call for a careful starting design from electromagnetic, mechanical, and thermal point of view.

There have been many review articles on couplers with valuable references [458–467]. The basic requirements for input couplers and fundamental design principles are also described in [1]. Here we cover advances in design, fabrication, and performance, and provide pertinent coupler examples. No single-coupler design suits all applications. A variety of coupler types have been explored and developed: coaxial and waveguide, one and two windows, cold and warm windows. We will discuss the advantages and disadvantages for these choices.

Couplers for superconducting cavities have been developed to deliver several hundred kilowatts of rf power spanning frequencies from 300 to 2000 MHz, and duty factors from 1 to 100%. Many versions have been tested off-line to MW power levels. Pioneering work at KEK, CERN, and DESY prior to 1995 on

RF Superconductivity: Science, Technology, and Applications. Hasan Padamsee
Copyright © 2009 WILEY-VCH Verlag GmbH & Co. KGaA, Weinheim
ISBN: 978-3-527-40572-5

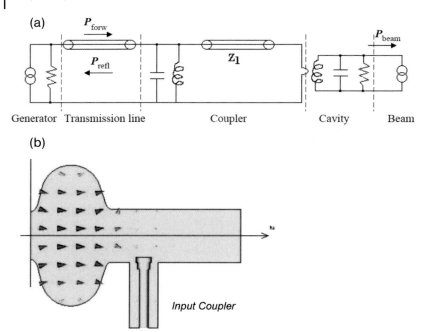

Fig. 8.1 (a) Equivalent circuit for an input coupler. (b) Coaxial input coupler at the beam pipe of a superconducting cavity.

the first generation of high-power cw couplers for TRISTAN, LEPII, and HERA helped establish a basic understanding of the design issues and performance limitation mechanisms in high-power applications at the 50–150 kW levels [468–472]. Since then, design efforts for the KEK-B Factory [473–475], the CESR upgrade at Cornell [476, 477], the LHC at CERN [478], the APT project at LANL [479], the TESLA Test Facility at DESY [480], the SNS at JLab [481], and ILC coupler variants at KEK [482–484] have significantly advanced input coupler technology. As a result there has been remarkable progress in power capability for both pulsed and CW operation: 300–500 kW of rf power in operating accelerators and up to 2 MW for prototype testing.

There are many reasons for progress. Extensive simulations take place in the design phase to predict electromagnetic, thermal, mechanical, and MP properties of coupler geometries. Commercial rf modeling codes (e.g., Microwave Studio [131], MAFIA [130], and HFSS [132]) are available for 3D simulations with high accuracy to predict and optimize rf transmission, voltage, current, and power densities. The goal of the rf design is to obtain good transmission properties (minimize reflections and insertion losses) over a workable bandwidth as well as over possible variations in temperature and assembly tolerances. The codes model electromagnetic field distributions over various elements in the coupler transmission line to establish best locations for cooling intercepts and window placement, and to determine the coupling strength, normally given in terms of the external Q (Q_{ext}).

The detailed electromagnetic field distribution is exported into commercial mechanical analysis codes (e.g., ANSYS [136], COSMOS [137]) to calculate stress, vibrations, and heating in regions that bridge ambient and liquid helium temperatures. The goal is to obtain a low cryogenic heat leak by introducing thermal intercepts at proper locations and temperatures. To minimize rf losses, stainless steel parts of the coupler must be coated with high conductivity copper of optimal thickness after taking into account the cryogenic heat leak due to conduction, with a goal to minimize both static and dynamic heat loads overall. In the warm sections of the coupler, the design should avoid a large temperature rise at the operating power level so as to keep the stresses manageable due to thermal expansion and contraction. Cooling designs should take into account the largest possible anticipated thermal load due to operation in traveling and standing wave modes.

The mechanical design of the coupler needs to be integrated with the cryomodule design taking into consideration assembly sequence issues as well as the movement of coupler parts due to cool down of the module. In some cases, these shifts (10–20 mm) are accommodated by bellows integrated into the mechanical design of the coupler [480].

Equally important is the implementation of clean practices during fabrication and assembly with high quality control of materials and platings to ensure reliability, high-power performance and preserve cavity cleanliness. Sharp edges should be eliminated in design and fabrication to avoid field emission. For coated parts, an excellent bond between a film and a substrate is essential to thermally stabilize the film, and to prevent particulate generation which can be dangerous for field emission. Use of a cold window is advisable for high-gradient applications to seal the cavity from the many coupler components. An additional warm window is necessary to avoid gas condensation on cold window surfaces. The second warm window also adds protection for vacuum integrity. The cold window should not be in such close proximity to the cavity that impact from field-emitted electrons from the cavity leads to window charging, arcing, and possible puncturing. Section 8.4 will discuss window topics.

Multipacting (*mult*iple im*pact* electron amplification) is a phenomenon of resonant multiplication of electrons under the influence of rf fields. We discussed MP in cavities in Chapter 5. Here we cover MP in couplers (Section 8.8). Secondary electrons created by a primary electron impinging on a metal surface are accelerated by the electric field, and impact the surface again. If the time for the trajectory is synchronous with the rf period, and if the impact energy is greater than unity, an avalanche of electrons is created. The synchronous condition depends on the details of the geometry, and on the local electric and magnetic field values. The secondary electron yield is larger than unity for most metals at a primary electron impact energy between 100 and 1000 eV. The secondary electron yield is particularly high for the ceramic surface, which must be coated with an antimultipacting thin film with minimal rf losses.

Codes are available to simulate MP in various regions of the coupler, and these help to make best choices for the geometry and thus to minimize the incidence of MP. While it has so far proven impossible to avoid MP through the entire power range a coupler will experience, it is possible to tailor design to avoid MP conditions around the operating power level, especially lower order MP bands which are more stable. This can be done by properly choosing the coupler diameter (and impedance of the coaxial line) for the operating power level. In cases where a MP band lies close to an operating point, voltage or magnetic biasing has been developed to disrupt MP resonance conditions for coaxial and waveguide input couplers, respectively. Degassing the coupler by baking keeps it free of surface contamination, decreasing the secondary emission, and thereby the time required to bring a coupler to the desired power level through the conditioning process (Section 8.9).

Additional good practices should be followed in coupler design and fabrication. For beam-tube couplers, the fields need to couple through a beam pipe that is below cut-off. The beam-pipe diameter and spacing between the end cell and the coupler port need to be sufficient that the antenna does not have to penetrate too far into the beamline, where it may become a source of strong wakefields or coupler kicks. The transverse electromagnetic field of the power coupler on the beam tube can create a small kick which increases beam emittance [485]. This effect is especially strong for a cavity at the low-energy end of an accelerator (the injector), where a high average rf power must be coupled to a vulnerable low-energy beam. In this case, a twin coaxial coupler design (Fig. 8.2) reduces harmful transverse kick fields, ideally to zero on-axis. The shape and location of the antenna tip can also be optimized to minimize penetration into the beam pipe. A good coupler design should also try to ensure that there are no significant rf fields from higher order modes that may cause anisotropic heating at the cold window to minimize thermal stresses.

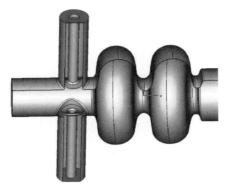

Fig. 8.2 Twin coaxial input coupler adopted for the Cornell ERL injector to eliminate the transverse kick to a low energy beam. For a 1-mm offset between the two antenna locations, the kick is reduced by more than a factor of 10 over the kick produced by a single coupler [485].

8.2 Power

The input power requirement, P_f, is determined by the operating cavity voltage, beam current and the overhead called for by peak microphonics detuning, and Lorentz-force detuning expected [486, 487]:

$$P_f = \frac{V_c^2}{4\frac{R}{Q}Q_{ext}} \left\{ \left[1 + \frac{I_b \frac{R}{Q} Q_{ext}}{V_c} \cos\phi_s \right]^2 + \left[\frac{2Q_{ext}\delta\omega_m}{\omega} \right]^2 \right\}$$

Here V_c is the cavity voltage, I_b is the average beam current, ϕ_s is the synchronous phase, $\delta\omega_m$ is the amplitude of the frequency detuning, and ω is the rf frequency. RF feedback loops provide cavity field stability reducing the microphonics influence on beam quality as well as the rf power overhead required to compensate for the microphonics detuning [486–490]. Environmental microphonic noise creates fluctuations in the cavity resonance frequency and thereby produces amplitude and phase modulations of the field, affecting both beam quality and rf system performance. This is especially true for high-Q superconducting cavities. The optimum Q_{ext} is determined by beam loading.

$$Q_{ext} = \frac{V_c^2}{\frac{R}{Q} I_b \cos\phi_s}$$

Typical loaded Q for beam-loaded applications range from 10^5 to several 10^6. In the case of near-zero beam loading, as for example for the ERL, the rf power required depends on the microphonics detuning level and choice of Q_{ext} (or loaded Q) as shown in Fig. 8.3 for 1.3 GHz, 7-cell cavity at 20 MV/m operating field level.

Although the required power decreases substantially with increasing Q_L, running cavities at Q_{ext} in the 10^8 range is challenging because the small cavity

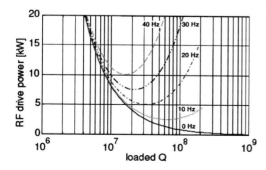

Fig. 8.3 Peak rf drive power as the function of Q_L for a 1.3-GHz, 7-cell cavity at 20 MV m^{-1} accelerating gradient. The power is determined by the peak microphonics cavity detuning during cavity operation [487].

bandwidth of a few Hz makes the rf field extremely sensitive to perturbations of the resonance frequency due to microphonics and Lorentz-force detuning. Operating at high Q_L makes it hard to meet amplitude and phase stability requirements which can be quite demanding for some applications, such as a high-current cw ERL-based light source, where the relative rms amplitude stability must be better than a few $\times 10^4$ and rms phase $< 0.1°$ in order to achieve the beam quality necessary for a good light source. In addition, Lorentz forces during filling detune the cavity by several hundred Hz making necessary precise compensation during turn-on. Consequently the highest loaded Q_s are presently limited to several 10^7; but advances are forthcoming [487] to lower the rf power requirements for cw accelerators.

8.3
Waveguide versus Coaxial Couplers

Many basic choices need to be made in selecting a power coupler design. Among the main factors governing these choices are the rf frequency, power level, ease of cooling, static heat leak, and coupling adjustability needed. Figure 8.4 compares the two primary varieties: waveguide and coaxial couplers. Waveguide couplers are not as widely used as coaxial couplers. Examples of waveguide couplers include CEBAF, CEBAF upgrade, and JLab FEL at 1500 MHz

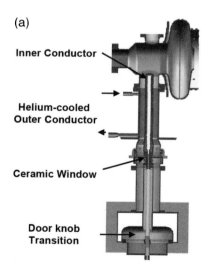

Fig. 8.4 (a) Coaxial coupler used for SNS 800 MHz cavities adapted from the KEK-B coupler. Gas flow or water flow through the inner conductor is used for cooling (courtesy of JLab and SNS).

(b) Waveguide coupler for the CESR 500 MHz cavity used in several storage rings. A planar ceramic disk-shaped window is incorporated in the warm waveguide, which is of reduced height [489, 490].

[491, 492], CESR at 500 MHz [489], as well as light sources such as SRRC, DIAMOND, and CLS, which have adopted the CESR SRF system.

Waveguide coupling is conceptually simpler since it does not require any rf transition between the output waveguide of the rf power source and the cavity interface. Coaxial couplers generally incorporate a transition from waveguide to coax, such as a door-knob-shaped element. Due to the existence of a cutoff frequency in waveguides, the size of the waveguide coupler is generally larger at a given operating frequency than for the coaxial case. Because of the larger cross section, the contribution to the infrared heat transfer to the cryogenic environment is also larger. The coupling strength depends on the size and shape of the coupling iris, the longitudinal location of the waveguide relative to the cavity's end cell, and the location of the terminating short (if any) of the waveguide on the opposite side of the beam tube. Some adjustable coupling is possible by deforming the gap distance. Coupling can also be adjusted using an external, 3-stub tuner waveguide on the air side [506, 507], though the additional field stress and heating due to the standing waves in the line can become problematic for heating and breakdown at high average power levels.

One of the main advantages of the waveguide coupler is the need for cooling only the outer wall. For one MW traveling wave power the peak electric field for a standard waveguide at 1.3 GHz is 400 kV/m, whereas for a coaxial line with an outer diameter (OD) equal to the small side of the waveguide is 800 kV/m [459]. The power density is lower for the waveguide, but the total longitudinal losses are the same in both cases, about 1 kW/m in copper. For the coaxial line, about 2/3 of this loss must be cooled from the inner conductor, which is not as readily accessible as the outer conductor. The losses at the waveguide wall can normally be intercepted at 70 K or 4.5 K, using straps or heat-exchanger piping. Waveguides also offer a higher pumping conductance over a coaxial line. MP electrons in the coax can be disrupted by an electrical bias of a few kV [493–496], whereas MP in the waveguide can be cleared with magnetic bias using a few gauss [497, 498]. However, this approach is not possible in the superconducting waveguide section due to persistent screening currents which exclude dc magnetic flux from the waveguide volume. Here grooving the waveguide wall is an option. The main disadvantage of waveguide couplers is their size, which increases the mechanical and thermal complexity of interfaces to the cavity and cryomodule. Plating and flanging are also harder for rectangular waveguides than round pipes in coax.

Not being limited by a cutoff frequency, coaxial couplers are more compact, especially for low-frequency systems. A variety of window geometries and arrangements are available to be discussed later. A large range of coupling values can be achieved by proper insertion of the center conductor into the line. Also variable coupling can be achieved with relatively simple adjustment of the inner conductor penetration via a bellows extension. The center conductor can be electrically isolated from the outer conductor using a kapton film to allow the use of bias voltage to disrupt MP. Larger diameter coax lines are useful to push MP bands to higher power levels [493]. But these have a higher thermal radiation heat leak and a larger interface to the beam tube. The sizing of the coax diameters should avoid azimuthal overmoding.

8.4
Windows: Warm and Cold

A window provides the physical barrier between the cavity vacuum and open waveguide of the power source, but the barrier must be transparent to microwaves at the operating frequency. Many designs use two windows. The main arguments for two windows are (i) to preserve the cleanliness of the cavity by sealing with a first, cold window, and (ii) vacuum safety provided by the second, usually warm window. Superconducting cavities must be handled and maintained under Class 10–100 clean-room conditions at all times to be dust free. It is therefore essential to seal the coupler opening of the cavity with a window at an early stage in the clean-room assembly of the input coupler. Placing a window near the cavity allows a compact cavity assembly for ease of handling after sealing in the clean room. Being near the cavity means the window is at 70 K or lower and therefore must have a vacuum on both sides. Hence the window can be cooled only by conduction, making high average power design more challenging. MP can occur in the vacuum on both sides of the cold window. If the cold window is too close to the cavity, field emission electrons from the cavity can charge it up and lead to arcing and eventually ceramic damage [499, 500].

The second window prevents gas condensation on the cold window, and serves as a backup to preserve the cavity vacuum in case the cold window develops a leak during operation. The vacuum between the windows must be pumped separately. The second window is normally incorporated into the transition from coaxial to waveguide. It can also be a planar waveguide window or coaxial disk window. The warm part of the coupler including the second window is generally assembled after placing the cavity string into the vacuum vessel, also under clean and dry conditions for faster processing to high power. Cooling designs for both windows should take into account the largest possible anticipated thermal load due to operating the window and coupler in a full standing wave condition swept through 180° phase change.

The cold window design is a must for applications aiming for the highest gradients (15–40 MV/m), to prevent dust contamination and field emission during subsequent assembly steps. For high (≥ 100 kW) average power applications at moderate to low gradients (5–15 MV/m), a single, warm window design is used with convection cooling or water cooling. A gas barrier serves as the second window to provide safety for the cavity vacuum in case the main window develops a leak. In this case, the cavity is exposed only to the dry, dust-free air in between the two windows. The warm window is located sufficiently far from the cavity cold mass to limit both conductive and radiation heat leaks into the liquid helium bath. The challenge for the single warm window design is that a large coupler assembly must be attached to the clean cavity in a clean room.

Several types of ceramic windows are in use. Coaxial couplers use either the cylindrical window [478, 480, 501–503] or disk window (Fig. 8.5) [468]. Waveguide couplers use either a planar disk [504] or planar rectangular ceramic incorporated within the rectangular waveguide [499], as shown in Figs. 8.6 and 8.7.

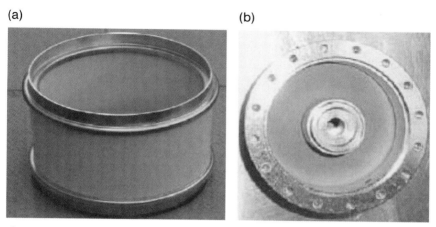

Fig. 8.5 Window geometries for coaxial couplers: (a) cylindrical ceramic (courtesy of DESY); (b) disk ceramic [463, 464] (courtesy of KEK).

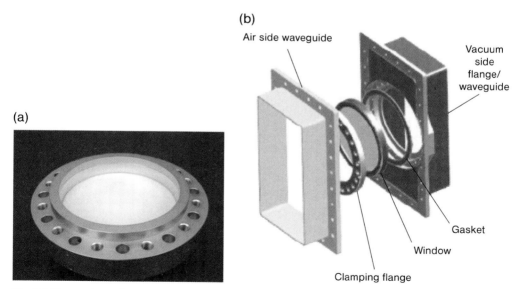

Fig. 8.6 Waveguide windows: (a) planar disk; (b) incorporation of planar disk into rectangular waveguide (courtesy of SLAC).

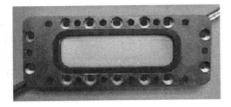

Fig. 8.7 Planar rectangular waveguide window used for JLab FEL, and tested to 50 kW cw [492, 500].

8.5
Coaxial Coupler Examples

Several generations of coaxial couplers have emerged for pulsed power applications above 200 kW. The performance of most types is summarized in Table 8.1 [464]. Figure 8.8 shows the 3D geometry of TTF-III (third generation), 1.3 GHz coupler developed at the TESLA Test Facility (TTF) to be used for the XFEL and possibly the ILC [480]. The coupler is designed for operation in the pulsed mode (1.3 ms) at several hundred kilowatts power and less than 5 kW average power. The HFSS-calculated electric field distribution (Fig. 8.9) shows electric field minima at the two windows in standing wave operation. The TTF-III cou-

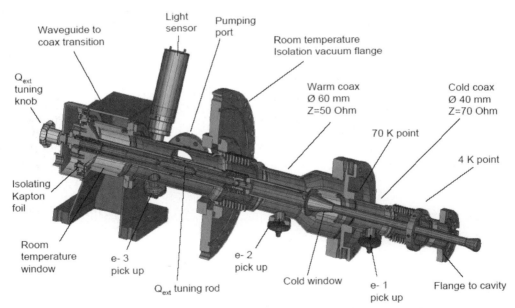

Fig. 8.8 The TTF-III coupler has two cylindrical ceramic windows, a "warm" window in the waveguide to coaxial interface and a "cold" window which seals the cavity after assembly in a clean room. The windows receive a nominal 10 nm coating of titanium-nitride on their vacuum surfaces to reduce their secondary electron emission coefficient [480] (courtesy of DESY).

Table 8.1 CW couplers [464].

Facility	Frequency	Coupler type	RF window	Q_{ext}	Maximum power	Comments
LEP2/SOLEIL	352 MHz	Coax fixed	Cylindrical	$2\times10^6/1\times10^5$	Test: 565 kW 380 kW	Traveling wave @$\Gamma=0.6$
LHC	400 MHz	Coax variable (60 mm stroke)	Cylindrical	2×10^4 to 3.5×10^5	Oper: 150 kW Test: 500 kW 300 kW	Traveling wave Standing wave
HERA	500 MHz	Coax fixed	Cylindrical	1.3×10^5	Test: 300 kW Oper: 65 kW	Traveling wave
CESR (Beam test)	500 MHz	WG fixed	WG, 3 disks	2×10^5	Test: 250 kW 125 kW Oper: 155 kW	Traveling wave Standing wave Beam test
CESR/3rd generation light sources	500 MHz	WG fixed	WG disk	2×10^5 nominal	Test: 450 kW Oper: 300 kW 360 kW	Traveling wave Forward power
TRISTAN/KEKB/BEPC-II	509 MHz	Coax fixed	Disk, coax	$1\times10^6/7\times10^4/$ 1.7×10^5	Test: 800 kW 300 kW Oper: 400 kW	Traveling wave Standing wave
APT	700 MHz	Coax variable (± 5 mm stroke)	Disk, coax	2×10^5 to 6×10^5	Test: 1 MW 850 kW	Traveling wave Standing wave
Cornell ERL injector	1300 MHz	Coax variable (15 mm stroke)	Cylindrical (cold and warm)	9×10^4 to 8×10^5	Test: 61 kW	Traveling wave
JLAB FEL	1500 MHz	WG fixed	WG planar	2×10^6	Test: 50 kW Oper: 35 kW	Traveling wave

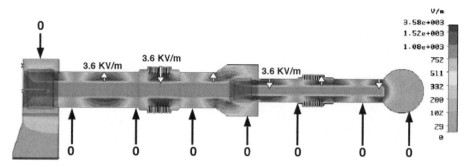

Fig. 8.9 RF simulation of TTF-III input coupler in standing wave operation. Windows are placed at the electric field minimum [508] (courtesy of DESY).

pler has adjustable coupling strength between 1×10^6 and 2×10^7 for 15 mm antenna movement. There are two cylindrical windows (97.5% Al_2O_3 with TiN coating), one at 70 K and the other warm near the door-knob transition. The cold part seals the cavity vacuum and is entirely inserted into the cryomodule. The warm part has its own separate vacuum. The cold coaxial line has 70 Ω impedance with 40-mm OD, and the warm line has 50 Ω impedance with 62 mm OD. All stainless steel parts are made of 1.44 mm thick tubes. Copper plating is 30 μm thick on inner conductor and 10 μm thick on outer conductor. There are two heat intercepts: at 4.2 K and 70 K. Dedicated tests show [509, 510] that 10 kW cw traveling wave and 5 kW cw standing wave operation is possible. Higher powers require cooling improvements discussed below.

Advances for alternate candidates continue [465, 482, 483]. A 4th generation coupler is under development [466, 518] with the aim of increasing the power handling capability and pushing the expected MP levels to higher powers, by increasing the line impedance for both the warm and cold parts of the coupler to 70 Ω. The coupler has an outside diameter of 80 mm.

Two new coupler designs are under development at KEK-STF for ILC at 1.3 GHz. One [482] is based on two disk windows, cold and warm, but the coupling is fixed (Fig. 8.10) by eliminating the inner conductor bellows for simplification. Tests exceeded 1 MW, 1.5 ms at 5 Hz (Table 8.2). The other development

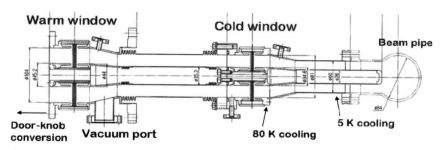

Fig. 8.10 KEK-STF two window disk ceramic coupler with fixed coupling [482].

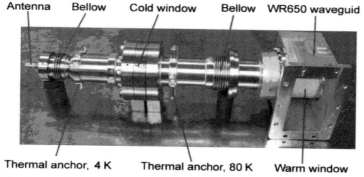

Fig. 8.11 Modular coupler at 1.3 Hz developed at KEK-STF [482] (courtesy of KEK).

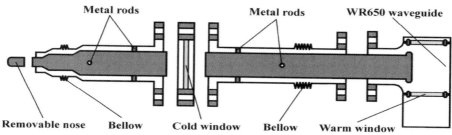

Fig. 8.12 Modular coupler schematic [483] (courtesy of KEK).

is a new modular design [483] for ease fabrication and assembly, with a capacitively coupled disk window for a low heat leak (Figs. 8.11 and 8.12). The complete coupler is divided into four parts: waveguide to coaxial transformer, coaxial line, rf window, and the antenna at cold side. Since the inner conductors are not attached rigidly to the waveguide, only two bellows in the outer conductor are needed to absorb the movement of the coaxial line due to thermal contraction and expansion. Pairs of rods mounted in the gap between the inner- and outer conductors are rotated by 90° from each other. High-power tests demonstrated successful operation in the pulsed mode at 1 MW × 1.5 ms × 5 Hz. Multipactor ceased above 200 kW.

The TTF-III input coupler has been improved [511, 512] for cw operation at 75 kW average power for the Cornell ERL injector (Figs. 8.13 and 8.14). Cooling of critical parts is enhanced to reduce the static and dynamic heat loads. The "cold" window is enlarged to the size of "warm" window. The cold part uses a 62 mm, 60 Ω coaxial line (instead of a 40 mm, 70 Ω line) for stronger coupling, better power handling and to avoid MP. The outer conductor bellows design in the warm and cold sections is improved for better cooling by introducing more heat intercepts. The warm inner conductor bellows is cooled by forced air flow. The antenna tip is enlarged and shaped for stronger coupling needed for high beam current. Q_{ext} can be varied from 9.2×10^4 to 8.2×10^5. The geometry is op-

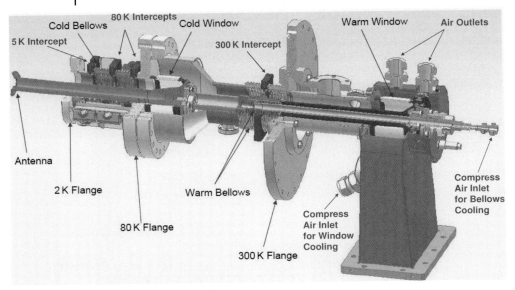

Fig. 8.13 3D CAD rendering of the variation of TTF-III coupler for 75 kW cw operation [511].

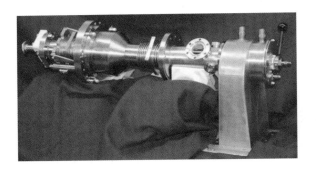

Fig. 8.14 Assembly of the Cornell ERL injector coupler developed for 75 kW cw operation [511].

timized for obtaining low reflection over the whole range of coupling. Simulations show a good match with broad bandwidth. Figure 8.15(a) shows the reflected power behavior (S_{11}) for different coupling values obtained by bellows extension and contraction.

The total (static plus dynamic) heat loads are about 0.2 W at 2 K, 3 W at 5 K, and 75 W at 80 K. The temperature profile along the inner and outer conductors is shown in Fig. 8.15(b) as results of ANSYS simulations. Figure 8.16 shows the integration of coupler with the cryomodule. Production couplers reached [512] 61 kW cw after pulsed rf conditioning up to 85 kW (see also Table 8.2).

Tables 8.1 and 8.2 summarize the performance of a variety of couplers developed for pulsed and CW applications [464].

The end faces of the ceramics are brazed to copper collars, which in turn are brazed to the warm and cold coupler parts. These parts are mainly fabricated

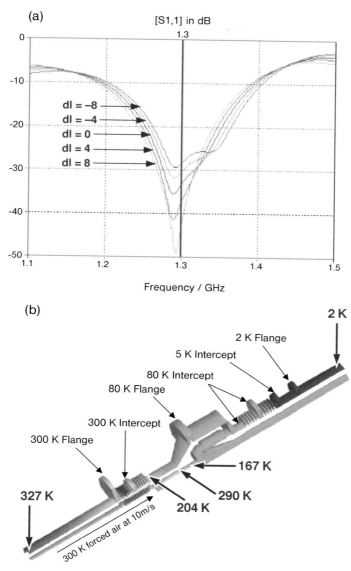

Fig. 8.15 (a) S_{11} parameter of the Cornell ERL injector coupler for a range of coupling values (due to different bellows' extension/compression). The value of *dl* corresponds to the antenna travel relative to the middle position. (b) Calculated temperature profile [511].

Table 8.2 Performance of pulsed power couplers [464].

Facility	Frequency	Coupler type	RF window	Q_{ext}	Maximum power		Pulse length and rep. rate
SNS	805 MHz	Coax fixed	Disk, coax	7×10^5	Test:	2 MW	1.3 ms, 60 Hz
					Oper:	250 kW	1.3 ms, 60 Hz
J-PARC	972 MHz	Coax fixed	Disk, coax	5×10^5	Test:	2.2 MW	0.6 ms, 25 Hz
						370 kW	3.0 ms, 25 Hz
FLASH	1300 MHz	Coax variable (FNAL)	Conical (cold), WG planar (warm)	1×10^6 to 1×10^7	Test:	250 kW	1.3 ms, 10 Hz
					Oper:	250 kW	1.3 ms, 10 Hz
FLASH	1300 MHz	Coax variable (TTF-II)	Conical (cold), WG planar (warm)	1×10^6 to 1×10^7	Test:	1 MW	1.3 ms, 10 Hz
					Oper:	250 kW	1.3 ms, 10 Hz
FLASH/XFEL/ILC	1300 MHz	Coax variable (TTF-III)	Cylindrical (cold and warm)	1×10^6 to 1×10^7	Test:	1.5 MW	1.3 ms, 2 Hz
						1 MW	1.3 ms, 10 Hz
					Oper:	250 kW	1.3 ms, 10 Hz
KEK STF	1300 MHz	Coax fixed (baseline ILC cavity)	Disks, coax (cold and warm)	2×10^6	Test:	1.9 MW	10 μs, 5 Hz
						1 MW	1.5 ms, 5 Hz
KEK STF	1300 MHz	Coax fixed (low loss ILC cavity)	Disk (cold), cylindrical (warm)	2×10^6	Test:	2 MW	1.5 ms, 3 Hz
						1 MW	1.5 ms, 5 Hz

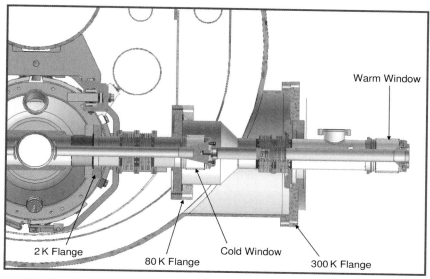

Fig. 8.16 3D CAD schematic showing interfaces between Cornell ERL injector coupler, cryomodule, and cavity [511].

from stainless steel with copper-coated rf surfaces, thickness a few electrical skin depths (~15 μm for the warm part and 5 μm for the cold part). The central antenna of the cold part is fabricated from bulk copper, and is moveable with the center bellows for adjustable coupling. An insulating kapton foil, under air, allows a dc bias voltage to the central antenna as an antimultipactor measure. There are substantial diagnostics installed for conditioning and operation. Electron detectors are installed in the warm and cold coaxial lines as well as a light detector on the vacuum side of the room temperature window. There is also a spark detector and an infrared temperature measurement on the airside of the room temperature window.

At the 50 kW average power level, the SNS 805 MHz coupler (Figs. 8.4a and 8.17) is a scaled down version of the single, warm, window KEK 508 MHz coupler which has operated successfully in KEK-B at 400 kW cw. The SNS coupler is designed to provide 550 kW pulsed power at about 8% duty factor (50 kW average power), and has been tested up to 2 MW in the pulsed mode.. A doorknob transition connects the coupler to WR975 rectangular waveguide. On the vacuum side of the main disk-type window, the outer conductor is helium gas cooled, and the inner conductor is conduction cooled. On the air side, the outer conductor of the coaxial segment is water cooled. A coaxial cylindrical capacitor using a Kapton insulator is provided between the inner extension and doorknob for dc biasing control of MP.

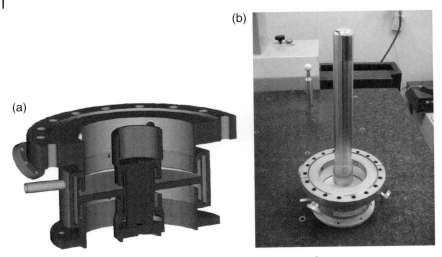

Fig. 8.17 SNS 805 MHz fundamental power coupler. (a) Warm window section; (b) coupler assembly [481] (courtesy of JLab and SNS).

8.5.1
Couplers for Low-β Resonators

Low-β applications usually require low power and therefore simpler coupler designs. Most of these are of the coaxial variety. Figure 8.18 shows the input coupler for a spoke resonator [513] and Fig. 8.19 shows the coupler for the ISAC QWR [514]. At 6 MV/m, the power required is 200 W with <0.5 W heat leak to the He bath.

8.6
Waveguide Coupler Examples

A prime example of the waveguide version is the CESR SRF waveguide coupler (Fig. 8.4b) [515]. It has a fixed coupling, $Q_{ext}=2 \times 10^5$, with a factor of 3 adjustability via a 3-stub WG transformer. Magnetic bias by solenoids wound around the normal conducting waveguide sections help to suppress MP (Section 8.8).

Waveguide couplers are used in CEBAF (Fig. 8.20) for 6 kW cw with cold and warm waveguide windows [342]. The dog-leg shields the cold window from the field-emitted electrons emerging from the cavity. The rectangular window and coupler has been upgraded to 13 kW for future energy upgrades to CEBAF [342]. JLab continues to develop waveguide couplers and windows for high-power FELs of the future where the demands could rise to several 100 kW cw.

8.6 Waveguide Coupler Examples | 281

(a)
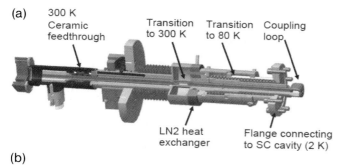

(b)

Fig. 8.18 (a) Schematic of input coupler for a spoke resonator; (b) completed coupler [513] (courtesy of ANL).

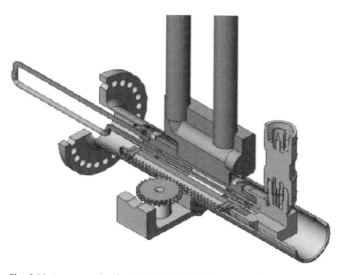

Fig. 8.19 Loop coupler for ISAC QWR [514] (courtesy of TRIUMF).

Fig. 8.20 Waveguide coupler examples from JLab. (a) 6 kW CW coupler for CEBAF with dogleg to protect the window from field emitted electrons emerging from the cavity [342]. (b) CEBAF upgrade waveguide coupler developed for 13 kW CW operation has a warm rectangular waveguide window [492, 500] (courtesy of JLab).

8.7
Materials and Fabrication

Great care must be exercised in fabrication and quality control of coupler parts to minimize coupler conditioning times and to protect the cavity from contamination during coupler installation and operation [516, 517]. Figure 8.21 shows typical inner and outer conductor parts including cylindrical ceramic windows in preparation for fabrication of a TTF-III coupler.

Maintaining tight tolerances is essential for good rf performance. The plating quality should be closely monitored to have the desired RRR (>30) and thickness uniformity without blisters [519]. Proper surface preparation is necessary for good plating adhesion. For the SNS couplers, the quality of the copper plating adhesion on the outer coaxial conductors was ensured by high-pressure water rinsing coated parts with 100 bar for 30 min. Vacuum firing at 300–400 °C is necessary for copper-plated parts. Ceramic parts should be protected from evaporated metal during brazing operations. Braze regions on the window should be shielded to decrease the impact of braze "blushing" onto the ceramic. All coupler components should be carefully cleaned to be free of hydrocarbons before final assembly. The final coupler assembly should be baked at 200 °C. During baking, an argon atmosphere should be provided for the outer surface

Fig. 8.21 Subassemblies for the TTF-III coupler (from left to right): cold coax, warm coax, tuning pushrod and waveguide box [508, 516, 517] (courtesy of Saclay and DESY).

Table 8.3 Properties of AL-300 alumina [521].

Alumina content	97.6%
Density	3.76 g/cm^3
Flexural strength	296 GPa
Dielectric constant	(1.43 GHz) 9.25–9.4
Dielectric loss tangent	(1.43 GHz) (1–3) $\times 10^{-4}$
Surface roughness	R_a 1.1 µm

to avoid oxidization of the rf contacts. After baking, the vacuum in the system should be better than 5×10^{-10} mbar.

Windows are usually made from Al_2O_3 ceramic of >95% purity with metallized ends for joining to other components of the coupler. Table 8.3 gives an example of the ceramic material specifications.

Properties of the KEK window and coupler materials can be found in [522].

Because of the high secondary electron emission coefficient of alumina (>6), a thin layer of Ti, TiN, or other antimultipactor coating is essential. Titanium vapor deposition in low-pressure (10^{-3} bar) ammonia has been successfully used for antimultipactor protection of windows, as well as various coupler components, such as waveguide sections and coax line segments. Figure 8.22 (a) shows a window-coating setup. After depositing 5–10 nm of Ti in the low-pressure ammonia vapor, the film is exposed to 100–300 mbar of ammonia vapor for about 10 h to convert any remaining Ti to TiN. Figure 8.22 (b) shows the room temperature and 70 K loss tangent measurements at 1428 MHz as a function of film thickness. The loss tangent is about three times lower at 70 K. Figure 8.22 (c) shows measurements of the secondary emission coefficient of TiN films as deposited, after 200 °C baking for 16 h, and after electron irradiation with 1100 eV. Baking reduces the yield from 2 to 1.7, and after a sufficient electron dose the secondary yield drops to 1.5.

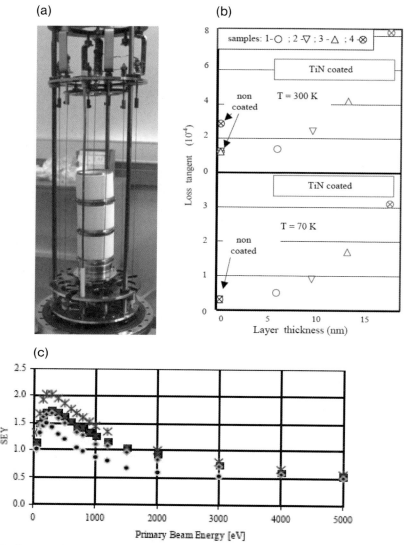

Fig. 8.22 (a) Plating set up for coating TiN on cylindrical window [508]. (b) Loss tangent measurements of ceramic coated with TiN. (c) Secondary emission yield of window ceramic coated with TiN. Electron bombardment lowers the yield [521] (courtesy of DESY).

8.8
Multipacting in Couplers

Chapter 4 discussed MP in cavities. MP in couplers is similarly an avalanche of electrons that absorbs energy from the rf field. Electrons desorb gas upon impacting the coupler surface, deteriorate the vacuum, and create conditions for an arc if the rf is continued. Two essential conditions must coexist for an electron avalanche. At certain power levels, there are resonant electron trajectories, which hit the same surface (one point MP) in multiples of rf cycles (Fig. 8.23), or trajectories across two opposite surfaces (two-point MP) in odd integer multiples of half an rf period. The number of rf periods (or half periods) is called the order of MP. High-order MP is less stable and easier to condition, because electrons have longer travel times and can get more easily out of the stable phase. In standing wave operation, MP sites remain fixed near the maxima of the electric field, but in traveling wave mode, MP electrons travel along with the wave.

If the secondary electron yield of the surface is larger than one, the number of electrons increases exponentially with increasing number of impacts. The secondary electron yield depends on impact energy and surface condition. Clean copper surfaces show a maximum secondary yield around 1.2, but contamination layers or absorbed gases increase the yield by typically 50%. In the cold part of couplers, cryopumping can collect several monolayers of gas and increase the secondary yield. Alumina rf windows have an extraordinary high yield value (~6) which can be reduced by coating with TiN and further reduced by electron bombardment.

Several numerical codes exist to predict MP trajectories and resonant conditions in coaxial lines as well as rectangular waveguides [210–214, 523–527]. The codes include both electric and magnetic fields. These should be used during the design stage of rf couplers to avoid geometries prone to MP. High computing power available recently allows calculations for complex rf structures in 3D and with a large number of particles. Simulation analysis with the code

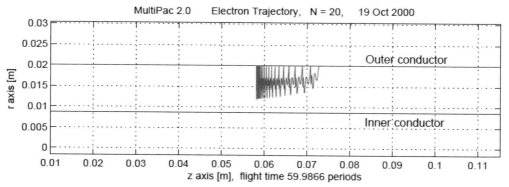

Fig. 8.23 One-side MP (third order) in a 1.3 GHz 50 Ω coax line with a standing wave [211, 523] (courtesy of University of Helsinki).

TRACK3P [525] includes traveling wave, standing wave, and any mixture of both configurations. Windows in coaxial lines have been included in some codes.

One important general outcome from the MP codes for coaxial lines is that the rf power at which a resonance occurs scales with the 4th power of the coax outer conductor dimension. Simple rules give the scaling of levels for one-point MP and two-point MP, as these vary with frequency (f), gap size (d), and coaxial line impedance (Z).

Power $\sim (fd)^4 Z$ (one-point MP)
Power $\sim (fd)^4 Z^2$ (two-point MP)

See also [1], Fig. 10.17 for a typical MP "susceptibility" chart.

The rules show that enlarging the gap between the inner and outer conductors will shift a given MP barrier to a higher power level. Another advantage from a larger outer diameter is an increase in pumping conductance which helps recover from gas burst events during MP conditioning, discussed below.

MP at windows can be particularly dangerous, as large amounts of power deposited in small areas of the dielectric can lead to large temperature gradients, dangerous stresses, and ceramic failure. Simulation studies show that resonant MP also occurs on the single ceramic surface or between the ceramic surface and surrounding metal surfaces [524] as shown in Fig. 8.24.

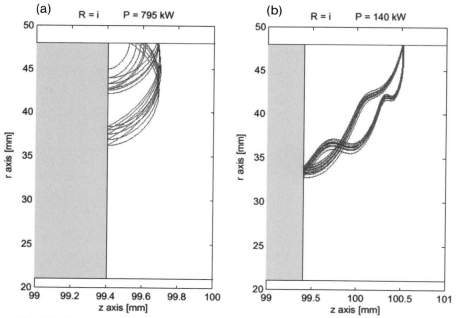

Fig. 8.24 Electron trajectories between the window surface and the outer wall of coaxial window during multipacting; (a) first order and (b) fourth order [524] (courtesy of University of Helsinki).

The avalanche of MP electrons can desorb sufficient quantities of gas to lead to an ionization discharge or rf breakdown. Bake-out of coupler components helps to reduce the processing time by reducing the adsorbed gas. Adequate pumping is essential for effective processing, and later, trip-free operation. This includes low-pumping impedance for the coupler geometry. When the stored gases on the surface layer are finally exhausted by rf processing due to electron impact, the incidents of rf breakdown decrease despite the persistence of MP electron trajectories.

Since resonance conditions for electron trajectories are closely coupled to the shape of the cavity wall and the resulting fields, a minor change in the field can disrupt the resonance condition. A standard MP remedy for power couplers uses the superposition of dc fields to disturb the resonance conditions. A dc bias of a few kV between the inner and outer conductors of a coaxial line will influence MP resonances [493–496]. A sufficiently thick insulation layer (e.g., 0.005" kapton) between the two conductors is important to avoid arcing. For full MP suppression, the bias voltage scales as

$$|V_{\text{bias}}| \sim fdZ$$

and has a typical value of 3 kV for most applications. Simulations predict and experiments prove that resonant trajectories can be suppressed by 2.5 kV bias voltage for the case of a TTF-III coupler [480, 501]. MP suppression can be realized with either polarity of the bias voltage. A negative bias voltage can also intensify one-sided MP on the outer conductor. This effect is sometimes used to better process input couplers before operations [459]. At an intermediate positive bias voltage one-sided MP on the inner conductor surface may occur. Biasing becomes more important for low-frequency applications because the power level for a given MP barrier decreases with frequency.

The dominant MP mode in a rectangular waveguide is two-sided MP between broad walls in the high electric field region [526], similar to MP across parallel plates. MP electrons traverse the gap in an odd-integer multiples of rf period. Figure 8.25 illustrates the trajectory of MP electrons in the midplane of a 500-MHz rectangular waveguide with a partially reflected wave. The CESR SRF system has a reduced-height waveguide (17"×4"), instead of a full height WR 1800 waveguide between the ceramic window and coupler iris. Moreover, a section of this guide is cooled by cryogens, resulting in an enhanced secondary emission coefficient due to condensed gas.

MP electrons travel along with the forward wave. The impact energy is linearly dependent on the forward power. Figure 8.26 shows MP bands, MP power levels, and impact energies for traveling wave conditions [528]. MP bands in the waveguide obey the following scaling law [208]:

$$P \sim (fb)^4$$

where f is the rf frequency, and b is the narrow dimension of the waveguide.

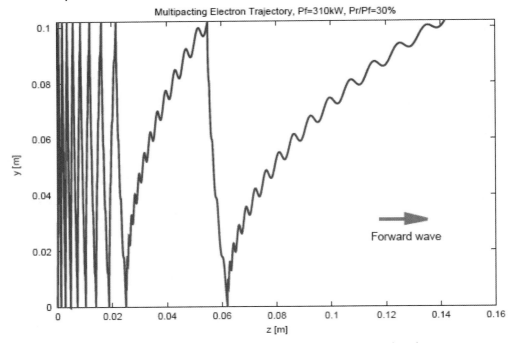

Fig. 8.25 High order two-sided MP in a 500 MHz rectangular waveguide with a partially reflected wave. The trajectory starts from the origin [208].

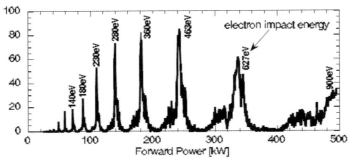

Fig. 8.26 Calculated MP bands and electron impact energy for an 18″×4″ reduced height WR 1800 waveguide [528].

Experiments with a room temperature copper waveguide (reduced height WR 1800, 18″×4″) showed that after rf processing, MP below 180 kW is completely eliminated due to reduction of the secondary electron emission coefficient from electron bombardment [528–532]. Calculations show the corresponding impact energy is about 360 eV for power levels below 180 kW. However, MP currents for higher power levels change little even after further aggressive processing [528].

8.8 Multipacting in Couplers

Two techniques have been used to suppress MP in rectangular waveguides: applying a dc magnetic field bias of about 10 G and placing grooves in the high electric field region. The latter technique is the only option for a superconducting waveguide section since the magnetic field cannot penetrate the superconductor due to persistent screening currents. The magnetic bias technique is effective up to 550 kW, for which a 10 G bias field suffices [528]. The suppression bias field is given by

$$B_{\text{bias}} \sim \sqrt{P/(fab^3)}$$

where P is the forward power, f the rf frequency, and a and b are the wide and narrow dimension of the waveguide, respectively. MP suppression coils providing 10 G bias have been in operation in the CESR cryomodule [528, 529]. Full suppression can be realized with either polarity of the bias field. Lower bias fields can enhance MP due to the oblique incidence of impacting electrons, whose orbit is bent due to the bias magnetic field.

MP in rectangular waveguides can also be suppressed by opening narrow slots on the broad waveguide wall. More than a factor of 2 reduction in MP current was obtained with a single 5-mm-wide slot at the maximum electric field location [529–531]. Processing barriers above 180 kW became much quicker. There was a reduction in light intensity during discharge at the groove location, as shown in Fig. 8.27. MP simulations with MAGIC show a corresponding reduction in electron intensity [530]. The code predicts a similar reduction for a ridge at the same location as the groove.

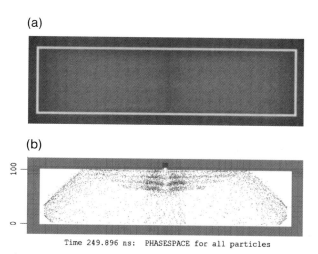

Fig. 8.27 (a) Multipacting-induced discharge in a waveguide with one groove on the broad wall shows attenuation of light intensity at the groove location due to reduction of electron density in the groove region. (b) MAGIC simulation of MP in a rectangular waveguide with a groove in the broad wall, after 250 ns of simulated run-time [530–533].

8.9
Conditioning

Before processing with rf power, baking an assembled coupler is necessary to degas surfaces and reduce conditioning times. The improved vacuum after bake reduces the amount of gas that condenses on cold parts of the coupler. Baking above 155 °C is needed to release water. Most labs bake couplers at 200 °C. Plated parts are often separately baked to 400 °C to clean the residues from plating and even to improve plating adhesion.

Before attaching to a cavity, couplers are generally preprocessed in pairs, with the common vacuum guarded by the two windows. Figure 8.28 shows a coupler processing test stand for the TTF-III coupler [465]. The joining element is a section of waveguide properly matched at the coupler's operating frequency, or by a room temperature, normal-conducting cavity. After many studies to improve design as well as fabrication and baking, the conditioning times of TTF-III couplers have been reduced from hundred's of hours to tens of hours [534]. Cold testing is necessary for higher average power (>50 kW) couplers. Figure 8.29 shows the assembly drawing of an LN nitrogen cooled test cryostat with a copper coupling cavity and two couplers [511]. One coupler is connected to the high-power cw klystron and the other coupler to a water-cooled load. Conditioning times to 60 kW are typically several tens of hours. Figure 8.30 shows two CESR rf planar waveguide windows assembled as a pair for high-power conditioning, typically for 50 h.

Diagnostics and interlocks with properly designed controls are necessary for safe processing to avoid coupler damage. In the presence of a cold surface with

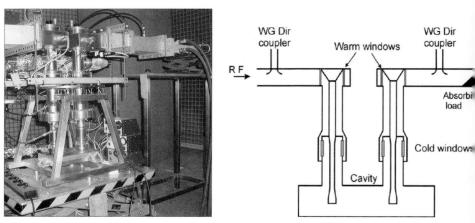

Figure. 8.28 A pair of TTF-III couplers mounted on their test bench at LAL-Orsay [465]. TTFIII coupler test stand (courtesy of Orsay). (b) Schematic of the coupler test stand. Two couplers are assembled with their antennas to a half height waveguide transition. The best prepared couplers need 10 to 40 h for processing [508] (courtesy of DESY).

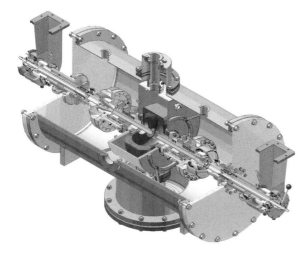

Fig. 8.29 3D CAD drawing of a pair of modified TTF-III couplers assembled to process while attached to a cold (77 K) cavity [511].

Fig. 8.30 CESR RF system waveguide window pair in high power test [463].

strong cryopumping, a vacuum monitor alone is not very sensitive. A fast (µs) interlock shuts off the rf at the first sign of reflected power, vacuum burst, or light. Photomultiplier tubes or diodes are helpful to detect early light from arcs. Coaxial rf pick-up probes detect electron or ion current. At high-power levels, infrared sensors keep track of the window temperature; platinum thermometer sensors placed at the window flange are also an inexpensive way to get information about temperature rise of the window assembly.

Typically, the initial processing is performed in a traveling wave mode with pulsed power and progressive increases of power levels and pulse lengths as the vacuum levels allow (Fig. 8.31). As the power level (or pulse length) rises, the vacuum in the coupler may suddenly degenerate and the reflected power jumps up, often accompanied by the emission of light and the appearance of electrons at rf pick-up probes. After pumping for some time till the vacuum recovers, the

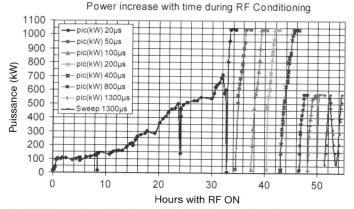

Fig. 8.31 A good conditioning run for the TTF-III coupler showing rate of power increase with different pulse lengths until the desire pulse length of 1300 μs is achieved [516, 517] (courtesy of Saclay).

power (or pulse length) may be increased beyond the level of the first event. The coupler is said to "condition" or "process." However, a similar event takes place again at a slightly higher level. Overall it takes many tens (and many hundreds of hours if fabrication or preparation is poor) for a coupler to reach its design power level and design pulse length through many such processing events. At some power levels, such as near MP barriers, the events are very frequent and the vacuum bursts more intense, so that it takes much longer to process through.

The presence of electrons and the sudden intensity of events at specific power levels shows that MP is involved. When the MP resonant conditions are satisfied, an avalanche of electrons bombards the surface and desorbs gas. With sufficient quantity of gas in the high electric field region, a gas discharge or spark occurs with the emission of light, and the rf power is fully reflected. The electron bombardment lowers the secondary emission coefficient most likely through the desorption of gas so that fewer electrons are emitted when the power is raised again. All regions of high secondary emission on metallic (or ceramic surfaces) need to be processed before significant progress can occur to the next MP power level. Residual gas analyzers show water and H and low levels of CO and CO_2 during processing events.

A large or sustained vacuum burst can permanently damage the coupler if the rf is not shut off in time (<1 μs). Metal can sputter and deposit on the ceramic surface, which increases losses, window heating, and mechanical stresses to eventually cause a crack. Successful conditioning requires a sufficient desorption rate without such harmful vacuum bursts. The trip level for pressure is usually set between 10^{-6} and 10^{-7} Torr. A high pumping speed through the coupler is therefore an important design feature. The choice of vacuum thresholds for power interruption and interlock-limit values is a compromise between establishing a safe procedure and avoiding lengthy conditioning. Optimizing these choices may allow a substantial gain in conditioning time. An insufficiently processed coupler will cause high rf trips rate which is undesirable for accelerator operations.

Standing wave conditions are also routinely used to locally enhance the fields at different locations of the couplers and to simulate the effects occurring during power or beam transients. After warm processing, the couplers need to be kept under vacuum or in a dry nitrogen atmosphere to prevent the recondensation of water vapor.

Conditioning couplers at warm test-stands is helpful for first processing, but more processing is generally also necessary when the coupler is operated with a cold cavity due to condensed gases. Coupler conditioning off resonance and without beam will establish standing wave conditions so that high electric field regions are processed. Sweeping through the cavity resonance or modulating the bias voltage helps to condition more (but not all) coupler surfaces. During pulsed operation the rf wave experiences a standing wave pattern during filling (typically several 100 µs), traveling wave during beam acceleration and a high-power transient after rf power switch off (by emptying the stored energy of the cavity). For long-term operation of a cold coupler system, regular processing may be necessary during maintenance periods. A warm-up cycle will degas the cold surfaces and improve operation.

9
Higher Mode Couplers and Absorbers

9.1
Overview of HOM Excitation and Damping

When passing through an accelerating cavity a charged particle beam excites a wide spectrum of higher order modes (HOM), depending on the impedances of the modes. The resulting electromagnetic field left behind by the beam, called the wake field, is illustrated in the simulation of Fig. 9.1. Thus the passage of the beam can deposit significant power in high impedance (i.e., high R/Q) monopole HOM. Unless properly extracted and damped, HOMs can also cause longitudinal beam instabilities and increase the energy spread. As discussed in [1], the energy lost by the passage, a single bunch, charge q, is given by

$$U_q = k_n q^2$$

$$k_n = \frac{\omega_n}{4} \frac{R_a}{Q_0}$$

where ω_n is the angular frequency of mode n, R_a/Q_0 is the geometric shunt impedance of the monopole, formally defined in [1]. k_n is also referred to as the loss factor of mode n. The total power deposited depends on the number of bunches per second, or the beam current.

Beam excited dipole, quadrupole, and sextupole modes with high impedance can deflect the beam to cause transverse instabilities. Among such deflecting modes, dipoles have the highest impedance. The energy lost by a charge to the dipole mode is given by

$$U_q = k_d q^2 \left(\frac{\rho}{a}\right)^2$$

$$k_d = a^2 \left(\frac{\omega_n}{c}\right)^2 \frac{\omega_n}{4} \frac{R_d}{Q_0}$$

where ρ is bunch displacement off-axis, a is the cavity aperture (radius), ω_n the angular HOM frequency of mode n. R_d/Q_0 is the dipole mode impedance, for-

RF Superconductivity: Science, Technology, and Applications. Hasan Padamsee
Copyright © 2009 WILEY-VCH Verlag GmbH & Co. KGaA, Weinheim
ISBN: 978-3-527-40572-5

Fig. 9.1 Excitation of higher order modes due to passage of a beam [535] (courtesy of DESY).

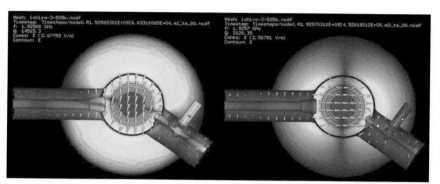

Fig. 9.2 Electric fields for the two polarizations of a dipole mode at 1.926 GHz. The input and HOM couplers are also included in the simulation. The frequency degeneracy of the two modes is split by about 50 kHz due to the presence of couplers. The calculations were made by the code OMEGA3P for an ILC cavity, Low-Loss cavity candidate [536] (courtesy of SLAC).

mally defined in [1]. Each dipole mode has two polarizations. Figure 9.2 shows the electric fields for the two polarizations of a dipole mode of a 9-cell cavity with couplers.

The main reasons for HOM couplers are to remove the beam-induced power in the monopole HOMs, and to damp the dangerous monopole and dipole modes to avoid energy spread, beam emittance degradation, and beam blow-up after multiple beam passages. Figure 9.2 shows the computed electromagnetic fields of a cavity dipole mode inside the body of the HOM couplers. Note that each dipole mode has two polarizations split with a small frequency difference due to perturbations, such as the presence of couplers. The beam-induced HOM power in monopole modes must be extracted from the cavity and deposited to higher temperatures to avoid cryogenic losses. Modes with high shunt impedance and high Q (i.e., high $R/Q*Q$) are of particular concern.

Damping requirements and basics of HOM coupler design are covered extensively in [1]. Here we will discuss some illustrative examples and advances in HOM coupler designs and their realization. There are three main types of HOM couplers: (a) antenna/loop couplers, (b) waveguide couplers, and (c) beam pipe couplers with absorbers. We will discuss examples of each case.

As a prime example for monopole mode excitation and power deposition, consider the European XFEL which plans to use the TESLA-shape cavity developed by TTF. Figure 9.3(a) shows the calculated R/Q for the low-frequency

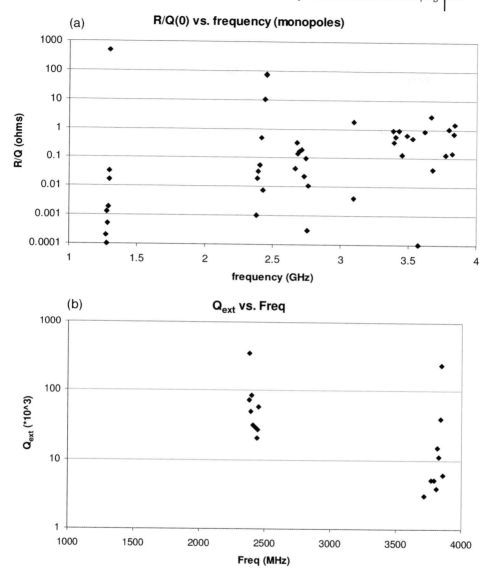

Fig. 9.3 (a) Shunt impedance of monopole modes below beam pipe cut-off frequency [537]. (b) Damping of dominant modes measured by two HOM couplers [538].

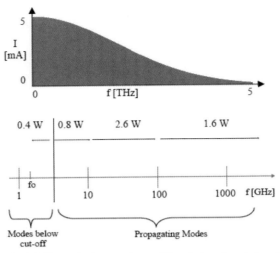

Fig. 9.4 The excitation spectrum of the electron bunch ($\sigma_z = 25$ μm) planned for the XFEL reaches frequencies up to 5 THz. The standard accelerator module has an integrated loss factor of 135 V/pC. The total power deposited by the nominal beam is 5.4 W per module [539] (courtesy of DESY).

monopole modes [537]. The dangerous modes are damped by two antenna/loop couplers placed just outside the end cells on either end of the cavity [25]. The measured damping of these modes is shown in Fig. 9.3 (b) [538].

The bunch length for XFEL will be 25 μm so that modes up to several THz will be excited, as shown by the bunch spectrum of Fig. 9.4 [539]. The nominal bunch charge will be 1 nC and the intrapulse repetition frequency will be 5 MHz. The nominal rf-pulse repetition rate is 10 Hz. The integrated longitudinal loss factor of the TTF type cryomodule (see Chapter 12) housing eight 9-cell cavities is 135 V/pC. The total deposited power by the nominal beam (40 000 bunches/s) is 5.4 W if no synchronous excitation takes place.

Note that most of the power (about 5 W) is generated in modes above the cut-off frequency of the cavity beam pipe. Antenna/loop couplers on the beam pipe optimized to damp the low frequency high impedance modes will not be effective against such a broad spectrum. Hence a broadband beam pipe absorber is needed for applications with a short bunch. Similarly for the ILC, with a bunch length of 0.3 mm, the total HOM power is 2 W/cavity of which 1.4 W will be above beam pipe cutoff out to frequencies of 400 GHz. We will discuss the broadband absorbers in Section 9.5.

Dipole modes with high transverse R/Q are harmful for emittance growth in XFEL and ILC. The impedances for these modes are shown in Fig. 9.5 [537, 540, 541] along with the damping (Q_{ext}) achieved by loop/antenna HOM couplers discussed in Section 9.3 [538]. Most of the dangerous dipole modes are well damped relative to the beam dynamics requirement of $Q_{ext} \sim 10^5$. Careful studies of HOM suppression have been carried out at the TTF linac by exciting

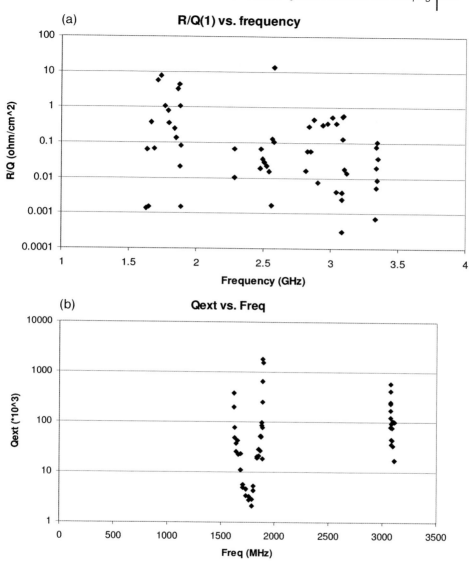

Fig. 9.5 Impedance and Q_{ext} for the TESLA 9-cell cavity dipole modes [537] and Q_{ext} for modes with high (R/Q) [538].

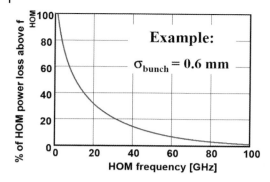

Fig. 9.6 HOM power distribution anticipated for an ERL with 100 mA beam current and 0.6 mm bunch length [535].

dipole HOMs with displaced beams to confirm dipole modes that are well damped and to excite any remaining problematic modes. To improve damping of a few modes with Q_{ext} above specification [542], the orientation of the two couplers relative to each other has been reoptimized [543].

A more demanding example with higher HOM power levels is a future ERL with a bunch charge of 77 pC, an average beam current of 100 mA, and bunch length of 0.6 mm. Here the total deposited power expected is 200 W at frequencies out to 100 GHz. Figure 9.6 shows the power distribution with frequency. For this application, dipole modes need to be damped to $Q_{ext} \sim 10^3 – 10^4$.

Codes, such as HFSS, MAFIA, GDFIDL, and OMEGA3P have been used to calculate R/Q and Q_{ext} values with reasonably good agreement between calculated and measured values as shown in Fig. 9.7 for TESLA-shape cavity.

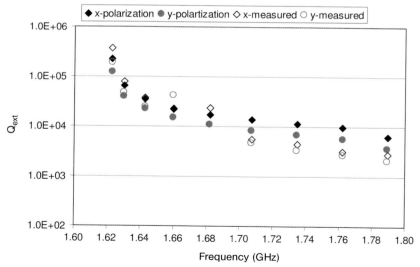

Fig. 9.7 Comparison of Q_{ext} between measurements and calculations using the 3D code, OMEGA3P [222, 546] (courtesy of SLAC).

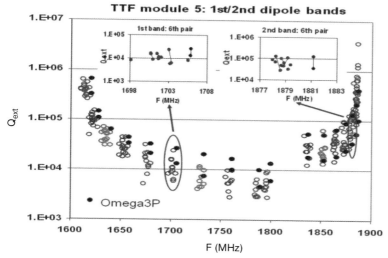

Fig. 9.8 Changes in frequency and Q_{ext} due to cell shape differences arising from manufacturing and other errors [547] (courtesy of SLAC).

Shape differences in the cells due to manufacturing tolerances can cause frequency shifts of several MHz, increases in mode splitting up to 0.5 MHz [547] causing significant differences between measured and predict Q_{ext} values as shown in Fig. 9.8.

9.2
Mode Trapping in Multicell Structures

As with input couplers, HOM couplers must also be placed outside the cells to avoid field enhancement in the cells which lead to premature quench or multipacting. Outside the end cell, the couplers are designed to intercept the evanescent energy available for modes with frequency below the beam pipe cut-off. However, some HOMs have very little stored energy in the end cells due to mode "trapping" within the central cells of a many-cell structure. Their suppression becomes difficult. The prime reasons for trapping are a small cell-to-cell coupling (which corresponds to a low-energy propagation velocity) and a large difference in mode frequency between the end cells and the inner cells. For example, in the TESLA-shape structure for XFEL, the 2.4 GHz monopole mode has very little stored energy in the end cells. The difference in frequency between end cells and inner cells is about 30 MHz as shown in Fig. 9.9. Since cell-to-cell coupling is 3%, the end cells cannot resonate "together" with the inner cells [548].

Reducing the number of cells and/or enlarging the iris diameter minimizes the likelihood of trapping by enhancing the cell-to-cell coupling. Another approach to minimize trapping is to match the end- and inner-cell frequencies

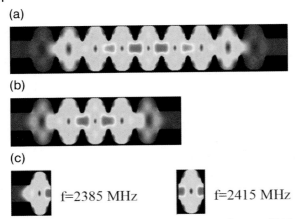

Fig. 9.9 (a) Example of mode trapping in a 9-cell cavity. (b) The stored energy in the end cells increases by reducing the number of cells to five. (c) The end cells and inner cells have different frequencies [548]. Bright regions represent higher stored energy (courtesy of DESY).

by adjusting the shape of the end cells, which yields a slightly asymmetric cavity. However, the impact on all dangerous modes needs to be considered for the best trade-off. The method has been applied to the TESLA structure as a remedy against trapping of the highest impedance mode in the third dipole passband. The modification of end-cell geometry resulted in increased stored energy in the dipole mode, improving its damping, but there is less suppression of the highest impedance monopole mode.

The 5-cell, 704 MHz structure for the electron cooling at RHIC (Fig. 9.10) uses all the above approaches to open the iris and beam pipe diameter to 17 cm, and successfully avoids trapping [549, 550]. The HOM energy radiates out of the cavity to beam line absorbers. Figure 9.10 also shows the computed Q_{ext} values for ferrite absorbers (discussed further in Section 9.5).

9.3
Coaxial Couplers

The development of HOM couplers was reviewed in [1]. For a recent review of HOM damping and power extraction see [548]. Coaxial HOM couplers were developed for HERA (500 MHz), LEP-II (352 MHz), and TRISTAN (508 MHz) cavities to operate in the corresponding high current storage rings. Subsequently, the HERA coupler was adapted for TESLA-shape 9-cell cavities [551] and scaled for application to SNS (800 MHz) and JLab upgrade (1500 MHz) cavities [552].

In general, the coaxial HOM coupler has an antenna or pick-up loop to extract HOM energy from the end cell via the electric or magnetic fields. Inductive and capacitive elements within the coupler body enhance coupling at the

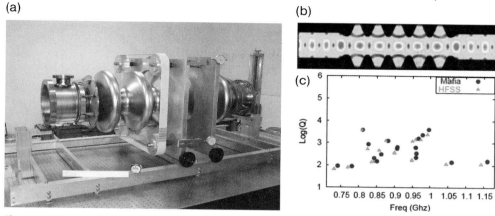

Fig. 9.10 (a) 704 MHz structure for the electron cooling at RHIC in its tuning fixture. (b) Free propagation of HOMs due to the large beam pipe. (c) Calculated Q_{ext} by two codes, MAFIA and HFSS, assuming a beam pipe damper discussed in Section 9.5 [550] (courtesy of BNL).

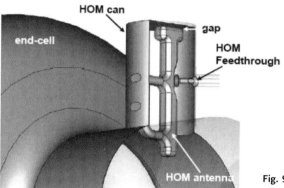

Fig. 9.11 The TESLA-type HOM coupler [25, 551] (courtesy of SNS).

desired frequencies of high R/Q modes, and suppress coupling to the fundamental mode via a rejection filter with a large rejection ratio, typically more than −70 db. The rejection filter must be carefully tuned prior to installation, and the transmission line must be terminated by a broadband load. For high average current application, the couplers are located inside the helium vessel for good cooling. The HERA couplers provide excellent suppression of dangerous modes to $Q_{ext} < 1000$, ensuring stable operation with electron and positron beams up to 45 mA. CW operation is routine to extract 100 W of HOM power.

For the 1.3 GHz, 9-cell TESLA-shape cavity, the rf design of the coupler has been scaled and simplified due to the moderate damping specification (Q_s of the order of 10^4–10^5) and the couplers are located outside the helium vessel because of minimal heating expected at 1% duty factor applications. Figure 9.11

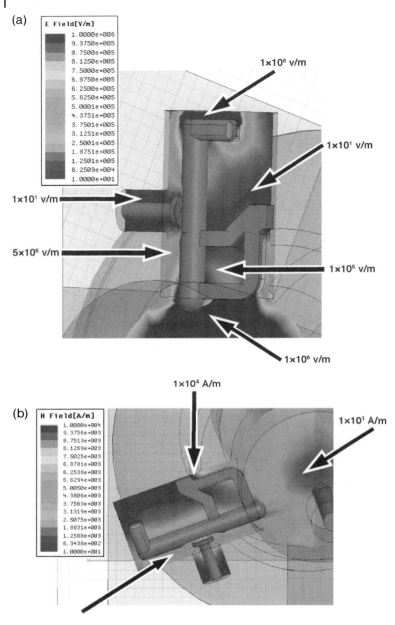

Fig. 9.12 Field distribution in the HOM coupler: (a) electric field; (b) magnetic field [555] (courtesy of KEK).

shows the layout of the popular TESLA style coupler developed for the linear collider and now used for the XFEL.

The TESLA HOM coupler has been scaled to 805 MHz for SNS at 6% duty cycle operation, to 1500 MHz for the CEBAF 12 GeV upgrade and cw operation, and also to 3.9 GHz for the third harmonic TTF injector cavity [553, 554]. These higher average power applications have met with some difficulties to be discussed so that some modifications have been developed.

Enabling higher duty factor operation or higher damping by bringing the coupler tip closer to the end cell requires improvements in cooling the TESLA HOM couplers, normally positioned outside the helium vessel. There is a significant amount of stored energy in the transmission line coupler. Figure 9.12 shows the electric and magnetic field distributions [555]. The problem arises due to the heating of the output antenna by the residual magnetic field of the fundamental mode (several percent of the field on equator) and the heat leak of the output line. The phenomenon was observed in cw cold tests of SNS and CEBAF 12 GeV upgrade cavities at JLab. In a cw test of the TESLA cavity there was a continuous decrease of the intrinsic Q due to the niobium antennae heating beginning at a moderate gradient of 5 MV/m. Abnormal heating of HOM couplers can detune the notch filter and couple out substantial amounts of power from the fundamental mode, leading to thermal runaway. The other causes for abnormal heating observed are multipacting in the HOM coupler (see next section) and heating from the impact of field emitted electrons emanating from the cavity, or from neighboring cavities.

Microwave analysis combined with thermal analysis using codes, such as HFSS and ANSYS have been used to analyze heating difficulties and devise solutions [556–558]. To keep the output antenna superconducting one approach has been to shorten the antenna probe tip, provided the HOM coupling loss can be tolerated. Another is to enhance the heat conduction at the output connector, for example by using a larger rf feedthrough with sapphire window and cooling copper blocks [559].

9.3.1
Multipacting in Coaxial Couplers

The high electric field regions of the loop coupler are susceptible to multipacting and associated heating. Multipacting in couplers has been simulated using the Omega3P (ANALYST) software [557, 558]. The troublesome regions are between the loop and the wall, in the small gap which defines the notch filter, between the coaxial post and the end wall of the can, and at several places between the post and the cylindrical wall, as shown in Fig. 9.13. For a 3.9 GHz cavity, the notch-gap barriers occur below 1 MV/m and the loop barriers between 10 and 20 MV/m. In the 3.9 GHz case, the heating was sufficient to cause severe stress at the junction of one of the welds and a fracture in the cold [554, 557]. Shortening the antenna tip reduced the fields and suppressed MP at the required field levels. Other geometries of HOM couplers for the low loss

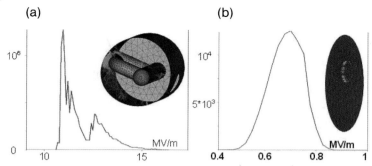

Fig. 9.13 Multipacting bands in a TESLA style HOM coupler for 3.9 GHz cavity. The graph shows the number of electrons generated. (a) MP in the 2 mm gap between the leg and the wall; (b) MP in the 0.6 mm gap of the notch filter [557, 558] (courtesy of FNAL).

cavity design have similarly been optimized to fully suppress multipacting barriers within an operating gradient range [560].

9.3.2
Coaxial Coupler Kicks

As with input couplers, antenna/loop-based HOM couplers introduce kicks which can spoil the beam emittance especially at low beam energy, right after the injector. These kicks arise from the wake fields introduced by the coupler geometry as well as by rf fields and the asymmetry of the coupler locations with respect to the beam axis. Figure 9.14 shows the standard and better orientations of the HOM couplers with respect to each other, and with respect to the input coupler antenna to reduce the kicks by a large factor. But the effect on the HOM damping of dipole modes needs to be evaluated with careful consideration for both polarizations of the dipoles [561].

9.4
Waveguide Couplers

As HOM couplers waveguides are natural high-pass filters to reject the fundamental mode and require no fine tuning as for the fundamental mode rejection filter essential to coaxial couplers. The waveguide coupler removes HOMs over a wide frequency range. It can handle high HOM power without heating difficulties. However, the bulkier waveguide adds to structure cost of the HOM endgroups.

The development of waveguide HOM couplers was reviewed in [1]. The first waveguide HOM couplers for superconducting cavities were designed at Cornell for 1.5 GHz muffin-tin cavities, and subsequently adopted for the 1.5 GHz Cornell/CEBAF elliptical 5-cell cavities (Fig. 9.15 a). The mA beam current of

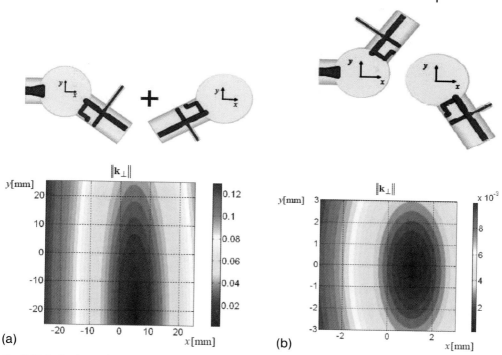

Fig. 9.14 Reduction and symmetrization of the kicks due to the wake field generated by the HOM couplers. The units of the kick are kV/nC. (a) With the arrangement of HOM couplers above, the magnitude of the kick voltages are high and the distribution asymmetric. (b) With the arrangement of HOM couplers above, the magnitude of the kick voltages are greatly reduced, and the distribution becomes more symmetric [561] (courtesy of DESY).

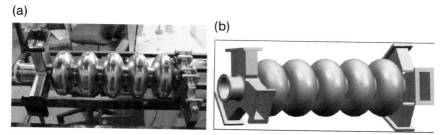

Fig. 9.15 Waveguide HOM coupler examples. (a) For the Cornell/CEBAF cavity. (b) Three-port couplers with strong damping for all HOMs under development for high current FEL applications [563, 564] (courtesy of JLab).

CEBAF results in a small amount of the HOM power to allow termination of the HOM couplers with waveguide loads inside cryomodules [562]. In case of higher beam current the terminating loads must be located outside the cryomodule, which adds to their mechanical complexity and introduces additional cryogenic heat leaks and shielding requirements.

Three-port HOM waveguide end-groups (Fig. 9.15 b) have been developed [563–565] to couple to any orientation of dipole and quadrupole modes as well as all monopole modes. These are under development for high current superconducting cavity for FEL applications as shown by the concept in Fig. 9.15 (b).

9.5
Beam Pipe Couplers and Absorbers

The cavity beam pipe can be viewed as a transmission line to couple out HOMs with frequencies above the beam pipe cut-off. The fundamental mode is below cut-off and does not propagate out, providing a natural rejection filter. As with the waveguide coupler, no tuning is needed. The cylindrical symmetry of the beam pipe avoids coupler kicks. The diameter of the beam pipe can be chosen to couple out all monopole and dipole modes. Extraction of the first two dipole modes demands the largest opening. These high impedance modes are especially dangerous. But a large diameter beam pipe reduces the rejection and R/Q of the fundamental mode, as well as enhances the peak surface fields for operation. These effects can be reduced by introducing a rounded step in the beam pipe at the iris of the end cell. In some cases as for KEK-B [566], the largest beam pipe is installed only at one end of the cavity for the extraction of the first two dipoles. In one case [567], flutes are added to the beam pipe (see Fig. 8.4) to extract the lowest frequency dipoles, but these add geometrical complexity and expense. Two sets of flutes are needed to remove both polarizations of each deflecting mode [567].

A section of the beam pipe lined with a microwave absorbing material serves as the load. This can be placed at room temperature outside the cryostat, or at ~80 K inside. But the presence of numerous absorbing sections along the beam line reduces the overall real estate gradient.

Beam line HOM couplers (Fig. 9.16) are especially suitable for high current, short bunch accelerators. The developments of these are reviewed in [1]. Several storage rings now use the approach: CESR at Cornell, KEK-B in Tsukuba, Taiwan Light Source, Canadian Light Source, DIAMOND light source, and BEP-II in Beijing [568]. The HOMs are damped to Q values between 100 and 1000. Ferrite absorbers lining the beam pipe dissipate several kilowatts of the HOM power at room temperature with water cooling. Similar absorbers will be used for the RHIC electron cooling 5-cell cavity [569].

KEK-B used the absorber to extract 25 kW to operate with 2 A beam current. At this power level, the inner HOM load surface temperature however reaches already 200 °C and the outgassing behavior needs further studies [570].

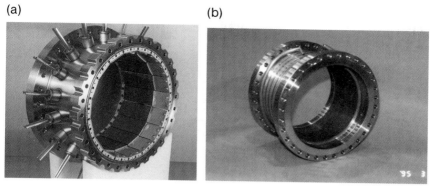

Fig. 9.16 Beam pipe absorbers lined with ferrite. (a) CESR load with 3 mm TT2-111R ferrite tiles bonded to a sintered copper/tungsten plate with Ag/Sn alloy [550]. (b) KEK load with 4 mm thick IB004 ferrite bonded to copper by HIP technique [566].

Several design, engineering, and technological challenges are being addressed for using beam pipe absorber loads over a wide frequency range to 40 GHz and at 80 K inside the cryomodule [571]. Figure 9.17 shows a prototype under development for a high current 100 mA ERL. The anticipated HOM power is 100–200 W at 80 K, and the required damping for beam stability is for Q between 100 and 10 000. The load is integrated with a bellows to join cavities.

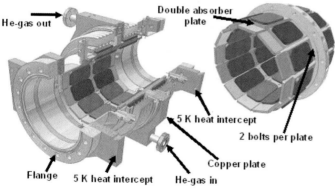

Fig. 9.17 Beam pipe absorber load for 80 K application over a wide frequency range. The load is integrated with the cavity joining bellows. HOMs generated in the bellows are absorbed by a second layer of tiles. A massive coupler block with helium gas flowing through channels intercepts the absorbed rf power. Tiles of TT2-111R and Co2Z ferrites [572] are soldered to plates made of Elkonite [574] chosen to match the thermal expansion coefficient of absorber plates. Ferrites are soldered by 90% Sn/10% Ag alloy in an argon atmosphere. Tiles of ceramic Ceralloy 137ZR10 (Ceradyne, Inc.) are brazed to the tungsten 18 alloy plates (95% W, 3.5% Ni, 1.5% Cu). The ceramic is brazed by TiCuSil foil (4.5% Ti, 26.7% Cu, 68.8% Ag) in vacuum [571].

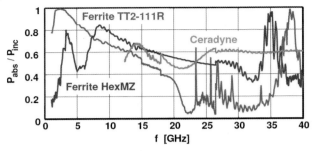

Fig. 9.18 Ratio of absorbed to incident power for three materials used in the HOM load under development for the Cornell ERL injector [571].

Several different kinds of ferrites or lossy ceramic are used. Measurements of electromagnetic properties of absorbing materials have been performed from 1 to 40 GHz. Figure 9.18 shows the absorption properties for examples in use [571]. The properties of the ferrite TT2-111R [572] have been measured at 110 and 300 K in the frequency range 1–15 GHz. In this frequency range TT2-111R shows higher magnetic losses at cryogenic temperatures than at room temperature and therefore is a promising material for 80 K. The Co2Z ferrites [572] provide high losses between 10 and 30 GHz. Tiles of ceramic Ceralloy, 137ZR10

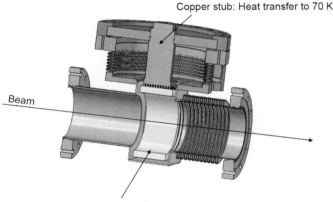

Fig. 9.19 Ceramic broadband beam line absorber. The absorption takes place in the ceramic ring hanging on a brazed copper stub. The dissipated energy is transferred to the 70 K cryostat shield via the stub and its external thermal connection. A stainless steel bellows serves as thermal barrier between the 70 K level of the flange and the 4 K cold vacuum chamber. The dielectric loss tangent of the ceramics has been measured at room temperature to be 0.1–0.4 between 5 and 40 GHz. The maximum expected temperature difference ΔT across the ceramics and copper stub is 110 K. The design absorber power capability is 100 W which will allow for acceleration of up to one million nominal bunches per second [539] (courtesy of DESY).

[573] provide broadband dielectric losses over a wide range out to 40 GHz. These absorbing materials must be installed and operated successfully in the demanding clean and high vacuum environment for superconducting cavities.

A different approach is under development for a lower power broadband beam pipe absorber for the XFEL [539]. For 25 µm bunches, a 5 W absorber is required over a frequency range out to several 100 GHz. Such a device will be installed between cryomodules to absorb the power and transfer the generated heat to 70 K. The proposed layout of the absorber is shown in Fig. 9.19. It is integrated into the vacuum chamber connecting two cryomodules. The ceramic material adopted for this design is used in HOM dampers installed in CEBAF at JLab.

10
Tuners

10.1
Overview

Frequency tuners are an essential component of acceleration systems. Both slow tuners and fast tuners fulfill important functions. The development of tuners is reviewed in [1]. Recent review talks can be found in [575–577]. Slow tuners bring a cavity resonance to operating frequency, compensating for a variety of effects: cavity-dimensional changes due to evacuation and cool down, or slow drifts in frequency due to pressure changes in the helium bath surrounding the cavity. Tuners also compensate the reactive effects of beam loading in high-current accelerators to minimize the reflected power. Occasionally cavities need to be detuned to bypass operation, or for diagnostic purposes.

Slow tuners must cover a wide tuning range (of up to several hundred kHz), while providing a resolution of the order of 1 Hz. Slow tuners are usually motor driven. Occasionally thermal tuners have been used [578].

Fast tuners provide a smaller tuning range of several cavity bandwidths but with a control bandwidth of several kHz and slew rates of 1 µm in 100 µs. Together with feedforward and feedback, these tuners compensate static and dynamic Lorentz-force detuning, especially at high-gradient operation [579]. Feedforward has worked effectively for Lorentz-force compensation. Fast tuners also have the potential to control microphonics, typically up to a several tens of Hz. Feedback has been used to a few ×10 Hz bandwidth. Fast tuners are important for cavities with little beam loading, when operation at high Q_{ext} is desirable to minimize rf power. At high Q_{ext} the bandwidth is sufficiently narrow that microphonic excitations disrupt the cavity resonance. Typical microphonic noise levels are of the order of few Hz to several tens of Hz with a frequency spectrum ranging up to a few hundred Hz. The observed spectrum is a result of a convolution of the spectrum of excitation and the coupling to the mechanical resonances of the cavities. Typical excitation sources of microphonics are vibrations from pumps and human activity. If the repetition frequency of the coarse tuner stepping motor matches a mechanical resonance of the cavity–cryostat system, strong mechanical vibrations can be excited. To avoid microphonic excitations in general, it is important to ensure that

RF Superconductivity: Science, Technology, and Applications. Hasan Padamsee
Copyright © 2009 WILEY-VCH Verlag GmbH & Co. KGaA, Weinheim
ISBN: 978-3-527-40572-5

the mechanical resonant frequencies of the structure do not coincide with the frequency of rf repetition rate (e.g., 5–10 Hz for XFEL or ILC and 60 Hz for SNS).

The fast tuner normally has a piezoelectric actuator which is integrated with the mechanical tuner mechanism. Magnetostrictive actuators have also been studied [580]. The actuators typically allow for a cavity length change of a few micrometers resulting in frequency changes of hundreds of Hz to kHz. In most designs, the fast actuator has been located inside the vacuum vessel for best Lorentz-force compensation. The fragile piezoelectric actuators must not be subject to tension, bending, and shear forces. To maximize their lifetime, a correct preload range must be maintained under all operating conditions. While fast tuner technology advances to compensate cavity detuning, important gains can also be made in both damping the sources of detuning and in stiffening cavity structure to reduce Lorentz-force detuning and to raise microphonic frequencies.

Tuner designs strive for compactness to avoid wasting beam-line space, avoid disrupting the field flatness from cell to cell, or tuning neighboring cavities. The tuner mechanical supports and operating motors should keep cryogenic heat

(a)
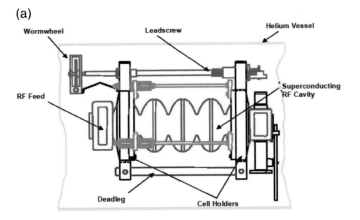

(b)

Fig. 10.1 The original CEBAF mechanical tuner: (a) principle [575]; (b) photograph of operating unit [581] (courtesy of JLab).

load to a minimum. Frequency tuners should be free of hysteresis. Presetting is required to avoid the neutral point between tension and compression over the entire range of expected operation. The hardware must be easy to maintain and repair, ideally without the need to warm up or disassemble a module, but in practice this has been realized only for a few tuner choices (the scissors tuner at JLab and the slide jack tuner at KEK, discussed later). Most designs have a cold drive motor inside the vacuum vessel and therefore require warm up to replace the fast actuator. Most cavity tuners and cavity piezotuners designs require opening the cryostat to effect repairs. The tuners must therefore be very reliable or redundant units installed. The desire for a long lifetime of operation (many years) calls for rigorous life-cycle testing, continuous monitoring, and periodic maintenance. Failure of the slow tuner will prevent operation of the cavity involved. Failure of the fast tuner may require operating at lower gradients.

10.2
Tuner Examples

There have been many advances in tuner design beyond the original concept of directly squeezing the cells via a yoke attached to the end cells of a multicell structure as in the original CEBAF tuner (Fig. 10.1). Tuner innovations have been stimulated by the need for higher compactness, reliability, and fast tuning. The original CEBAF tuners attach to the end cells of the cavities inside the helium vessels and are actuated by external motors via reduction gears and rotary

Table 10.1 Comparison of tuner system properties for 9-cell TESLA-style cavities [576, 582] with bandwidth 520 Hz, LF detuning coefficient ~1 Hz/MV/m², and tuning sensitivity 315 Hz/μm.

	Saclay original	Saclay modified	INFN blade tuner	TJLAF upgrade	KEK slide jack tuner	KEK coaxial ball screw
Coarse range (kHz)	440	500	500	500	1100	>4000
Coarse res. (Hz)	<1	<1	<1	<1	<100	<120
Fast actuator	Piezo	Piezo	Piezo	Piezo/Magnetostr.	Piezo	Piezo
Number of fast actuators	(1–2)	2	2	2	1	1
Fast range (Hz)	<500	1000	1200	1000/30000	1900	2500
Position of fast actuator	5 K, vacuum	5 K, vacuum	5 K, vacuum	5 K, vacuum	5 K, vacuum	80 K, vacuum
Position of motor	5 K, vacuum	5 K, vacuum	5 K, vacuum	5 K, vacuum	Warm, outside	80 K, vacuum

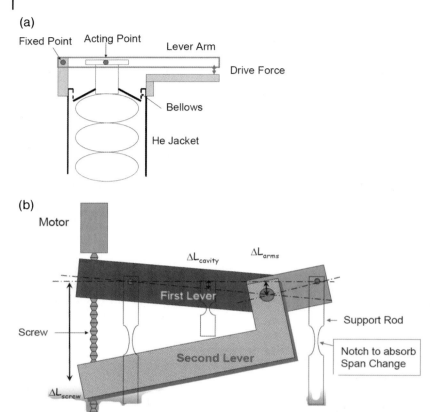

Fig. 10.2 (a) Basic principle of most slow tuners [577].
(b) Principle of the Saclay tuner used in TTF [584] (courtesy of KEK).

feedthroughs. The original installations had significant hysteresis due to backlash in couplings and finite service life due to the feedthroughs.

A variety of tuner designs are now available as a result of inventive efforts at a number of laboratories. Table 10.1 [582] compares the general properties of the examples discussed here. The first generation Saclay tuner (Figs. 10.2 and 10.3) has been in use at TTF for several years for the TESLA 9-cell cavities [583]. It is activated by a stepping motor with a harmonic drive gear box. Special lubricants are used for smooth operation in vacuum and at low temperatures (balzers Balinit C coating for the screw-nut and gear box, and Lamcoat for ball bearings). The measured stiffness is 100 kN/mm. The tuner has a double lever system which boosts the force by factor of 25. The maximum coarse tuning range ±400 kHz for a 9-cell cavity. The cavity tuning sensitivity is 315 kHz/μm. The theoretical resolution is 1.5 nm (0.75 Hz). The force applied always keeps the cavity in a state of compression to avoid the neutral point between compression and tension, where there may be a dead zone for the cavity response to the

Fig. 10.3 (a) Photograph of the Saclay/TTF tuner acting on the cavity at right. (b) Incorporation of two piezo elements for fast tuner tests on TTF [575] (courtesy of Saclay and DESY).

tuner force. The original design did not incorporate fine tuners, but one version was modified for proof-of-principle tests on Lorentz-force detuning compensation to provide 500 Hz tuning range with the piezo (Fig. 10.3 b) [582].

10.2.1
Saclay TTF Tuner

10.2.1.1 Modified Saclay Tuner
One disadvantage of the original Saclay tuner is the occupation of valuable beam-line space, about 7 cm for TESLA-type cavities and cryomodules. The modified Saclay tuner (Fig. 10.4) [584] is more compact, and incorporates piezo-actuators (about 1 kHz tuning range). The neutral point is avoided by design which also ensures that the piezos are always under the correct preload necessary for a long lifetime.

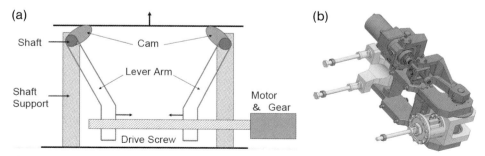

Fig. 10.4 (a) Principle of lever-cam design of the improved Saclay tuner [577] (courtesy of KEK); (b) 3D CAD layout [585] (courtesy of Saclay).

10.2.2
INFN Coaxial Blade Tuner

The coaxial blade tuner [586] minimizes the intercavity distance, allowing about a 5% reduction over the first generation Saclay tuner for the overall accelerating structure length for a TESLA-like cryomodule. The tuner evolved from a new approach [587]. The piezoblade tuner system (Figs. 10.5 and 10.6) has three main components: the ring-blade assembly, the lever system to actuate the slow tuning, and a piezo system for fast tuning. The main tuner mechanism consists of a three-ring bending system connected by welded blades at an angle (with a slant of $15°$ with respect to the central axis). The twisted blades transform the azimuthal rotation (in opposite directions) of the two halves of the central ring

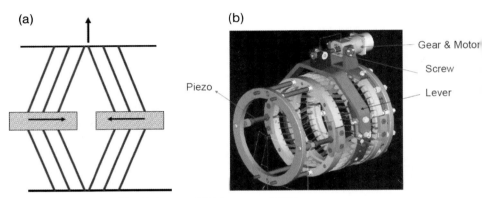

Fig. 10.5 (a) Principle of the blade tuner [577] (courtesy of KEK); (b) 3D CAD version. Each half blade-ring assembly has an array of 23 "packs" of two blades on each side, for a total of 184 flexural elements (blades). The blade packs for the two halves of the blade-ring assembly are welded [586] (courtesy of INFN).

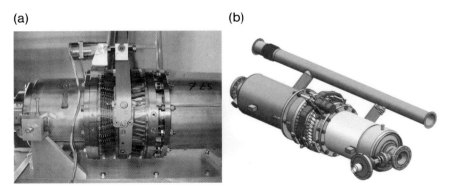

Fig. 10.6 (a) First generation of the coaxial blade tuner (without piezos) assembled to the helium vessel of a 9-cell cavity [586]; (b) improvement with piezo [588] (courtesy of INFN).

into a variation of the distance between the end rings, producing elastic change of the cavity length. Replacing a set of gears with flexures reduces backlash, and increases reliability and reproducibility. Titanium as the main tuner material is compatible with the titanium helium vessel and niobium cavity for thermal shrinkage. The matched elastic and yield strengths reduce the stresses in the flexural elements with respect to the choice of cheaper structural materials, such as stainless steel. The azimuthal rotation is provided by a lever system connected to a cold stepper motor. The lever system amplifies the torque of the motor, increasing the tuning sensitivity. In the standard design, the two end rings are rigidly connected to the cavity He tank, which is split in two sections joined by a bellow in order to allow the variation of the cavity length. The maximum axial displacement foreseen for the blade tuner is ±1.5 mm. Although the ring-blade mechanism is extremely stiff (~100 kN/mm when fixing the central ring rotation), the overall combined tuner stiffness is limited to about 25 kN/mm by the slack in the leverage mechanism.

Figure 10.5 (b) shows two piezoelectric elements acting between the ring-blade assembly and the outer ring welded on the He tank [588]. These provide a maximum stroke of ~4 µm at 2 K, corresponding to a fast tuning range of 1.2 kHz. For a prototype case of two piezos, with a cross section of 10 mm×10 mm, a blocking force of 4 kN (each) is expected. With a 12 kN maximum force generated by the slow tuner, the available tuning force is 8 kN to allow a slow tuning action of 1.6 kHz, which is four times the frequency reproducibility from cavity to cavity or warm-cold frequency changes. Longer piezo-tuning stacks with greater stroke and better reliability are also possible.

The location of the piezo stacks avoids stresses due to the deflection and vibration of the helium tank. The four threaded bars, parallel to the piezo elements are tightly bolted at both ends to provide stiffness to the system during transportation, handling and assembly, and to avoid large deformations of the bellows. Under the operating condition, the inner bolts are properly loosened and the bars act as safety devices in case of piezo failure, mechanical failure, or overpressure conditions inside the helium tank.

The cavity pretuning procedure avoids the neutral point of the lever system and chooses a correct range of preload values for the piezoelectric elements over the entire slow tuning range. The blade tuner operation always acts to stretch the cavity and to compress the piezoelectric elements.

Prototypes of the coaxial blade tuner have been tested on individual cavities inside a horizontal test cryostat and on the superstructure in 2002 [589]. An improvement under development reduces the number of blades and the mass from 21 to 12 kg, lowering the cost of Ti material [590].

10.2.3
KEK Coaxial Ball Screw Tuner

One main objective of this design [591, 592] is to provide a large tuning range while maintaining a high stiffness of the tuner system in the longitudinal direction. The full tuning range of 4 MHz available is not usable of course since that would imply cavity deformation. Another aim is to operate a 9-cell ILC cavity at 45 MV/m for which the Lorentz detuning of the cavity is 3 kHz. With the static tuning characteristics of 368 kHz/mm, a 10 µm fast tuning stroke is required. The prototype ball screw tuner achieved a large dynamic tuning range at room and liquid nitrogen temperatures.

The longitudinal movement of the cavity (Figs. 10.7 and 10.8) is realized by circumferential movement on a large wheel attached on a male screw (lead/

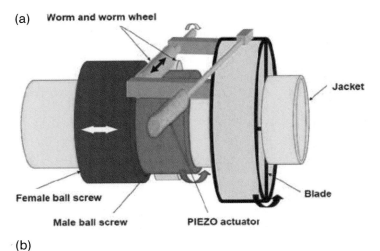

Fig. 10.7 (a) Component layout of the coaxial ball-screw tuner with piezo. (b) View of the motor, gear, ball-screw, and helium vessel [577, 591] (courtesy of KEK).

Fig. 10.8 Overview of the ball-screw tuner and helium vessel [577] (courtesy of KEK).

diameter 1:7). The slow tuning is performed by a worm gear driven by a pulse motor, mounted on a ring loosely coupled to helium vessel via 12 thin blades. The slow tuner as a whole can be pushed fast by a piezo device mounted on the helium vessel. Both the motor and the piezo are placed at intermediate temperature inside of the vacuum vessel.

10.2.4
KEK Slide Jack Tuner

The slide jack tuner is one of the few versions where the motor is outside the vacuum vessel and so can be less expensive and more easily replaceable in case of failure. However, a rotary feedthrough is needed to penetrate the vacuum vessel and the shields. Figure 10.9 demonstrates the principle and Fig. 10.10 the hardware [577, 593].

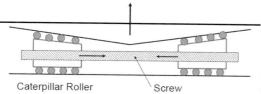

Caterpillar Roller   Screw

Fig. 10.9 Principle of the KEK slide jack tuner [577] (courtesy of KEK).

10.2.5
JLab Tuners

JLab has developed several tuner designs for SNS, the CEBAF upgrade, and the high-current FEL cryomodule. The SNS tuner (Fig. 10.11) is scaled from the original Saclay tuner design to be used with 6-cell 800 MHz SNS cavities [124]. It has a stainless steel frame which attaches via flexures and threaded studs to the helium vessel head. The bearings and drive screw are coated with dichronite for lubrication. The cavity is tuned in compression only. Each tuner is preloaded in the warm state. The stepper motor and harmonic drive are rated for UHV, cryogenic, and radiation environment.

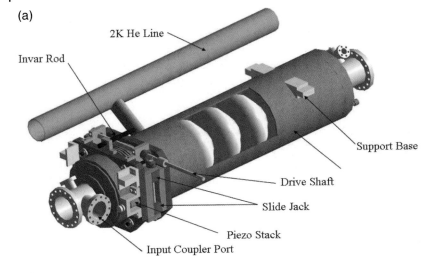

Fig. 10.10 KEK slide jack tuner. (a) 3D CAD layout; (b) hardware details [577] (courtesy of KEK).

The CEBAF upgrade cavities will use a "scissor-jack" tuner (Fig. 10.12), which is very rigid and has practically no hysteresis and has a warm external motor [576, 581]. The jack mechanism uses Ti-6Al-4V cold flexures and fulcrum bars, and attaches to hubs on the cavity. The stepper motor, harmonic drive, piezo, and ball screw are all mounted on top of the cryomodule with openings required in shielding and vacuum tank. Each tuner is preloaded and offset while warm. The tuner has proven rugged and serviceable.

A "zero-length" rocker type (Fig. 10.13) has also been developed for possible use with the high-current FEL module [576, 581]. A 316-L stainless steel frame attaches to the helium vessel head. The tuner consists of the mechanical linkage, gear reduction, stepper motor, and two piezoelectric actuators (Fig. 10.13). All components including the motor are in vacuum and cold. The mechanism provides a 30:1 mechanical advantage and is preloaded at room temperature.

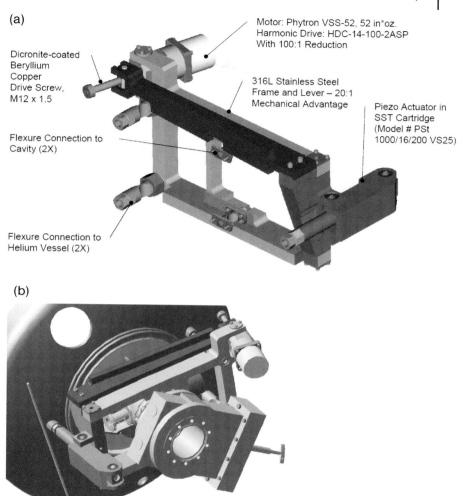

Fig. 10.11 SNS tuner assembly with piezoactuator.
(a) Tuner layout; (b) attached to cavity [119] (courtesy of JLab).

The cavities are tuned (in tension only) by the actuation of a stepper motor. The harmonic drive provides a gear reduction of 100:1 between the motor and BeCu drive screw coated with Dicronite™ for lubrication. The drive screw moves the primary lever relative to the motor. Force and displacement are distributed evenly via the secondary lever to a pair of struts. The struts, acting in compression, rotate through an angle of 6° about the end closest to the motor and push on the helium vessel shell and head (not shown) to stretch the cavity. This portion of the mechanism can apply a large force over a relatively short distance.

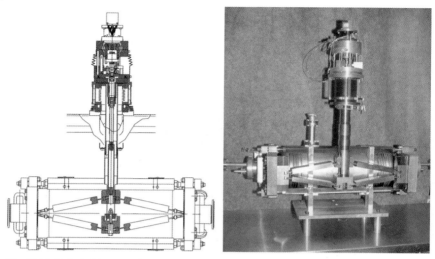

Fig. 10.12 JLab CEBAF upgrade scissor-jack tuner [576, 581] (courtesy of JLab).

The helium vessel heads are reinforced to limit distortion under high tuning loads. Two high-voltage piezoelectric actuators are loaded compressively during tuning and provide fine tuning capabilities for fast turn-on or microphonics control.

A flanged port on the vacuum tank provides access to the tuner motor and harmonic drive for repair or replacement, if necessary. One of the two piezoelectric actuators can be removed and replaced through this port in the event of failure of both the primary and secondary actuators. The coarse tuner range exceeds 1 MHz, the resolution is <2 Hz per step, and the tuning sensitivities are <1 Hz per step.

10.3
Fast Tuning

Fast tuners are mainly used to compensate for Lorentz-force detuning, as discussed in the remainder of this section [119, 596]. Their application to counteract microphonics detuning is under development. The Lorentz force between the rf magnetic field and the induced surface currents causes a slight deformation of the cells (~µm) and a shift in resonance frequency, which is proportional to E_{acc}^2. In pulsed operation, the detuning causes a mismatch between klystron and cavity, corresponding to a time-dependent reduction of the accelerating field, and a strong variation of the cavity phase with respect to the rf frequency.

For TESLA 9-cell cavities the LF detuning coefficient is 2–4 Hz/(MV/m)^2, which leads to sizable 5 kHz detuning at 35 MV/m, an order of magnitude higher than the cavity bandwidth of 400 Hz at a beam loaded Q of 3×10^6. The accel-

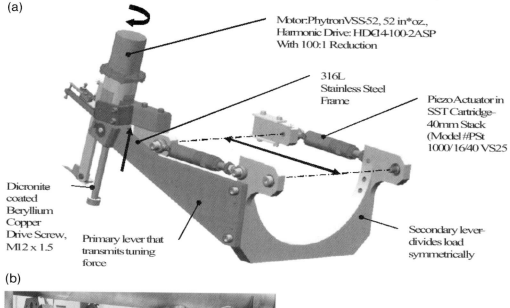

(a) labels:
- Motor: Phytron VSS-52, 52 in*oz., Harmonic Drive: HDCl4-100-2ASP With 100:1 Reduction
- 316L Stainless Steel Frame
- Piezo Actuator in SST Cartridge– 40mm Stack (Model #PSt 1000/16/40 VS25)
- Dicronite coated Beryllium Copper Drive Screw, M12 x 1.5
- Primary lever that transmits tuning force
- Secondary lever divides load symmetrically

Fig. 10.13 (a) Layout of JLab rocker tuner design for the high current FEL module. (b) Rocker tuner attached to helium vessel [576, 591] (courtesy of JLab).

erating structure is therefore reinforced by stiffening rings which are welded between neighboring cells to reduce the detuning by a factor of 2. The rf control system can handle the stiffened cavity detuning (~250 Hz) adequately up to about 23 MV/m by using additional rf power. But the demand for rf control power increases as the square of the detuning. For 35 MV/m pulsed operation the rf control power still becomes prohibitive unless the stiffening is further enhanced, but this greatly increases the forces required for coarse tuning. Hence the cavity detuning must be compensated with a piezoelectric tuner using feedforward in the rf control system. The LF detuning is repetitive with the pulse rate, and hence can be predicted. A 1 μm change in cavity length shifts the frequency by about 300 Hz. The piezoactuator changes the cavity length dynamically by a few μm, stabilizing the resonance frequency to better than 100 Hz during the flat top time at 35 MV/m. Corrections for beam loading can be accomplished with digital signal processing. Figure 10.14 shows a high-power pulsed test of a TESLA-style cavity at 35 MV/m [294]. The compensation of the cavity detuning by a piezoelectric actuator leads to a constant accelerating field and a reduced drift in the relative phase.

A significant spread in the dynamic Lorentz-force detuning constant has been seen (as much as a factor 2 at TTF). Also, a significant difference can exist between the static range and the dynamic range (maximal frequency shift within the RF pulse length).

A larger fast tuning range is helpful to deal with these factors.

As discussed in Chapter 8, for low-beam loading applications, such as heavy ion linacs or the ERL, the rf power demand can be decreased if high Q_{ext} ($>10^7$) can be used. The extent of detuning due to microphonics sets the upper limit on Q_{ext}. Microphonics degrades the phase and amplitude stability of the rf field. The application of fast tuners to reduce the frequency excursion of the cavity would allow higher Q_{ext}. However, the strategy of microphonics compensation has to take into account that the action of cancelling one mechanical mode can lead to the possible excitation of other mechanical modes.

Typical microphonic noise level of a few Hz to 10 Hz can limit the permitted Q_{ext} to a few $\times 10^7$. Using seismometers, the microphonics spectrum of the helium vessels and quadrupoles in the TTF cryomodules have been measured (Fig. 10.15 a) [599]. Vibration measurements have also been carried out with the wire position monitors [600]. The success of microphonics compensation by piezotuners has been demonstrated with the Saclay tuner on TESLA cavity tested in 1-cavity cryomodule [601]. Figure 10.15 (b) shows the reduction with feedback and with feedback plus feedforward.

Fast tuners are generally integrated with the slow mechanical tuner and mostly based on piezoelectric elements [119]. The typical static tuning range is 1 kHz or less. To guarantee long (~10 years) lifetime of piezostacks the preload force need to be set around 30% of the blocking force [602, 603], typically 1.2 kN for piezos in use. The piezo preload varies with the slow tuner action; therefore a suitable pretuning strategy is needed to ensure that the preload remains within valid operation limits. Also, the piezo preload has to be lower than its blocking force, which is the force above which a piezo no longer provides stroke. Diagnostics, such as strain gauges, can help to keep track of preloads.

Piezo elements have been tested to operate for 10^9 cycles under proper operating conditions [604]. The characteristic behavior of piezos at cryogenic temperatures, in terms of stroke and preload needs to be measured before selection and application. The stroke at liquid helium temperature is about 15% of stroke at room temperature, with some uncertainties on cryogenic characteristics and

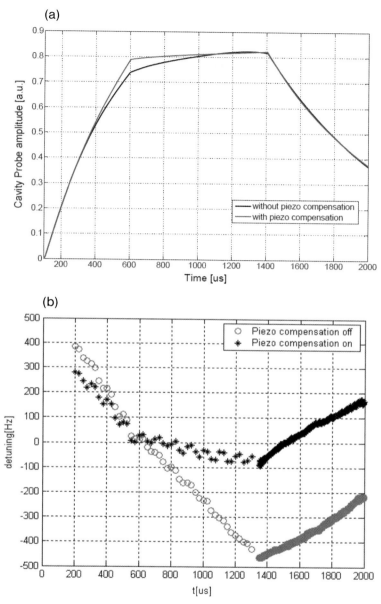

Fig. 10.14 High-power pulsed test of a TESLA-style cavity at 35 MV/m.
(a) The compensation of the cavity detuning by a piezoelectric actuator leads to a constant accelerating field and a reduced drift in the relative phase. (b) The measured detuning at 35 MV/m with and without piezoelectric compensation [294] (courtesy of DESY).

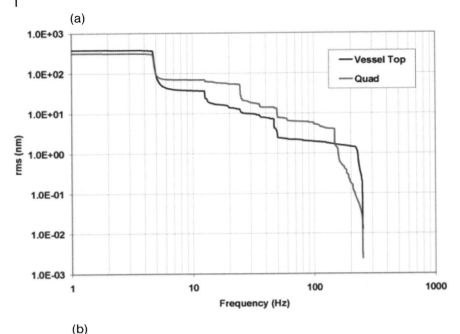

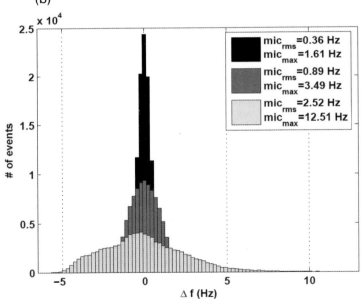

Fig. 10.15 (a) Amplitude of helium vessel and quad vibrations vs. frequency for the TTF modules in FLASH [599] (courtesy of DESY). (b) Microphonics compensation achieved by piezo with the Saclay tuner on TESLA cavity [601] (courtesy of BESSY).

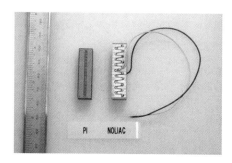

Fig. 10.16 Example of piezo tuning element from Noliac and Pi (left) [603] (courtesy of Orsay).

piezo operation. Piezos have shown moderate radiation hardness in cryogenic environment and under pulsed conditions. A redundant design for the fast actuator is important for reliability.

Piezos are readily available from several manufacturers with good fabrication reproducibility from batch to batch. Figure 10.16 shows two examples. A common material is PZT (lead zirconate titanate). Typical values for piezo specifications at 2 K are: blocking force 4 kN, stiffness 25 kN/m, control speed 0.01 µm/µs, 10 kN load limit to avoid damage during assembly, cooling and operating voltage < 200 V to limit self-heating in cryo environment. (Other piezos can have different properties.) A 40 mm stack will have a stroke of 60 µm at room temperature and stiffness greater than 100 N/µm. Typical cross sections are 7–10 mm^2. Figure 10.17 shows the familiar hysteresis at 300 K with a maximum stroke of 40 µm at 120 V. This hysteresis becomes negligible at liquid helium temperature.

Fast tuners based on magnetostrictive actuators are under development. These have significantly larger stroke than piezos at 5 K, produce less heat and may have a higher lifetime and higher tolerance for preload change than piezos. The actuator needs a high drive current, so that residual magnetic field is a concern. The actuators are similar size to the piezos, and can therefore be used as a replacement in tuner designs. First tests at cryogenic temperatures have been done and detailed characterization is underway [605, 606].

10.4
Low-β Cavity Tuners

Low-β resonators require continuous tuning to compensate slow frequency drifts due to the same sources as for high-β cavities. The cryogenic system should be carefully designed with sufficient pressure stability. The typical pressure sensitivity values are from a few to a few tens of Hz per mbar. A good mechanical design can substantially reduce these values. There are also innovative "self-compensating" resonator designs, where every displacement in high electric field region is associated with an equivalent one in a high magnetic field region, to substantially reduce the tuning sensitivity.

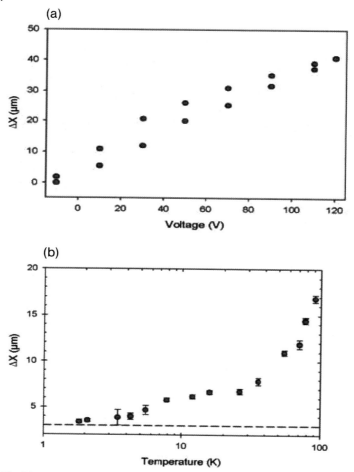

Fig. 10.17 Piezo stack actuator PICMA#1 characteristics. (a) Displacement vs. voltage; (b) displacement vs. temperature [603] (courtesy of Orsay).

Various tuning devices have been developed in different laboratories and their good performance allows reliable resonator phase locking. Mechanical tuners are used for slow frequency compensation in low-β resonators. We show some examples of slow tuners. Figure 10.18(a) shows a pneumatically actuated mechanical slow tuner which compresses the ATLAS QWR along the beam axis [607]. The tuning range is 40 kHz. Another example (Fig. 10.18b) is for the ISAC-II cavities at TRIUMF [608]. The tuning plate is actuated by a vertically mounted permanent magnet linear servomotor, at the top of the cryostat, using a "zero backlash" lever and push rod configuration through a bellows feedthrough. The tuning plate is radially slotted and formed with an "oil can" undulation to increase the flexibility. The demonstrated dynamic and coarse range of

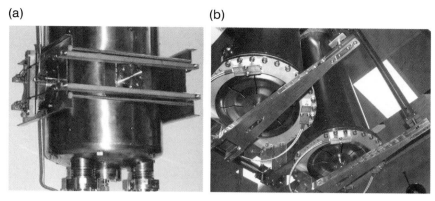

Fig. 10.18 Slow tuners for low-β resonators. (a) For the QWR at Argonne [607]; (b) the ISAC-II quarter-wave resonators [608] (courtesy of TRIUMF).

the tuner are ±4 kHz and 33 kHz, respectively. The system resolution at the tuner plate center is about 0.055 µm (0.3 Hz). The demonstrated mechanical response bandwidth is 30 Hz. Amplitude and phase regulation can be maintained for eigenfrequency changes of up to 60 Hz/s. The ISAC cavities are also equipped with a mechanical damper which limits microphonics to less than a few Hz rms. A demountable flange on the high field end supports the tuning plate. Similar tuners have been developed for ALPI [609]. The HWR for SARAF has a mechanical slow tuner with 0.2 Hz per step resolution and piezo-based fast tuners with a 900 Hz tuning range corresponding to a 6 µm stroke.

Low-β resonators also need fast tuners to compensate fast frequency changes due to mechanical resonances excited by microphonics. Variable reactance tuning has been successful here. The development of mechanical dampers, which reduce the cavity sensitivity to mechanical vibration, also allowed the locking of resonators without using fast tuner devices [610]. Continuous tuning is not required in Nb-sputtered cavities due to the thick copper backing. They can be reliably locked at high accelerating fields also without any fast tuner devices, simplifying control systems. Voltage-controlled reactive fast tuners have proved to

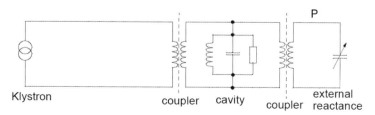

Fig. 10.19 Circuit diagram for voltage control reactance as a fast tuner [575] (courtesy of DESY).

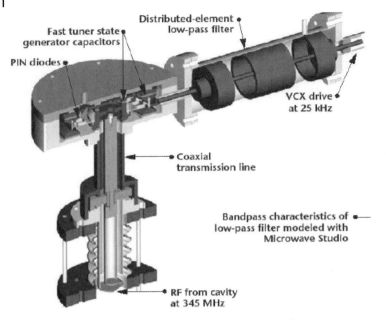

Fig. 10.20 Component layout for voltage-controlled reactance fast tuner designed for the RIA cavity [575] (courtesy of LANL and LLNL).

be reliable and effective for fast tuning low β, heavy ion resonators with stored energies at design gradient of < 10 J. Figure 10.19 illustrates the basic principle and Fig. 10.20 its application to a cavity. The reactive power required is

$$P = 8 \times \pi \times \Delta f \times U_0$$

where Δf is the frequency shift and U_0 ist the resonator stored energy.

Part III
Applications

11
Applications and Operations

11.1
Storage Rings

Superconducting cavities are used in storage rings worldwide because they operate cw at high gradients (5–10 MV/m) minimizing the number of cavities needed. Almost all the available rf power can be transferred to the beam. With large beam holes superconducting cavity designs have low impedance in higher order modes (HOMs), which is important for stable operation at high currents (100–2000 mA) and for preserving good beam quality through the linac. High-current storage rings for light sources and electron–positron colliders for high energy physics achieve high intensity beams with a very large number of bunches, spaced very closely together (10–100 ns). The high bunch charge, short bunch length, and tight bunch spacing raise serious issues with strong wakefields, higher mode power, and control of multibunch instabilities. With a small number of cells and open-beam hole designs, superconducting cavities offer attractive options for high-current beams by greatly reducing beam–cavity interaction and multibunch instabilities. Another issue for high-current storage rings is transient beam loading. In the fundamental mode the beam-induced voltage is nearly all reactive and must be compensated by rapid detuning. Gaps in the bunch train, such as abort or injection gaps, cause excessive phase modulation of the beam, which leads to capture losses during injection. Superconducting cavities reduce such transient effects by virtue of their low R/Q and high cell voltage.

For the applications in the mA to 100 mA regime, the late 1980s and early 1990s saw pioneering work by KEK [611] for the TRISTAN e+e− collider (508 MHz), DESY [612] for the electron ring of the HERA electron–proton collider (500 MHz), and CERN [613] for the LEP storage ring (352 MHz). Table 11.1 lists some key aspects of these installations.

11.1.1
LEP-II and LHC

CERN installed the largest SRF system in the late 1990s to double the energy of LEP [613]. The system is now decommissioned for the large hadron collider

Table 11.1 Cavity installations in TRISTAN, HERA, and LEP.

	TRISTAN	HERA	LEP
Number of cavities	32 5-cell cavities	16 4-cell cavities	288 4-cell cavities
Frequency (MHz)	509	500	352
Cavities per vacuum vessel	2	2	4
Time of installation	1988 and 1989	1991	1996–1999
Provided voltage (MV)	200	30	3600
Technology	Bulk Nb	Bulk Nb	Nb/Cu

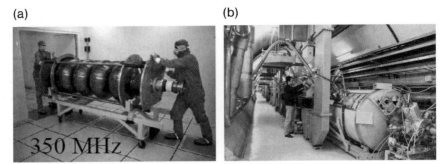

Fig. 11.1 (a) Copper 4-cell, 350 MHz cavity sputtered with Nb [614].
(b) Four-cavity cryomodule in the LEP tunnel with rf distribution system
[613, 615] (courtesy of CERN).

(LHC), which will become the frontier hadron collider for high energy physics (HEP). As discussed in Chapter 6, the goal to avoid quenches drove the cavity technology choice for LEP-II for sputtered Nb on Cu due to the high thermal stability provided by the copper substrate. Figure 11.1(a) shows the 4-cell copper cavities sputter coated with niobium, and Fig. 11.1(b) shows one of the 100 cryomodules in the LEP tunnel. Figure 11.2 shows the steady growth in LEP-II energy with installed voltage as 288 cavities were completed and commissioned over 5 years.

The emphasis of the LEP upgrade program was to reach the highest possible energy to increase the discovery reach of LEP-II. Over the operating period of LEP-II the average gradient rose from 6 MV/m (design) to 7.5 MV/m (Fig. 11.3) by optimizing the rf distribution over the large number of cavities and by rf conditioning at high power [117]. Both continuous (cw) and pulse conditioning with pulse length varying between 10 and 100 ms and duty cycle between 1 and 10% were successfully used to reduce field emission. The ultimate LEP-II energy reached was 104.5 GeV. For stable operation with beam, the total gradient was set about 5% lower than the maximum achieved during conditioning.

The operating experience was also dominated by the push to the highest gradients to extend the high energy reach with the available rf. The trip rate was about 2/h at 98 GeV rising to about 4/h at 100 GeV due to high gradient operation. The

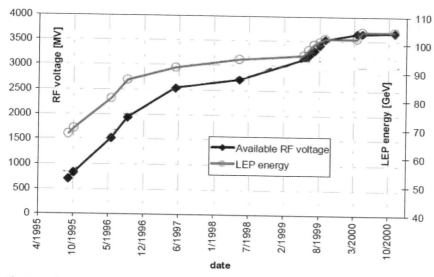

Fig. 11.2 The growth in LEP energy with available voltage [257] (courtesy of CERN).

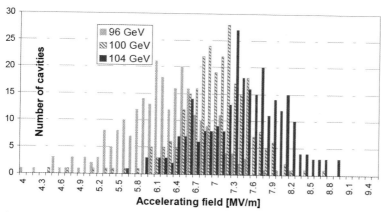

Fig. 11.3 Histograms of increasing cavity gradient distribution for three different maximum beam energies in 1999 and 2000. Gradient increase came from optimizing the rf power distribution and high-power rf processing to suppress field emission [257] (courtesy of CERN).

trip rate also rose with beam currents above 5 mA. Most trips occurred mainly due to field emission so that *in situ* processing played an important role. Each trip meant the loss of one klystron, or 100 MV out of a total installed voltage of about 3500 MV. This resulted in a loss of beam energy of 0.8 GeV. A trip of a HV supply (two klystrons) lost 200 MV and 1.6 GeV. But the beam generally suffered no serious performance degradation since the operating strategy was based on running with a voltage margin provided by 2.5 rf units.

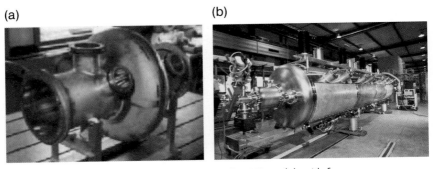

Fig. 11.4 (a) LHC single-cell Nb–Cu cavity [386]; (b) LHC module with four cavities [618] (courtesy of CERN).

CERN continued to develop LEP technology of sputtered niobium on copper for application to the LHC which will collide 7 TeV high current proton beams [616]. A serious issue for ampere size currents is transient beam loading. Since the induced beam voltage is nearly all reactive and compensated by detuning, the equilibrium is disturbed by the presence of injection and abort gaps in the bunch train. The transmitter must compensate rapidly until the slow tuner settles to its new position. The resulting phase modulation of the beam causes capture losses during injection. A low R/Q superconducting cavity running with high stored energy minimizes the periodic transient beam loading.

A single-cell cavity allows the strong HOM damping needed for beam stability at the close bunch spacing (< 10 m). The cavities (Fig. 11.4a) were fabricated in industry including the sputtering. Each cell has a large number of HOM ports for strong damping [386]. Figure 11.4(b) shows one LHC SRF module housing four single-cell cavities to provide a total of 8 MV per beam at 5 MV/m. Four modules are installed in the LHC (two modules per beam). After assembly by CERN, modules were tested to more than 3 MV per cavity.

11.1.2
CESR and KEK-B

The high current (≥ 1 A), high luminosity machines with SRF cavities are CESR in the US, and KEK-B in Japan, both operating for copious production of C and B quark mesons, respectively. CESR is an e+e− storage ring operating in two regimes: as a collider for HEP experiments and as a second generation synchrotron light source (CHESS). Both CESR (500 MHz) and KEK (508 MHz) designs are based on a single-cell niobium cavity with HOMs strongly damped by room temperature, ferrite-lined absorbers on the beam pipe. The round beam pipe on one side of the CESR cavity allows all but the two lowest dipole modes to escape from the cell [567, 619]. The beam pipe on the opposite side has additional "flutes," which lower the TE_{11} cutoff frequency to permit the lowest dipole modes to propagate out of the cavity. In the KEK version, the round beam pipe

Table 11.2 Operating parameters for the CESR and KEK SRF system [622] (consolidate).

	CESR	KEK-B (HER)
Number of cavities	4	8
Beam current	0.8 A	1.4 A
Bunch length	1.4 cm	0.7 cm
RF voltage with beam per cavity	1.6–2 MV	1.2–2 MV
Q	1×10^9 at 2 MV	0.5–2×10^9 at 2 MV
Maximum RF power to beam per cavity	300 kW	380 kW
Maximum HOM power extracted	5.7 kW	15 kW

on one side is further enlarged to propagate the dipole modes [620]. An iris between the cell and the large beam pipe avoids a severe drop in the fundamental mode shunt impedance. The CESR cavity is equipped with a waveguide input coupler while the KEK cavity has a coaxial coupler. Figure 11.5 shows the two cryomodule layouts. Table 11.2 summarizes the operating parameters for the two systems [568].

After the new B-factories surpassed CESR luminosity for B-physics, CESR switched operation in 2003 to a Tau-Charm Factory (1.55–2.5 GeV) mode [568], and continued the light source operation. The emphasis changed from delivering high power to providing high voltage for short bunch length and high synchrotron tune [568]. The rf power required for CESR-c is a very moderate 40–160 kW depending on the energy of the experiments, whereas the voltage requirement range is higher: 1.9–3 MV. Two CESR cavities were replaced with higher gradient cavities provided by industry [622].

Running with four SRF cavities, the availability of CESR operating at high currents from 1998 to 2003 was between 84 and 95% of the scheduled operating time. Over the same period CESR's beam current increased from 300 to 780 mA. The average trip rate was about 1 per day. Some of the trouble areas were condensed gas in the cavity and the waveguide coupler. Periodic high power processing and occasional warm up (one per year) to release condensed gas and window baking also improved the trip rate. Minor troubles were water leaks at HOM couplers and slippage in the tuner linkage.

At KEK-B, the cavities have been operating stably at high beam current and high power allowing luminosity records far beyond design values [624]. Beam currents have reached 1.4 A. RF power delivered to the beam is 350–400 kW per cavity. Absorbed HOM power is 14–16 kW per cavity. No coupled-bunch instability due to HOMs has been observed due to the excellent HOM damping provided by ferrite absorbers. The cooling for HOM dampers has been increased to allow 20–40 kW per cavity in anticipation of higher beam currents. The input coupler is routinely conditioned with a bias voltage during maintenance days.

The trip rate of eight cavities in total is 1–2 per week [623]. All cavities have kept the performance levels needed for routine operation, although there has been some Q degradation at higher accelerating voltage. After 7 years of opera-

340 | *11 Applications and Operations*

(a)
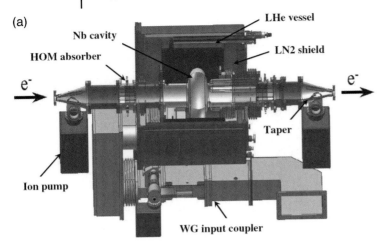

(b)
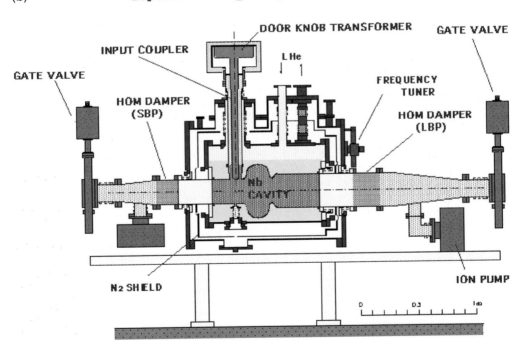

Fig. 11.5 (a) CESR module [619]; (b) KEK-B module [620].

tion, the Q at the operating voltage (1.4 MV) remains greater than 1×10^9, and all cavities can provide more than 2 MV (8 MV/m) each. However, the maximum possible voltage capability has degraded due to vacuum difficulties, and a change of input coupler, as discussed below. At 2 MV the Q has gradually degraded to $(3-5) \times 10^8$. The most likely cause is the significant outgassing from the hot isostatic press (HIP) bonded ferrite dampers. There have been a few additional trouble spots, but all were adequately resolved.

The Beijing Electron Positron Collider (BEPC)-II upgraded their beam energy from 1.55 to 1.89 GeV to run in the C-quark regime. The beam current increased from 26 to 250 mA (for the SR mode). They adopted the KEK-B module (Fig. 11.6a) with a slight modification to adjust rf frequency to 500 MHz. Cavity performance in vertical tests is shown in Fig. 11.6(b). Two cryomodules produced by KEK and MELCO, reached 2 MV during horizontal tests. Upon instal-

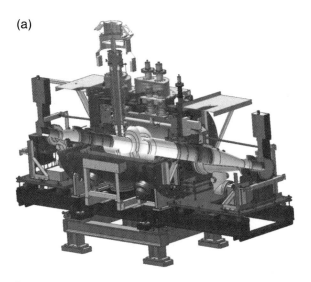

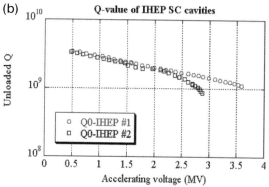

Fig. 11.6 (a) KEK-B module for BEPCII [625]; (b) vertical test result for KEK cavity [622].

lation in the ring and commissioning, the cavities are operating stably and reliably with the beam current of 180 mA during the user run. 250 mA beam current was stored in a test run [625].

11.1.3
Storage Ring Light Sources

Electron storage rings as x-radiation sources have an enormous impact on materials and biological science. Molecular and electronic structure determination, elemental analysis, imaging, and microtomography are among the many applications. The worldwide growth in storage-ring-based synchrotron radiation (SR) sources has been phenomenal, from just a few machines in the late 1960s to more than 70 machines in 2008.

With the successful and reliable operation of HOM-damped cavities at CESR and KEK-B, the Nb–Cu cavity developments at CERN, and the transfer of technology to industry, SRF has become a readily available technology for third generation storage-ring-based light sources. These are specialized storage rings operating in the energy range of 1.5–3.5 GeV, where x-ray beams are generated in insertion devices (wigglers and undulators).

By providing the needed few MV with one or two cavities, the SRF option frees up premium space for installing additional insertion devices and photon beam lines. With high-current SRF systems delivered by industry, operation has been carried out at institutions without an extended background in SRF technology.

At CESR, CHESS has been operating as a prolific light source for two decades. After replacing the CESR copper rf system of 20 cells by a superconducting system with 4 cells, the beam current could be increased from 300 to 780 mA, and accordingly the SR flux. CESR was the first storage ring to run (1999) entirely on SRF cavities [568].

Cornell transferred cavity and SRF technology to ACCEL Corporation [622] to provide turnkey modules which are used at the Taiwan Light Source [626], Canadian Light Source [627], Diamond Light Source [628], and the Shanghai Light Source [629]. Figure 11.7(a) shows a cavity under clean room assembly at Accel Vertical test results of eight cavities tested at Cornell by Accel are summarized in Fig. 11.7(b). The typical specified voltage of 2 MV at $Q > 5 \times 10^8$ was reached in all cases with considerable margin. Table 11.3 summarizes some of the main parameters for SRF-based storage ring light sources.

At SRRC (Taiwan) the replacement of copper DORIS-style cavities with the CESR-style SRF cavity (Fig. 11.8a) cured longitudinal coupled-bunch instabilities to raise the maximum beam current to 400 mA. Routine operation in top-up mode is at 300 mA. The average trip rate during user shifts has improved steadily to 0.5 trips/wk as shown in Fig. 11.8(b).

The Canadian Light Source storage ring operates routinely with beam current of 250 mA, accelerating voltage up to 2.4 MV, and rf power up to 225 kW. During an early test, the beam current of 300 mA was achieved, with 270 kW of rf power. The operation of the SRF system is robust and generally trouble free.

(a)

(b)
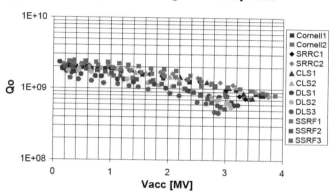

Fig. 11.7 (a) CESR SRF cavity fabricated and assembled at ACCEL Corporation [630]. (b) Vertical test results. The cell voltage of 2 MV corresponds to 6.7 MV/m [622] (courtesy of ACCEL).

Three cryomodules were fabricated for the light source DIAMOND. Two met specifications during commissioning without beam. During operation one module developed a leak from the helium vessel to insulation vacuum and was later repaired. The storage ring was subsequently commissioned at 3 GeV with one cavity. DIAMOND has accumulated up to 300 mA. A second module has been installed. The trip rate during initial commissioning stages was high while a number of difficulties were addressed with cryogenics, coupler processing, and tuner operation. At the Shangai Light Source under construction, two modules have been delivered and tested. One has been installed in the storage ring tunnel.

With the KEK-B cavity installed, BEPC-II operates well in the collider and light-source modes. SOLEIL in France adopted the 352 MHz LEP-based Nb–Cu technology. Saclay and CERN collaborated to develop a single module (Fig. 11.9a) containing two single-cell cavities to reduce the beam line space [631,

Table 11.3 Main parameters of SRF systems for storage-ring-based light sources [568].

Machine	CESR-CHESS	TLS	CLS	DIAMOND	SSRF	BEPC-II	SOLEIL	NSLS-II	TPS
Cryomodule	CESR-type					KEKB-type	SOLEIL	CESR or KEKB	CESR or KEKB
Beam energy (GeV)	5.3	1.5	2.5/2.9	3.0	3.5	2.5	2.75	3.0	3.0–3.3
Beam current (mA)	500	500	250 (500)	300 (500)	200 (300)	250	500	500	400
Frequency (MHz)	500	500	500	500	500	500	352	500	500
R/Q (Ω)	89	89	89	89	89	93	90	—	—
Q_{ext}	1.4×10^5	2.2×10^5	2×10^5	2×10^5	1.7×10^5	1.7×10^5	1×10^5	1.2×10^5 (6.7×10^4)	—
Cavity voltage (MV)	1.3	1.6	2.4	2.0 (1.3)	1.3 (2.0)	1.5	1.1	1.7 (2.5)	0.9–1.2
Number of cavities	4	1	1 (2)	2(3)	3	2	4	4 (2)	4
RF input coupler type	Waveguide	Waveguide	Waveguide	Waveguide	Waveguide	Coaxial antenna	Coaxial antenna	—	—
Power per coupler (kW)	160	82	245	270 (300)	200	96	150	225 (500)	180
HOM damper type	Ferrite beam line	Ferrite beam line	Ferrite beam line	Ferrite beam line	Ferrite beam line	Ferrite beam line	Loop	Ferrite beam line	Ferrite beam line
Status	Operational	Operational	Operational	Operational	Construction	Operational	Operational	Planned	Planned

(a)

(b)

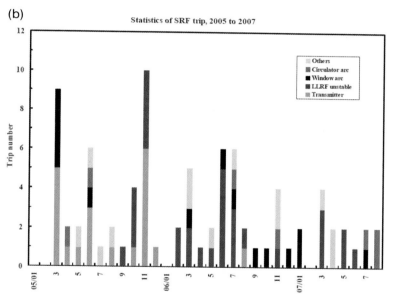

Fig. 11.8 (a) CESR cavity installed at SRRC in Taiwan. (b) Decrease in trip rate over time [568].

632]. In contrast to the KEK-B and the CESR ferrite lined beam pipe, damping of the HOMs is done by several loop couplers situated on a large diameter beam pipe in between the two cavities. Two couplers are optimized for longitudinal modes and two for transverse modes. The individual performance of cavities is seen in Fig 11.9(b). The completed module was tested with high power at CERN, where both cavities reached 2.5 MV, and the input coupler was operated at 120 kW at full reflection (standing wave operation). In 2001, a beam test at the ESRF took place during which the module provided in total 3 MV, and

(a)

(b)

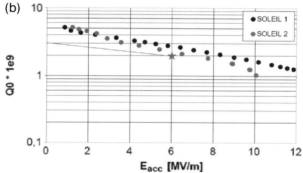

Fig. 11.9 (a) Two-cavity Nb–Cu module for SOLEIL prepared by a joint CERN–Saclay collaboration. (b) Cavity performance in vertical tests [631, 632].

each coupler transferred 190 kW rf power to the beam. Some weak points observed during this test, such as high static cryogenic losses and HOM coupler performance, were overcome in 2003 after introduction of a new thermal shield and an improved HOM coupler design. In a second high power test at CERN (2005), each cavity reached 2.5 MV and each coupler operated at 200 kW rf power at full reflection.

Once installed in SOLEIL (Fig. 11.9a) conditioning in the presence of beam was smooth. There were a few coupler vacuum trips while processing to 150 kW per cavity, which is sufficient for beam currents up to 300 mA [631, 632]. There was no serious HOM excitation. SOLEIL produced first light in September 2006. The stored current was raised to 300 mA in one month For the eventual goal of 500 mA, two cryomodules are foreseen. The second module is under construction by ACCEL. Installation and commissioning in SOLEIL is planned for 2008.

Future projects anticipating the use of similar systems are NSLS-II [633], the Taiwan Photon Source [629] and damping rings for ILC (at 650 MHz) [634], as well as third harmonic systems (1500 MHz). The NSLS-II at BNL will be a 3 GeV, 500 mA x-ray storage ring which will need over 5 MV and 1 MW of rf power. Four CESR-type cryomodules with two third harmonic cavities for bunch lengthening can fulfill the requirements. KEKB-type crymodule is also an option. Further development of high-power couplers would reduce the number of cavities needed from four to two.

Besides the main accelerating structures, higher harmonic superconducting cavities have been developed to increase beam lifetime in storage rings. Large-angle intrabeam scattering often limits beam lifetime (Touscheck effect). Addition of a third harmonic rf system lengthens bunches by flattening the rf waveform (Fig. 11.10) to increase lifetime by reducing the charge density. In addition, Landau damping suppresses coupled bunch instabilities. The typical accelerating voltage needed is 0.5 MV per cavity.

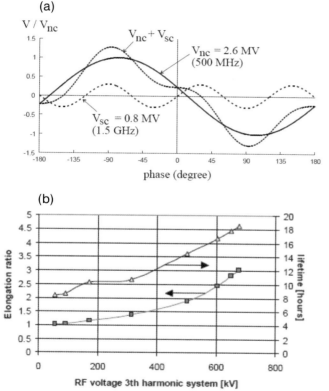

Fig. 11.10 (a) Addition of third harmonic rf flattens the rf waveform [635]. (b) Average bunch lengthening ratio and lifetime versus voltage due to operation with third harmonic cavity [393].

A collaboration of CEA (Saclay), SLS, ELETTRA and CERN designed, installed and operated a 1.5 GHz system at the SLS and at ELETTRA [393]. The modules increase the bunch length by a factor of 3, and the beam lifetime by a factor of 2 (Fig. 11.10b). Longitudinal-coupled bunch instabilities were also suppressed.

11.1.4
Crab Cavities

At KEK-B, two beams intersect at a finite angle (11 mrad). The crossing arrangement simplifies beam optics at the colliding region and reduces background, but hurts luminosity because of the reduced geometrical overlap (Fig. 11.11a). Angle crossing can also give rise to synchrotron–betatron instabilities at high currents. To circumvent instability and raise luminosity, rf deflectors called crab cavities can be used to tilt bunches and bring them to head-on collision. The tilted motion of bunches after deflection resembles the crawling motion of crabs; hence, the name "crab crossing" [636, 637]. The fields in the crab cavity installed near the collision point kick the heads and the tails of bunches in opposite directions starting an oscillation which results in a head-on collision at the interaction point. After collision another set of crab cavities kick bunches back to their original direction to re-enter the ring. Thus four crab cavities are generally required for a crab crossing. As a more economical first option, KEK adopted a modified crabbing scheme with just one crab cavity in each ring [638, 639]. The electron and positron bunches kicked by crab cavities installed in the high energy ring (HER) and low energy ring (LER) wiggle around whole ring and make a crab crossing at the collision point. The design kick is 1.44 MV, corresponding to peak surface fields of 21 MV/m and 60 mT. Ultimately the crab crossing is expected to boost the luminosity by a factor of 4 [638, 639].

The 508 MHz deflecting mode (TM110) cavity (Fig. 11.11b) has a nonaxially symmetric "squashed shape," with a racetrack cross-section, to push the frequency of the undesired polarization of the deflecting mode (700 MHz) above the cutoff frequency of large beam pipe. Three other modes below the beam pipe cutoff frequency also need to be removed. These are often called lower order modes (LOM). A beam line coaxial coupler (Fig. 11.12) extracts the lowest (430 MHz) fundamental TM010 mode, as well as the two polarizations of the TE dipole mode. The long hollow center conductor is made from niobium, and supported at the midpoint by a stub structure which also provides mechanical support and cooling. A notch filter set at the end of coaxial coupler rejects the main crab mode flow out from the cavity. The other HOMs which propagate out from the large beam pipe as well as from the coaxial coupler are damped by ferrite lined absorbers at room temperature.

An antenna type input coupler is connected horizontally to the large beam pipe to excite the crab mode. The external Q of the input coupler is set to about 10^5 to tolerate about 1 mm offset of beam position during operation and to handle more than 100 kW. The inner conductor is cooled by water and outer conductor is cooled by cold helium gas from helium vessel.

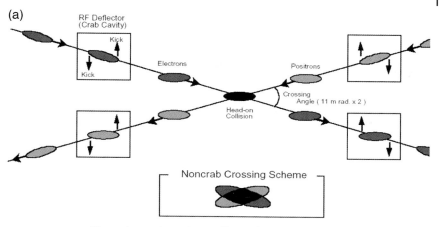

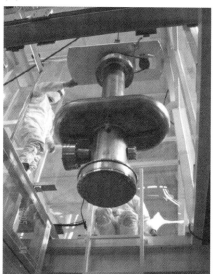

Fig. 11.11 (a) The crab-crossing scheme tilts bunches to provide head-on collisions. (b) High pressure water rinsing of the crab cavity. The cavity is 1 m by 0.5 m in cross-section. The cavity iris is reinforced by four ribs against stresses due to nonsymmetric cell shape [638, 639] (courtesy of KEK).

The coaxial LOM coupler also serves as the tuner for the crab mode by adjusting its insertion into the cell. The penetration of the LOM coupler is controlled by two driving rods connected to the stub support and driven by piezo- and motor-drive mechanical actuators set outside the cryostat. The tuning sensitivity is 38 kHz/mm. Figure 11.13 shows the layout of various components together with the helium jackets, which cool the cavity and the stub support.

Two crab cavities have been installed in KEK-B after commissioning off-line to reach kick voltages of 1.4 and 1.5 MV. Figure 11.14(a) shows the crab module

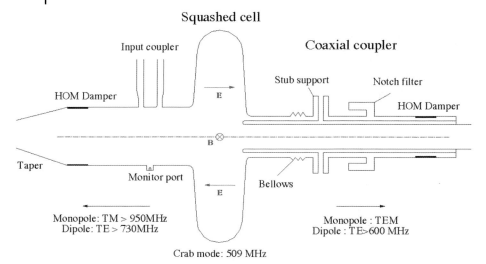

Fig. 11.12 The coaxial coupler removes the fundamental and TE modes. The stub support helps to set it accurately in the center of the cavity to prevent the crab mode from propagating out to ferrite absorber. A notch filter on the outer conductor rejects the crab mode [638, 639] (courtesy of KEK).

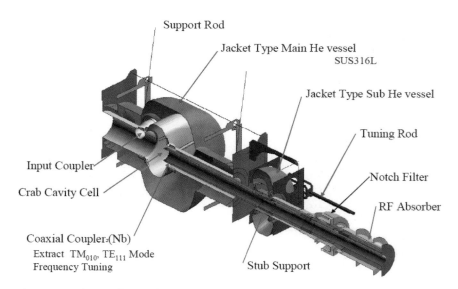

Fig. 11.13 3D layout of the crab cavity in its helium vessel [638, 639] (courtesy of KEK).

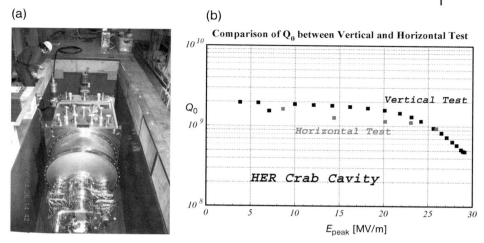

Fig. 11.14 (a) The crab cavity module situated in the pit for horizontal tests. (b) Comparison of vertical and horizontal test results [638, 639] (courtesy of KEK).

in the pit for commissioning, and Fig. 11.14(b) compares the commissioning test result with the vertical test qualifying result for one (HER) cavity. While operating in KEK-B, effective head-on collision of electron and positron beams was achieved successfully with a luminosity above $10^{34}/\text{cm}^2/\text{s}$ at beam currents of 1.3 A in the low-energy positron ring and 0.7 A in high-energy electron ring. With crab cavities detuned, the currents were increased to 1.7 A for LER and 1.35 A for HER. At these currents, the HOM dampers attached to beam pipe absorbed about 10 kW and 2 kW. During operation, the crab cavities were tripped by a sudden degradation of vacuum caused by release of condensed gas on the cold surface of the cavity and coupler part. The trip rate could be reduced by warm up to 80 K.

Crab cavities are under consideration for a variety of future developments, in particular the luminosity upgrade of the LHC, the International Linear Collider (ILC), and for ultrafast x-ray sources. A small (<1 mrad) angle crab approach is being considered for a Phase I, LHC luminosity upgrade [641]. The crab compensation in the LHC can be accomplished using only two sets of deflecting cavities, placed in collision-free straight sections of the LHC. This global crab compensation scheme is similar to the one used at KEK-B. It requires half the number of cavities, and allows freedom in their location, which is essential, the large transverse size of the 400 MHz deflecting elliptical cavities. However, a later Phase II upgrade calls for large crossing angles (>2 mrad), and the more challenging local crab compensation in the interaction region.

Before the change to head-on collisions, ILC was based on a 20 mrad crossing angle, which required 11.4 MV at 1.3 GHz and 3.8 MV at 3.9 GHz [642]. Choosing a higher frequency gives a shorter system. Any phase error between the bunch and the cavity leads to a deflection and rotation of the bunch. Any differential phase error between the positron crab cavity and the electron crab cavity leads to differing deflections and consequential loss in luminosity. Therefore the phase error tolerances are stringent. A variant of the FNAL 3.9 GHz CKM multicell cavity [643] was considered as an acceptable choice. Figure 11.15 shows a candidate deflecting cavity under development at Fermilab for beam diagnostics application and the performance of the cavity [644–646]. The TM110 mode of the 13-cell cavity (effective length 0.5 m) is at 3.9 GHz. The cavity will be used as a beam-slicing device to measure the longitudinal profile of a short bunch. For a kick of 5 MV/m the surface fields are 18.6 MV/m and 80 mT.

A large segment of the light-source user community is interested in time-resolved experiments, such as for photoemission and magnetism [647]. But the

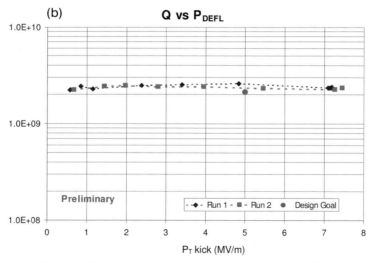

Fig. 11.15 (a) Multicell crab cavity under development by Fermilab [645].
(b) Cavity performance in a vertical test. The maximum kick is about 7 MV/m corresponding to a peak surface electric field of 26 MV/m [644].

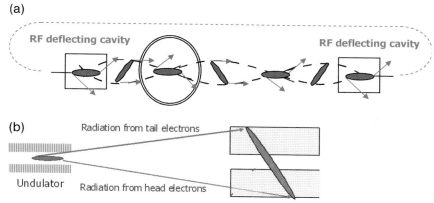

Fig. 11.16 (a) Bunch deflection with crab cavity before and after passing through an undulator. (b) Correlated distribution of the radiation emitted by tilted electron bunch in an undulator [647] (courtesy of LBNL).

resolution of interest (sub-ps to ps) is well below the typical 10–100 ps bunch durations available from high-current storage rings. Conventional storage rings such as ANL and LBNL are considering delivery of ultrafast x-rays using cw SRF deflecting mode cavities [648–650]. Electron bunches receive a head–tail kick from the dipole mode (TM110) of the deflecting cavity, phased such that the beam centroid passes through the cavity at zero phase, and receives no perturbation (Fig. 11.16). The head and tail receive transverse kicks in opposite directions. In a downstream-radiating undulator, the time-correlated transverse kick results in an angular or spatial distribution of the radiation emitted by electrons along the length of the bunch. The resulting extended x-ray pulse with correlated distribution is then treated by special x-ray optics, to result in an x-ray pulse of approximately picosecond duration. After the bunch radiates, the kick introduced is cancelled by another downstream deflecting cavity. Repetition rates as high as the full storage ring bunch rate, e.g., 500 MHz, or even up to the rf frequency may be achievable.

11.2 Electron Linacs

11.2.1 CEBAF and Upgrade

CEBAF is the largest SRF accelerator in operation as of 2008 [651]. It has two 0.6 GeV linacs with 20 cryomodules each and another 2.25 cryomodules in the injector making a total of 42.25 cryomodules. Each cryomodule has eight 5-cell cavities, installed as four pairs making a total of 338 cavities. Figure 11.17 shows a cavity pair under assembly in the clean room. The cavities were pre-

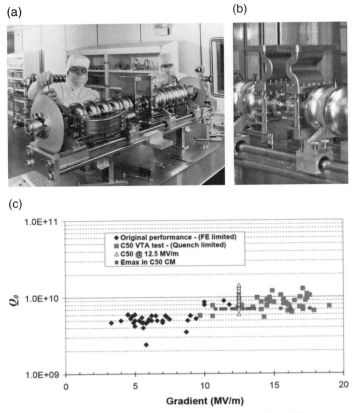

Fig. 11.17 (a) CEBAF cavity pair under clean room assembly [652]. (b) The new input coupler has a dogleg waveguide installed to eliminate window charging and arc trips [581]. (c) Improvement in performance of CEBAF cavities after reprocessing with BCP and HPR [342] (courtesy of JLab).

pared by SRF technology prevailing in the mid-1990s, based on BCP chemical preparation, ultrapure water rinsing, and clean room assembly, but with no high-pressure rinsing. Each cavity is powered cw by an individual klystron which offers enormous operational flexibility by operating cavities at their highest performance. The maximum beam power is 1 MW (typically >0.5 MW). Compared to the original cavity design gradient of 5 MV/m, the achieved gradient is 5–8 MV/m. Most CEBAF cavities are limited by field emission, only a few by quench. Initial performance of all cavities was improved by helium processing. The maximum operating gradient is set to 1 MeV/m below quench gradient limit. The overall trip rate is about 10/h. Recovery from each trip takes <30 s. The leading cause of downtime from the SRF part of the linac is due to field-emission-induced cold window arc trips. As expected from the Fowler–Nordheim field emission behavior, the arc trip rate increases exponentially with gradient. Field emission also causes radiation near the beam line. The limit is set for one rad. Operating at higher limits risks viton seals on the beamline vacuum which creates a later problem to remove cryomodules from the tunnel, in case replacement is needed.

The CEBAF main linac has been operating with high reliability. Figure 11.18(a) shows the trip-rate distribution by system in the entire accelerator for one year, and Fig. 11.18(b) further breaks down the distribution for SRF related trips. Other than window arc trips, the SRF system directly contributed less than 0.1% of unscheduled downtime. The field-emission-induced arc trips were minimized by developing Fowler–Nordheim-based trip rate models for each cavity, and adjusting gradients for a minimum overall trip rate. The cause of window arcs has been understood and eliminated by introducing dogleg waveguides (Fig. 11.17b) in modules which are being refurbished for the upgrade.

With more than a decade of operation, CEBAF has accumulated over five cryomodule-centuries and 40 cavity-centuries experience at cryogenic temperature with three cryomodule-centuries and 25 cavity-centuries under power.

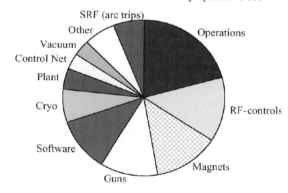

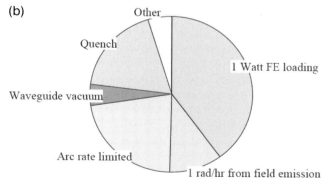

Fig. 11.18 (a) CEBAF trip rate distribution for 1999.
(b) Breakdown of the SRF trip rate [653, 654] (courtesy of JLab).

There were no cryomodule failures for 7 years of operation. After this period three cryomodules were damaged during warm-up. Two cryomodules degraded severely in performance. All five were early production cryomodules. Tuners malfunctioned in six cavities.

In September 18, 2003, Hurricane Isabel caused 4-day power outage. The Central Helium Liquefier was shutdown and the helium inventory evaporated. This released a decade's worth of adsorbed gas and added an additional thermal cycle to all of the modules. Two cryomodules were damaged by uncontrolled warm-up [655, 656]. Loss of insulating vacuum caused thermal gradients which flexed internal indium beam line vacuum joints, causing helium leaks into the cavities. Emergency generator power for vacuum pumps and controls is installed as a future counter measure. The hurricane event shows that future machines should have emergency power capacity to maintain cryomodules at 4 K indefinitely. All the cavity trip rate models had to be rebuilt because of a strong randomization in field emission, but overall the loss in capability was surprisingly small. After the hurricane, some cavities continue to show degradation due to new electron emission sites, lowering the overall operating gradient by 65 MV/year.

11.2.1.1 CEBAF Upgrade

CEBAF will upgrade its energy from 6 to 12 GeV by adding 0.5 GV in each of its two linacs, so that each linac will deliver 1.1 GeV. The beam power will remain unchanged at 1 MW. The plan is to install a total of 10 new 100 MV cryomodules with 7-cell cavities operating at 18 MV/m. This should be compared to the original 30 MV per CEBAF cryomodule with cavities operating at 7.5 MV/m. The size of the cryoplant will be doubled to provide 10 kW total at about 2 K [651, 652].

Figure 11.19 shows the upgrade cavity design and one Nb cavity unit [657]. The cavity will be of 7 cells of the low-loss shape with lower cryogenic losses as discussed in Chapter 2. Specifications call <250 W of dynamic heat load at 2.07 K for an eight-cavity module. Improvements over the original cryomodule design include increased filling factor by eliminating bellows, valves, and other beam-line components. The new cavities occupy almost the same length as the original 5-cell ones. Waveguide input couplers are upgraded to 13 kW instead of 5 kW, and more compact coaxial TESLA style HOM dampers are planned instead of waveguide couplers. Hence the helium vessel size can be greatly reduced. Because the end-group components are outside the helium vessel and cooled only by conduction, the heat transfer of HOM output feedthrough was enhanced over the TTF version. Several prototype cryomodules have been built and operated in the CEBAF and FEL accelerators during the prototype development phase.

CEBAF has scheduled a regular refurbishing of ten of the original cryomodules [342]. Several hardware improvements have been developed. A dogleg section of waveguide is added to remove a direct line of sight from the beamline to

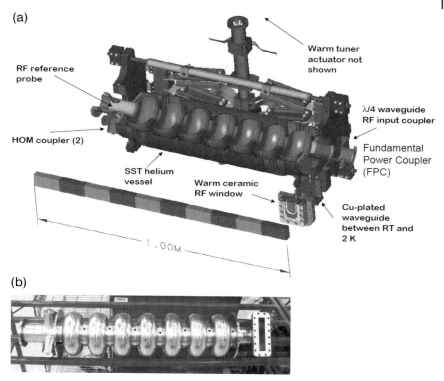

Fig. 11.19 (a) CEBAF upgrade cavity layout [657]. (b) The low-loss shape 7-cell cavity with stiffeners, TESLA-style HOM couplers outside the He vessel and a waveguide fundamental power coupler [658, 659] (courtesy of JLab).

the cold ceramic rf window, which reduces window charging and arcing. The polyethylene warm rf window is replaced with a ceramic window. The cavity tuner mechanical linkage is stiffened to reduce backlash in the tuner operations. For the cavities, the BCP treatment is maintained, followed by 600 °C hydrogen degassing, and final high-pressure rinsing to keep field emission low. After reprocessing more than 40 cavities, performance routinely goes out to quench field levels with rare field emission. Figure 11.17(c) compares the gradient before and after rework. The average increases from 7 to 15 MV/m.

11.2.2 Free Electron Lasers (FEL)

Free electron lasers (FELs) are sources of tunable, coherent radiation at wavelengths covering a wide range from mm, IR, vacuum UV, and soft x-rays [661, 662]. An FEL consists of an electron accelerator and a "wiggler" magnet. The magnetic field of the wiggler causes the electrons to oscillate transversely and radiate. When the electromagnetic waves in turn interact with the beam, the electrons bunch together and start to radiate coherently near a resonant wavelength. In the oscillator configuration (Fig. 11.20a), the laser light reflects back-and-forth between mirrors, gaining strength on each pass through the wiggler. The rf is run in the pulsed mode, and the beam is dumped after extracting the power needed for lasing. As the electron beam power rises with demand for photon flux, dumping the beam becomes expensive. Hence the energy recovery configuration (Fig. 11.20b), where the rf runs cw. After passing through the wiggler, the beam recirculates through the main linac in opposite phase. The superconducting linac recovers the energy from the beam and restores it in the rf fields to accelerate another train of bunches from the injector. The decelerated beam which now has low power can be dumped.

FELs offer many desirable characteristics over conventional lasers: wavelength tunability, high average power, and high efficiency of conversion of ac to laser power. High peak power and high average power infrared and ultraviolet FELs

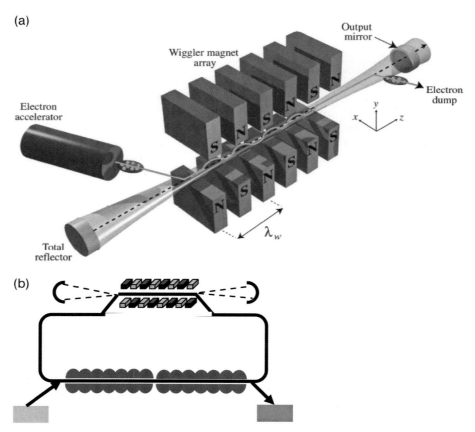

Fig. 11.20 (a) Principle of the free electron laser in resonator configuration [662]. (b) Free electron laser with energy recovery [663] (courtesy of JLab).

serve as valuable research tools in solid state physics, chemistry, biology, and medicine. Picosecond pulse trains allow fast time domain studies, such as vibrational dynamics in condensed matter systems. High-power FELs offer a variety of applications in materials processing, surface processing, micromachining, surgery, high-power microwaves, and defense.

To achieve lasing, it is necessary to focus the electron beam so that there is adequate spatial overlap with the laser beam. Therefore good beam quality (i.e., low energy spread and low emittance) is essential for FEL operation. Linacs, in general, and SRF linacs, in particular, can deliver beams which satisfy these requirements. The injector determines the emittance and energy spread. High-gradient, low-impedance SRF structures allow the preservation of exceptional beam quality required for short wavelength FELs. Linacs offer exceptional amplitude and phase stability of the rf fields, thereby ensuring minimum contribution to the energy spread. Linacs also have operational flexibility so that changes in beam energy, bunch length, and pulse patterns are all possible. SRF-driven FELs have reached unprecedented values of wavelength and average output power. Sub-ps bunches are possible, whereas in storage-ring-based light sources the typical rms bunch lengths are not shorter than about 10 ps.

Table 11.4 shows some of the main parameters of SRF-linac-based FELs. The first FEL beam was demonstrated in the 1980s with a 50-MeV beam from the SCA at Stanford [664]. The SCA experiment at 1.6 μm wavelength converted more than 1% of beam energy to laser energy and made the first demonstration of energy recovery. Decelerating a 50-MeV beam to 5 MeV by recirculating through the linac at the appropriate phase required only 10% of the power in the recovery mode. At the follow-on Stanford Picosecond FEL Center, the linac provided a 200 μA electron beam of high quality at energies from 15 to 45 MeV. The beam-powered FELs covered wavelengths from 3 to 13 μm and 15 to 65 μm [665].

Table 11.4 Main parameters of operating FELs [661].

	Frequency (MHz)	Energy (MeV)	Current (mA)	Wavelength (μm)	Type
JAEA-FEL Japan	500	17	5.2–40 (pulsed)	22	FEL, ERL
JLab-FEL USA	1500	120	5–10	1–6 + UV upgrade	FEL, ERL
ELBE Germany	1300	12–40	1 mA	2–10	FEL
S-DALINAC Germany	3000	30	0.06	7	FEL, ERL
SCA, USA	1300	40–50	0.15	1–2	FEL, ERL
PKU FEL (under constr.)	1300	30	1–5	5–10	FEL
FLASH Germany	1300	1000	mA (pulsed)	>6.5 nm	SASE-FEL

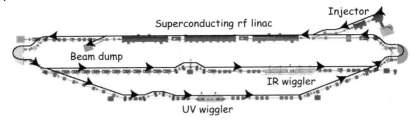

Fig. 11.21 Layout of the JLab IR FEL with UV upgrade. The FEL runs in the energy recovery mode. It comprises a 10 MeV injector, a linac consisting of three Jefferson Lab cryomodules generating a total of 80–160 MeV of energy gain, and a recirculator [663] (courtesy of JLab).

The Jefferson Lab (JLab) IR FEL [666] (Fig. 11.21) has lased in the 1–6 μm wavelength range and reached average IR output power of 14 kW at 1.6 μm, the highest cw average laser power ever to be achieved. JLab runs in the energy recovery mode with more than 99.8% efficiency. More than 1.3 MW beam power was recovered with a beam of 9.1 mA at 150 MeV [667]. This is an important milestone toward high-beam power ERLs of the future (see Chapter 12). To deal with high currents some of the principal modifications to the CEBAF cryomodule were to locate the waveguide HOM loads at higher temperature and to use higher power waveguide windows in the injector modules. The upgraded module demonstrated 82 MV of continuous acceleration surpassing all previous such systems before suffering a rf window failure. The accelerator has been configured with another recirculation for the UV upgrade.

Other cw FEL projects underway are at JAERI in Japan [668, 669], Darmstadt in Germany [670], and ELBE at Rossendorf [671]. The Japan Atomic Energy Agency (JAEA) FEL (Fig. 11.22) in JAERI, Japan is based on a 500 MHz SRF system with four cavities operating at $E_{acc}=5$ MV/m (4.2 K). The 17 MeV linac operates in a pulsed mode (1% duty factor) with a beam power of about 140 kW in the energy recovery mode. It lases between 1 and 22 μm with a laser power of 700 W at 22 μm.

The S-DALINAC at Darmstadt drives an FEL with a cw recirculating linac operating at 3 GHz using 20-cell SRF cavities accelerating a 50 MeV beam. When

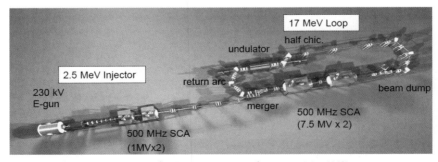

Fig. 11.22 The JAEA FEL runs with 500 MHz superconducting cavities [668].

diverted through a wiggler and mirror section, lasing takes place at 7.0 μm. The radiation source ELBE at Rossendorf runs a superconducting 1300 MHz linear accelerator using TESLA cavities that accelerate a 1 mA electron beam to energies of 12–40 MeV. Two undulators allow access to a wide range of wavelengths in the mid- and far-infrared. The facility also provides different beams for neutron production, positron production, bremstrahlung, and x-radiation.

11.2.3
TTF SASE FEL (FLASH)

At ultrashort wavelengths, less than 100 nm, mirrors are not available for FELs. In this case, coherent bunching of the electron beam in a single-pass through a long wiggler opens an exciting alternative. As a short bunch of high intensity interacts with the undulator field, microbunches develop which emit photons coherently (Fig. 11.23). In the "high-gain mode" the radiation field amplitude grows exponen-

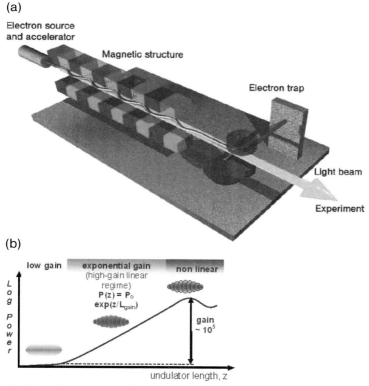

Fig. 11.23 The SASE principle: (a) a short bunch of high intensity interacts with the undulator field of a long wiggler and (b) microbunches develop which emit photons coherently. The gain in power is exponential with undulator length until it reaches saturation at full bunching [675] (courtesy of DESY).

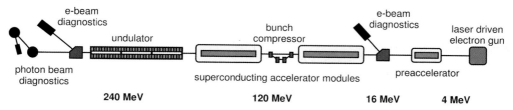

Fig. 11.24 Component layout for the first phase of the TESLA Test Facility (TTF) [677] (courtesy of DESY).

tially with distance along the undulator. The power increases as the square of the number of particles per bunch. This process is called self-amplified-spontaneous-emission (SASE) [673, 674]. SASE FELs are the most attractive candidates for extremely high brilliance coherent light with wavelength in the angstrom regime.

Self-amplified stimulated emission of photons requires electron peak currents of kA, e.g., from bunches with 1 nC shorter than 100 fs. Since the minimum bunch length provided by a photocathode rf gun is restricted by collective effects, a multistage bunch compression scheme is needed to reach the required peak current. Compression is based on accelerating the beam off the rf field crest and passing through a transverse magnetic chicane which converts the energy modulation into spatial bunching. For satisfactory performance, high stability in field amplitude and phase is essential.

The TESLA Test Facility (TTF) at DESY advanced the science of SASE using a superconducting linac, while providing UV to x-ray beams to a user facility [676–680]. From inception, the TTF also served a dual purpose to demonstrate a high-gradient SRF linac in preparation for a future superconducting linear collider, now called ILC.

In the first phase (Fig. 11.24), TTF installed three modules of eight 9-cell cavities to accelerate the electron beam to more than 270 MeV. Figure 11.25 shows a third generation TTF module [681]. The cavities were all prepared by BCP and HPR; hence the gradient was limited to 25 MV/m. The SASE lased over a wavelength range from 80 to 180 nm, corresponding to beam energy between 181 and 272 MeV, demonstrating saturation at 98 nm. Over 13 000 h of operation, the average availability was 84%.

The TTF Linac was operated 7 days per week, at 24 h/day. Approximately 50% of the time was allocated to FEL operation including a large percentage of user time [681]. In the second phase (Fig. 11.26), TTF was upgraded to 1 GeV and renamed FLASH. A total of six modules installed have 48 cavities operating above 20 MV/m [680]. A large fraction of these cavities were prepared by advanced methods of EP and baking to demonstrate higher gradients ($E_{\mathrm{acc}} < 30$ MV/m). The installation of an additional bunch compressor and an improved injector allow 2.5 kA peak current and a normalized emittance of 2 mm mrad. First lasing was achieved at 30 nm and since then saturation at 13 nm. The expected minimum wavelength is 6 nm, with the use of a 30 m

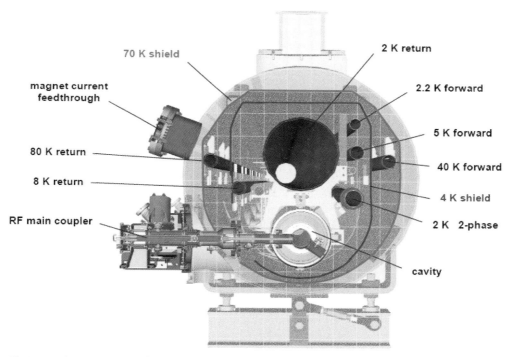

Fig. 11.25 The TTF-III cryomodule. A typical 1.3 GHz module consists of eight cavities, a quadrupole magnet, a beam position monitor, and beam steering doublet. After two design iterations, the "type three" cryostat design is – with small modifications – the baseline for XFEL, and serves as a starting point for an ILC module [682, 688] (courtesy of DESY).

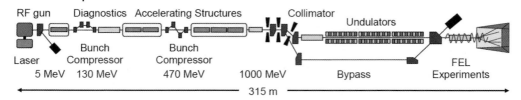

Fig. 11.26 The second phase of TTF [683] (courtesy of DESY).

undulator. Figure 11.27 shows the evolution of FEL wavelength with electron beam energy.

To maintain high gradients, the TTF linac was assembled in dust-free conditions from the beginning. There has been no degradation of gradient over time, and there have been no vacuum accidents. Over 13 months of operation starting July 2005 there have been 10 kh of operation with an average downtime of about 15%. The operation relies on one maintenance day (8 h) per week. The major contributors to down time were rf (klystrons and modulator troubles) (70%), LLRF (7%), laser (6%), photon lines (5%), and controls (3%) [675, 684].

The successful completion of the first and second phases of TTF demonstrated that superconducting 9-cell Nb cavities can reliably perform at 25 MV/m, as demonstrated by the results of modules 5–7 in Fig. 11.28. Stable beam acceleration near this gradient was also demonstrated. TTF/FLASH has proven to be an important testbed for future SRF-based accelerators.

The TESLA Technology Collaboration (TTC) has now adopted electropolishing and baking for higher gradients toward a future linear collider. To establish proof-of-principle, a 9-cell, EP-baked cavity equipped with input couplers, HOM couplers and tuners has been operated inside a single-cavity test cryomodule with a high power klystron to reach gradients between 35 and 36 MV/m. The cavity operated stably for more than 1100 h at 35 MV/m at a Q value of 7×10^9.

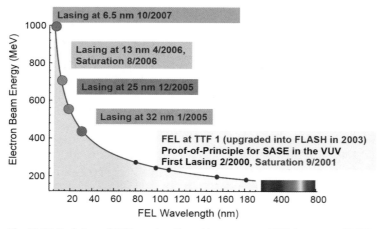

Fig. 11.27 Evolution of TTF wavelength and beam energy [680] (courtesy of DESY).

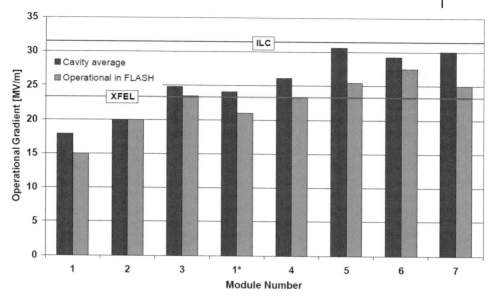

Fig. 11.28 Evolution of cryomodule gradients over installation history. The bar chart compares linac and cavity acceptance test results. The average degradation is a few MV/m. Module 6 has only cavities prepared by EP [680] (courtesy of DESY).

Operation was without quench or trips originating from the cavity–coupler system [685]. There was no significant field emission below 35 MV/m as judged from the absence of x-rays. A second cavity reached 35 MV/m at a Q of 6×10^9 with no x-rays detected up to 34 MV/m. For operation at 35 MV/m, Lorenz-force detuning compensated by fast piezotuners has been demonstrated with the 9-cell cavity operating at 35 MV/m [294].

In addition to gradient and Q performance, dark current, alignment, and vibration are important issues for future applications of TESLA technology, such as XFEL and ILC. The TTF experience gave valuable data in all these arenas. Dark current values in the final module are acceptable. Only one cavity in module number 5 produced a mentionable dark current. Detuning this cavity left an integrated dark current of the order of 20–25 nA at 25 MV/m average gradient [686]. (For ILC, the captured dark current must remain below 50 nA per cavity.) The final alignment of the cavities in the module came within 0.3 mm of the expected position after evacuation and cool down. The alignment was repeatable over several thermal cycles. The vibration of the quads was not amplified over the vibration level outside the cryomodule. The quad vibration level was not affected by rf operation or by the refrigeration system with no difference between warm and cold operation except for a large amplitude of ~30 Hz oscillation (plus harmonics) that built up in the cold. This was determined not to be a mechanical resonance of the cold mass/quad structure.

11.2.4
European XFEL

X-rays have played a crucial role in the study of structural and electronic properties of matter on an atomic scale for many decades. Some of the most fascinating applications have come from the life sciences. An exponentially increasing number of biological structures have been solved and deposited with the protein data bank. The growth trend is likely to continue, but there remain many challenges in structural biology, especially resolving systems that are difficult to crystallize.

Linear-accelerator-driven FELs using the SASE principle provide one of the most promising approaches to produce x-radiation with unprecedented quality as well as a path for extension to the angstrom wavelength regime. With the ultrahigh-brilliance coherent radiation and ultrashort (10–100 fs) pulse length from FEL x-ray sources, the research will enter a new era, impacting the full span of the materials and biological sciences. The femtosecond time scale opens the possibility for novel time exposure experiments in biological, chemical, and physical processes to investigate structural changes, and multiphoton processes in atoms and molecules. Ultimately, it will become possible to acquire holographic snapshots with atomic resolution in space and time on the scale of chemical bond formation and breaking. Figure 11.29 compares the peak brilliance of the SASE light sources with the third generation storage ring light sources, showing ultimate gains of ten orders of magnitude.

The first high brilliance short wavelength facility [687] is under commissioning, using 20 GeV electrons from the existing SLAC normal conducting 2.86 GHz linac. At the European XFEL to be built at DESY in Germany [683, 688], a superconducting driver linac will deliver short bunches ($\sigma_z = 25$ μm) at maximum energy of 20 GeV to long undulators The nominal charge will be 1 nC and the intrapulse repetition frequency will be 5 MHz. Operation in this extremely short bunch-length regime presents considerable technical and beam dynamics challenges. The x-ray pulses will have a duration in the range of 10 to a few 100 fs, with a corresponding peak power of tens of GW.

Figure 11.30 shows the overall layout of the XFEL, and Table 11.5 gives an overview of the beam and accelerator parameters. The main linac will have 116 twelve meter long accelerator modules with eight superconducting cavities each, grouped into 29 rf stations. The linac will be housed in a tunnel (Fig. 11.29 b) 15–30 m underground. The 10 MW klystrons are in the same tunnel and connected to the modulators in an easily accessible surface building on the DESY site by the high voltage pulse cables. The required klystron power per station is 4.8 MW, well below the maximum power capability of the multibeam klystrons.

To optimize availability, there will be two parallel injectors to produce and accelerate the electron beam before combining the beam lines at roughly 100 MeV. A short accelerator section at the third harmonic rf frequency will be used for linearization of longitudinal phase space. This section is followed by a booster linac increasing the beam energy to 500 MeV. The booster will consist

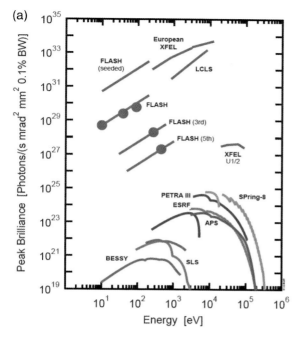

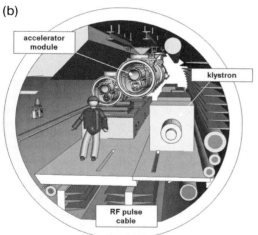

Fig. 11.29 (a) Comparison of peak brilliance from a variety of operating and planned light sources [675]. (b) The European XFEL main linac tunnel cross-section, 5.2-m diameter tunnel. The accelerator modules will be suspended from the ceiling [683] (courtesy of DESY).

Table 11.5 Main parameters for the European XFEL [683, 684].

Performance goals for the electron beam	
Beam energy	10–20 GeV
Emittance (norm.)	1.4 mrad · mm
Bunch charge	1 nC
Bunch length	80 fs
Performance goals for SASE FEL radiation	
Photon energy	15–0.2 keV
Wavelength	0.08–6.4 nm
Peak power	24–135 GW
Average power	66–800 W
Number of photon per pulse	$1.1–430 \times 10^{12}$
Peak brilliance	$5.4–0.06 \times 10^{33\,a)}$ *
Average brilliance	$1.6–0.03 \times 10^{25\,a)}$
Accelerator parameters	
Energy for 0.1 nm wavelength (maximum design energy)	17.5 GeV (20 GeV)
No. of installed accelerator modules	116
No. of cavities	928
Acc. gradient (104 active modules) at 20 GeV	23.6 MV/m
No. of installed RF stations	29
Klystron peak power (26 active stations)	5.2 MW
Loaded quality factor, Q_{ext}	4.6×10^6
RF pulse length	1.4 ms
Beam pulse length	0.65 ms
Repetition rate	10 Hz
Maximum average beam power	600 kW
Unloaded cavity quality factor, Q_o	10^{10}
2K cryogenic load (including transfer line losses)	1.7 kW
Maximum no. of bunches per pulse (at 20 GeV)	3250 (3000)
Minimum bunch spacing	200 ns
Bunch charge	1 nC
Bunch peak current	5 kA

a) In units of photons/(s mrad² mm² 0.1% b.w.).

of four accelerator modules with 32 cavities at the design gradient of 12.5 MV/m. At this energy the electron bunches are compressed by about a factor of 100 down to σ_z of 22 µm corresponding to about 5 kA peak current for 1 nC charge. The linac will accelerate more than 3000 bunches per rf pulse. In total there will be about 1000 cavities operating at a gradient of about 23 MV/m (for the main linac) and housed in 120 modules with about 30 rf stations. Industrialization of linac components is one of the key tasks toward construction of future large accelerators such as the XFEL and the ILC.

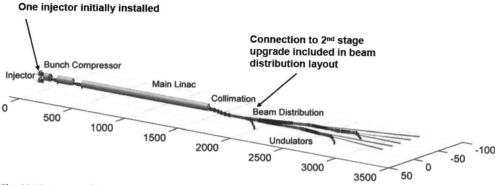

Fig. 11.30 Layout of the European XFEL driven by a 20 GeV superconducting linac [688] (courtesy of DESY).

11.3 Heavy-Ion Accelerators

Superconducting linacs providing precision beams of heavy ions have consistently been one of the most successful applications of SRF [22]. Heavy ions, from helium to uranium, are accelerated to energies from a few to 20 MeV/nucleon and used to bombard other nuclei. Above 5 MeV/nucleon, ions have sufficient energy to overcome the Coulomb barrier and penetrate the nucleus. The collisions cause energy, mass, and angular momentum to be transferred between the projectile and target nuclei, enabling structure research on the evolution of nuclear shape as a function of excitation energy and other aspects, such as spin.

At Argonne, ATLAS with 68 cavities installed has been operating for nearly three decades as a national user facility for heavy ion, nuclear and atomic physics research, logging well over 100 000 h of beam-on-target operation. The split ring resonators have built up a track record in excess of 3 million unit-hours of operation with better than 99% availability. ATLAS will be upgraded in energy by 30–50% (depending on ion species) by adding additional accelerating structures to the end of the linac. The structures will be housed in a single 8-cavity cryomodule [607] containing seven quarter-wave (115 MHz, $\beta=0.15$) and one half-wave cavity (176 MHz, $\beta=0.26$). The new cavities exhibit improved performance relative to existing ATLAS cavities due to a combination of optimized geometry and the implementation of the latest clean handling and processing techniques. The new cryomodule will also have separate systems for the beam and insulation vacuum which will help to reduced field emission and multipacting by reducing particulates and condensed gases.

Heavy-ion facilities that continue operation are at Florida State University, Kansas State University, TRIUMF (Canada), JAERI (Tokai, Japan), INFN (Legnaro, Italy), and Australian National University. New heavy-ion accelerator facil-

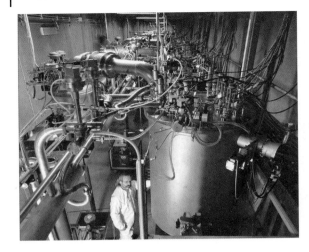

Fig. 11.31 An overview of the heavy-ion accelerator ALPI at INFN Legnaro [689].

ities utilizing superconducting structures are running at Sao Paolo (Brazil), New Delhi, and Mumbai. We will discuss just a few of these many examples.

The SC linac ALPI at INFN Legnaro (Fig. 11.31) has been serving as Tandem booster since 1994, accelerating species from 28Si to 197Au [689]. Figure 11.32 shows the evolution of ALPI energy with the continuous development of resonator technology from Pb–Cu to Nb–Cu, and solid Nb. In all, 64 quarter-wave resonators (QWRs) at 160 MHz are in the main beam line delivering a total of 49 MV. 44 cavities are medium-β (0.11, 160 MHz) and 8 are high-β (0.13, 160 MHz). 12 low-β (0.055) resonators at 80 MHz are made from solid Nb. Both Nb and Nb–Cu resonators reached 6–8 MV/m at 7 W in vertical tests. Table 11.6 shows their off-line and operating performances.

Off-line values of the peak surface electric field range between 22 and 32 MV/m for both types of resonators. Operational field values are close to off-line values in Cu-based QWRs, while they are significantly lower for the full Nb QWRs (45–65%), which is expected to improve after an upgrade of the tuners. The upgrade will boost low-β voltage from 10 to 20 MV.

The main reason for the better operational performance is that the thick Cu housing makes the Nb–Cu resonators stable to both mechanical vibrations and changes of the liquid He pressure. The sensitivity of the Cu-based resonator frequency to variations of pressure is as low as ~0.01 Hz/mbar. Even large varia-

Table 11.6 On-line and off-line resonator performance at ALPI [395, 689].

	Nb/Cu QWR	Full Nb QWR
$\bar{E}_a$ (MV/m)	4.5/6.5	6.5
$\bar{E}_p$ (MV/m)	22.5/32	32
$\bar{E}_{a,\,op}/\bar{E}_a$ (%)	98/92	45–65

(a)

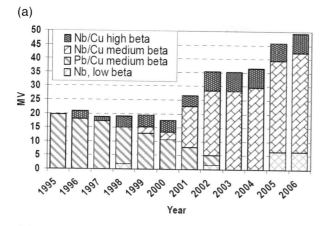

(b)

Fig. 11.32 (a) Evolution of ALPI energy with technology development [689].
(b) ALPI medium-β (0.11) Nb–Cu QWR [395, 689] (courtesy of INFN).

tions do not need to be controlled via the mechanical tuners. Operation of full Nb resonators is more critical. The main challenge is to ensure the resonator phase stability with respect to slow changes of the liquid He pressure (time scale of seconds or slower) and to mechanical vibrations (10–100 ms). The frequency sensitivity to pressure changes in full Nb QWRs is ~100 times larger so that mechanical vibration modes as low as 22 and 42 Hz were seen in operation. The phase stability issue was addressed with an improved mechanical tuner and development of mechanical dampers.

A new positive ion injector upgrade (PIAVE) has been completed to extend the mass range of ion beams available from the tandem up to uranium delivered by an ECR ion source, as well as to extend the energy range to 6 MeV per nucleon. PIAVE features two superconducting radio frequency quadrupoles (SRFQs) and eight low-β (0.05) solid Nb QWRs.

The SRFQs (Fig. 11.33) were discussed in Chapter 2. Built with solid Nb within a stiffening Ti jacket, the SRFQs are 0.8 m in diameter and 1.4 and

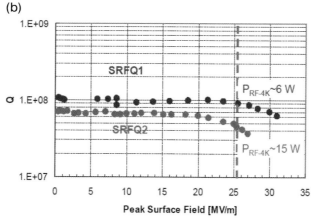

Fig. 11.33 (a) The SRFQ housed in its module. (b) Off-line performance of two SRFQs for PIAVE [689] (courtesy of INFN).

0.8 m long, respectively, each with a resonant frequency of 80 MHz. Since the RFQ is mostly a focusing structure, with a small on-axis component of the electric field, the ratio E_{pk}/E_{acc} is particularly high, i.e., 7.33 and 10 for the two resonators, respectively, as compared to about 3 for the QWR. At the maximum peak surface field of 25.5 MV/m the interelectrode voltages of SRFQ1 and SRFQ2 are 148 and 280 kV, respectively, which are significantly higher than those achieved by typical normal conducting RFQs. This is one of the main attractive features of a superconducting RFQ. The stored energy is acceptably low

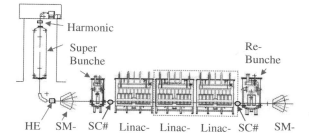

Fig. 11.34 Pelletron and linac booster at the Inter University Accelerator Centre, New Delhi, India [90].

for phase-amplitude stabilization circuits. The design values of the accelerating field are limited to 2.4 and 3.4 MV/m with a dissipated power of ~10 W at 4.5 K. After conditioning MP and field emission for several days, PIAVE has been successfully commissioned and fully operating with noble gases up to 132 (Xe) in its first stage [689].

Superconducting linac boosters for increasing the energy of heavy ions from the Pelletron accelerators at Inter-University Accelerator Centre (IUAC), New Delhi and Tata Institute of Fundamental Research (TIFR), Mumbai are nearing completion [90]. The accelerating structures for both linacs are QWRs.

At IUAC, the linac (Fig. 11.34) requires 27 QWRs, of which 13 were built in collaboration with Argonne National Lab, and the rest in Delhi after a full suite of new SRF infrastructure was installed. The QWR (Fig. 11.35) is made of bulk niobium operating at 97 MHz and optimized for $\beta=0.08$. One module with eight cavities (Fig. 11.36a) has been operated with beam. Figure 11.36(b) shows the resonator performance. Problems encountered with the drive coupler and slow tuner have been resolved. Two additional cryostats are expected to be operational in 2008.

At TIFR, Mumbai, the QWRs are made of lead-plated copper operating at 150 MHz, optimized for $\beta=0.1$. Each module houses four QWRs. Seven modules have been constructed and installed in the beam line. In the first phase, three modules were used for acceleration followed by a cycle of experiments. Subsequently all seven modules and 23 out of the 28 resonators were used for acceleration of Si^{13+} ion beams to provide energy gain over 100 MeV.

11.4
Heavy-Ion Accelerators for Rare Isotope Beams (RIBs)

Several heavy-ion accelerator facilities are preparing superconducting linacs to accelerate rare isotope beams (RIBs) up to energies of at least 6.5 MeV per nucleon [690]. Among other fundamental questions, experiments with the radioactive beams will provide basic insight into the origin of the heavy elements in supernovae. An ultimate Rare Isotope Accelerator (RIA) facility of this type will be discussed in Chapter 12. As with most heavy-ion boosters, the choice of short superconducting cavities, exhibiting wide velocity acceptance allows opti-

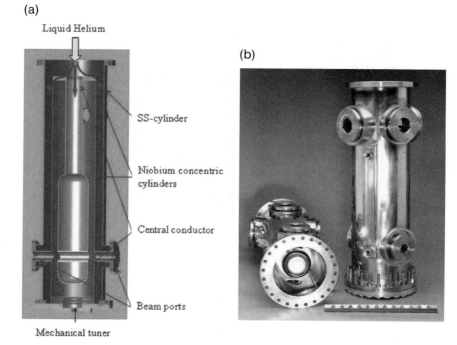

Fig. 11.35 (a) Design of the QWR for the Delhi booster at 97 MHz with $\beta = 0.08$. (b) Completed resonator made from solid Nb and electron beam welding [90] (courtesy of IUAC).

mization of the output energy for each ion species by adjusting individual rf phases. Beam loading is an important consideration in these driver linacs and beam halo must be understood and controlled.

11.4.1
ISAC-II

TRIUMF in Canada is operating an Isotope Separation On-Line (ISOL)-based radioactive beam facility, ISAC, supplying both low-energy (60 kV, mass $A = 238$) and high-energy (0.15–1.5 MeV/u, $A = 30$) experimental areas with exotic beams. To extend the final energy to at least 6.5 MeV/u and mass range up to 150, TRIUMF is installing ISAC-II [92, 514]. The expansion includes the addition of a superconducting heavy-ion linac supplying a total of 42.7 MV of acceleration. The linac consists of three cavity geometries, all bulk niobium, two-gap QWRs with β values of 0.042, 0.072, and 0.105 and rf frequencies of 70.7, 106.1, and 141.4 MHz, respectively. There are 20 bulk niobium quarter wave cavities at 106 MHz housed in five cryomodules. Another 20 cavities at $\beta = 0.11$ (141 MHz) are planned for the end of 2009 to raise the final energy to 6.5 MeV/u. A final low-β stage is foreseen after 2010.

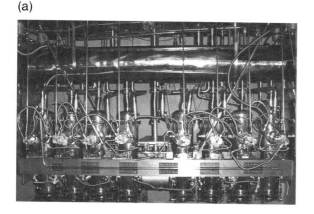

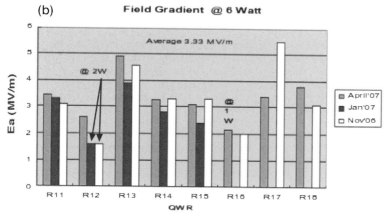

Fig. 11.36 (a) Eight-resonator module for the Delhi Pelletron booster. The dimensions of the cryostat are 2.3 m in length, 1.3 m in height, and 1.1 m in width. (b) Field gradient achieved by eight resonators during three online tests. The average is 3.3 MV/m [90] (courtesy of IUAC).

Prototype Nb cavities have been completed successfully in collaboration with INFN–LNL. At 4.5 K, these achieved a gradient of 6.7 MV/m at 7 W dissipated power [514]. The medium-β cavity design goal is to operate cw at peak fields of 30 MV/m and 60 mT with <7 W dissipation or operating Q of 5×10^8. At 18 cm effective length, the gradient corresponds to an acceleration voltage of 1.1 MV, at a stored energy of 3.2 J.

The initial stage completed consists of five cryomodules, each module housing four 106 MHz QWRs ($\beta=0.071$) and one 9 T superconducting solenoid, all operating at 4.2 K (Fig. 11.37b). On-line performance has confirmed cw cavity operation at gradient of 7.6 MV/m corresponding to a peak surface field of 38 MV/m. Performance after 18 months of operation and a full thermal cycle during the annual shutdown shows very little (5%) degradation.

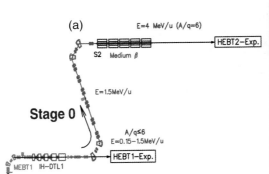
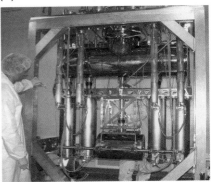

Fig. 11.37 (a) ISAC-II, Phase 0 with medium-β section. (b) A medium-β cryomodule for ISAC-II consists of four QWRs and a central superconducting focusing magnet. The 9 T solenoid is not shielded but is equipped with bucking coils to reduce the magnetic field in the neighboring rf cavities. Each module has two main assemblies, the top assembly and the resonator tank assembly. The top assembly includes the vacuum tank lid, the lid mu-metal and liquid nitrogen shield, the cold mass, and the cold mass support [92, 514] (courtesy of TRIUMF).

11.4.2
SPIRAL2

With the SPIRAL2 project [691, 692], GANIL in France aims to make an intermediate step between existing RIB facilities and future RIB projects like EURISOL or RIA (see Chapter 12). GANIL is a stable beam and RIB facility using a cascade of three cyclotrons to accelerate the primary heavy-ion beam. Energies up to 100 MeV/u and intensities up to 6 kW are available. SPIRAL2 will be composed of a room-temperature injector to 1.5 MeV, followed by superconducting linac driver, to deliver a 5 mA deuteron beam with energy up to 20 MeV/u, and heavy ions with 14.5 MeV/u. Figure 11.38 shows the planned layout. Beam commissioning is expected in 2011.

Most of the acceleration is planned with twenty-six 88 MHz QWRs divided into two β-families ($\beta = 0.07$ and $\beta = 0.12$). This design results in an identical frequency for all rf power sources, a lower total number of cavities, a larger cavity aperture, no frequency jump which would require longitudinal matching. The first SRF linac section is made of twelve $\beta = 0.07$ QWRs, and the high-energy section comprises fourteen $\beta = 0.12$ QWRs. The transition β is 0.11. Room temperature quadrupoles provide transverse focusing so that higher alignment tolerances can be achieved compared to superconducting solenoids.

The QWR structure design has $E_{pk}/E_{acc} = 5$ and $B_{pk}/E_{acc} = 10$ mT/(MV/m), which allows for a qualifying gradient of 8 MV/m with a realistic operating gradient of 6.5 MV/m. A strict cleanliness requirement for high-field reliable operation is being followed including separate vacuum systems for the cavity and the cryostat insulation.

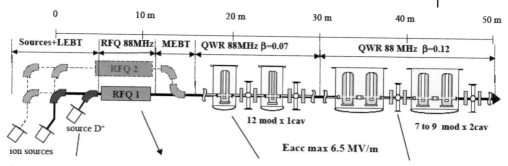

Fig. 11.38 Planned layout of the SPIRAL2 at GANIL [691].

The low-β cavities and cryomodules (single resonator) will be developed at CEA-Saclay [77–79], while the high-β ones (two resonators per module) at IPN-Orsay [693, 694]. Figure 11.39 shows the low-β cavity under fabrication and a high-β cryomodule under assembly. Each team has carried out the integration study of the QWRs in their respective cryomodules, taking into account tight longitudinal space requirements. Prototypes of each cavity type have been built, prepared, and tested in vertical cryostats at Saclay and Orsay. The resonators have demonstrated gradients of 9 and 11 MV/m in qualifying tests. The prototypes, built by ACCEL ($\beta=0.07$) and Zanon ($\beta=0.12$), have both reached fields at least 30% higher than required. Cryostats are under fabrication. Both types of cavities will be equipped with the same power couplers (operation at 12 kW) with cylindrical and disk ceramic windows under development at a third French laboratory, LPSC/Grenoble. The specified maximum power is 20 kW. Several units have been tested to 30 kW.

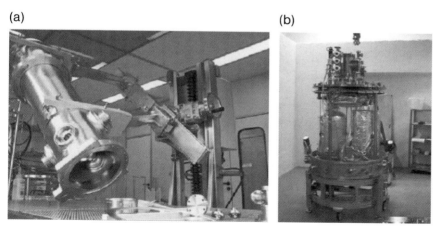

Fig. 11.39 (a) Low-β resonator by Saclay for Spiral-II [692]. (b) High-β cryomodule by Orsay. Tuning is accomplished by squeezing at the beam ports in low-β cavity and a unique cold plunger in high-B field region of the high-β cavity [692, 694].

11.4.3
MSU Reaccelerator

Rare isotope beams (RIBs) are also available at the National Superconducting Cyclotron Laboratory (NSCL) at MSU [695]. A novel system stops the RIBs in helium gas and reaccelerates them to provide opportunities for experiments ranging from low-energy Coulomb excitation to high-energy studies of astrophysical reactions. The beam from the gas stopper is first brought into a electron beam ion trap (EBIT) charge breeder on a high-voltage platform to increase its charge state, and then reaccelerated from 0.3 to about 3 MeV/u by a system consisting of an external multiharmonic buncher and a RFQ followed by a superconducting linac. The superconducting linac uses QWRs (Fig. 11.40) with β of 0.041 and 0.085 together with superconducting solenoid magnets for transverse focusing and dipole coils to provide alignment error corrections. There is a possibility to upgrade the output energy to 12 MeV/u. The reaccelerator will be a testbed for a future RIB facility. Its design derives significantly from the past Rare Isotope Accelerator (RIA) driver linac R&D efforts at NSCL. A total of three cryomodules consisting of fifteen 80.5 MHz QWR SRF cavities will be used.

In rf tests, the 0.041β QWR exceeded a surface field of 50 MV/m at 4.2 K compared to the design field of 16.5 MV/m. The 0.085β QWR exceeded 30 MV/m compared to the design field of 20 MV/m. A prototype cryomodule has been designed and fabricated (Fig. 11.41). The module consists of an 80.5 MHz, 0.085β, QWR cavity, a 322 MHz, 0.285β HWR cavity, a superconducting dipole

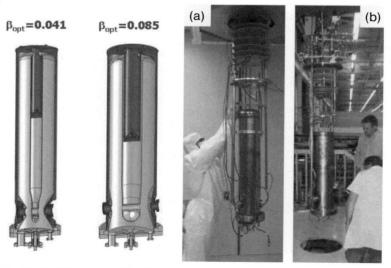

Fig. 11.40 QWRs designed for the NSCL reaccelerator at MSU. (a, b) Prototypes in preparation for testing [695].

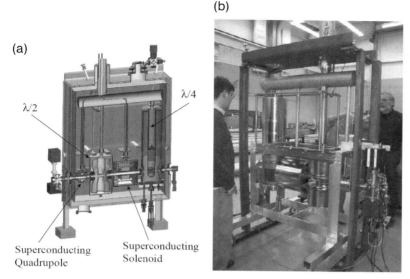

Fig. 11.41 (a) Design of a prototype cryomodule.
(b) Assembly of the prototype module [695] (courtesy of MSU).

solenoid and quadrupole. Both resonators have titanium vessels welded around them using Nb–Ti alloy transitions. The superconducting focusing magnets inside the cryomodule consist of a 9 T solenoid with an integrated steering dipole and a 31 T/m quadrupole.

The SOREQ Institute in Yafne Israel has launched the SARAF project to install a linac for protons and deuterons to energies of 1.5 MeV/u as an alternative to cyclotrons. The final energy will be 40 MeV and the cw current will be 4 mA. It will have 48 half-wave resonators at 176 MHz built by industry [93]. The main application will be radioisotope production for medical use, together with applications to RIB production, nuclear physics, and neutron physics. Operation is expected to start in 2008. The design is based on two families, one with optimized $\beta=0.09$, and the second with optimized $\beta=0.15$. In the first phase, protons/deuterons starting from an ECR ion source, followed by 176 MHz RFQ will be accelerated to about 7 MeV by six half-wave $\beta=0.09$ resonators. Resonators built in series production at ACCEL reached surface fields of 28–30 MV/m ($E_{pk}/E_a \sim 3$) at 10 W dissipation. Multipacting could be processed within seconds up to few minutes. Figure 11.42 shows the layout of a cryomodule populated with six half-wave resonators and three superconducting solenoids.

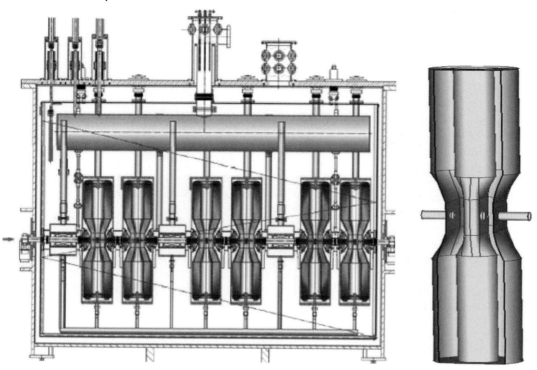

Fig. 11.42 (a) SARAF module with six HWRs at 176 MHz [93]. (b) HWR schematic. The $\beta = 0.09$ resonator built from 3 mm thick Nb has a height of 85 cm (courtesy of ACCEL).

11.5
Neutron Source

11.5.1
SNS

Neutron scattering is an important tool for material science, chemistry, and life science providing the choice wavelength to probe a large variety of materials at a broad range of length scales. By measuring the velocity of atoms, neutrons follow catalysts in action and transport through biological membranes. By direct interaction with nuclei, neutrons offer greater contrast than x-rays (especially for H). Isotopic contrasting makes possible the study of hydrogen bonds important to chemistry and biology. In polymers, hydrogen atoms can be located precisely with neutrons, but not with x-rays. Neutrons can detect light atoms next to heavy ones, as for example to determine crucial oxygen positions in high-T_c superconductors. Neutrons penetrate deep into matter to allow exploration of material properties over significant depths allowing characterization of deep

11.5 Neutron Source

welds and their associated stresses. With sensitivity to magnetic moments, neutrons are useful for studying the magnetic structure of materials and developing advanced magnetic materials.

To increase the capability for such studies, higher neutron flux is desired. For many years, nuclear reactors have been the main source of neutrons. Since neutrons emerging from reactors must be slowed down to useful energies in a hydrogen-rich moderator, a large fraction is lost in the moderator. A serious obstacle to advanced-reactor-based sources is environmental unpopularity, adding to the large inventory of fissile material already built up. Therefore, there is a strong incentive to push accelerator-based neutron sources.

Although lower in flux, accelerator-based pulsed neutron sources have begun to compete with reactors. High-energy (1-GeV) protons produce neutrons by hitting a heavy metal target and exciting nuclei to energies where neutrons are "evaporated" in a process referred to as spallation. Accelerator-based neutron sources provide high-peak intensity and very short (ms) pulses at rep rates of the order of 50 Hz. The advantage of short pulses is time-of-flight measurement to determine the incident neutron energy, eliminating the need for monochromatization and the accompanying waste of neutrons.

The highest intensity accelerator-based neutron sources in operation today are LANSCE at Los Alamos and ISIS at Rutherford Lab. The flux of these pulsed sources is still a factor of 30 below what reactors can provide. The SNS at Oak Ridge National Lab will be an advanced pulsed spallation source with 1.4 MW of beam power at the target, eventually upgradeable to 3 MW [696]. This would correspond to the average flux of the highest flux reactor at Grenoble. SNS comprises a 1.5-MW, 60-Hz, 7.8% duty factor, 1-GeV linac, an accumulator ring, associated beam lines, and a spallation neutron target. Construction began in 1999 and the accelerator was commissioned in 2006 with beam delivered to the spallation target. At 1.4 MW SNS will have eight times the ISIS beam power to become the world's leading pulsed spallation source.

For the high-energy segments of the linac, the choice of superconducting structures over the previously selected normal conducting technology reduces the overall rf installation requirement for SNS [697]. Superconducting cavities also offer the advantage of a larger aperture that reduces beam losses which activate radioactivity in the structures. At the proton beam power frontier the central challenge is to control beam loss to minimize activation. This demands careful control of beam injection and extraction. For 1 nA protons at 1 GeV, a 1 W beam activates stainless steel to 80 mrem/h at 1 ft after 4 h.

Another benefit is operational flexibility possible with individually powered cavities. Each cavity can be run at its full performance capability.

Figure 11.43 shows the SNS layout. From 186 to 1000 MeV, 800 MHz superconducting cavities drive the SNS linac. As one of the partner labs, JLab was responsible for linac design and construction. There are total of 81 (one meter-long) 6-cell cavities operating at 804 MHz. Eleven cryomodules have three medium-β (0.61) cavities each and 12 cryomodules have four high-β (0.81) cavities each. Figure 11.44 shows a cavity string and cryomodules installed in SNS.

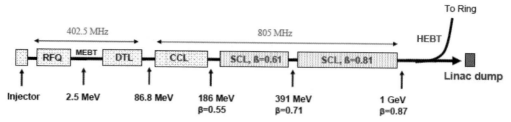

Fig. 11.43 Layout for SNS. The superconducting linac starts at 186 MeV [697].

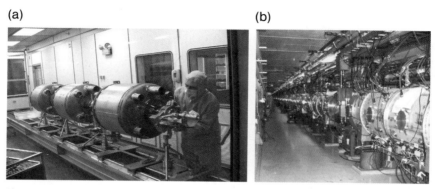

Fig. 11.44 (a) Medium-β cavity string for SNS [698] (courtesy of JLab). (b) Cryomodule installation in SNS [699].

The cavity properties have been discussed in Chapter 2. The design gradients are 10.2 and 15.6 MV/m for the medium- and high-β cavities, respectively, with a design Q above 5×10^9. Operation at 2.1 K allows a highest Q of 1.7×10^{10}. Each of the 81 cavities is powered by a pulsed 550 kW klystron via a high-power coupler, previously processed in test stands at JLab or at SNS. The cryogenic system is designed for 2.4 kW of power at 2.1 K, which provides a comfortable margin of operation for the static and dynamic losses of all the cavities leaving sufficient reserve for a possible power upgrade in the future. The upgrade would be based on an additional 36 high-β cavities (nine cryomodules).

The performance of bare cavities during qualifying tests at JLab is shown in Fig. 11.45 [700, 701]. Field emission was responsible for the wide scatter in performance. After adopting improved preparation procedures to reduce field emission, the average gradient improved significantly with experience. At the specified Q the average reached 15.5 MV/m (design 10.2), for $\beta=0.61$. The average number of tests needed to qualify a cavity dropped from 1.9 to 1.1. Continuous (cw) rf processing was used routinely to destroy early onset emitters. Figure 11.46 shows the yield of the medium-β cavities for various Q values. The yield at the specified gradient and specified Q is about 80%.

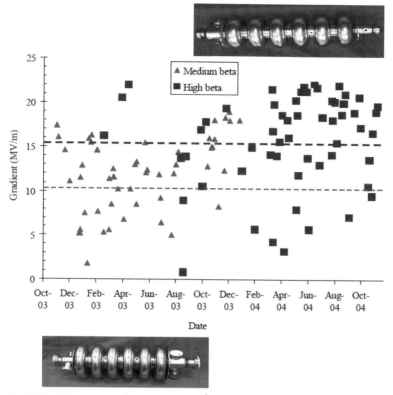

Fig. 11.45 SNS cavity gradients at $Q=5\times10^9$ measured in vertical tests. The specified gradients are indicated with dashed lines [700, 701] (courtesy of JLab). Insets are medium and high-β cavities.

The average gradient for all high-$\beta=0.81$ cavities at the specified Q was 15.8 MV/m (design 15.6 MV/m). Poor cavities were reprocessed and retested after which the average gradient for all passed cavities rose to 17.7 MV/m. The field emission onset was low (6 MV/m) even after improved rinsing procedures. Again rf processing was used routinely to improve the cavity yield. Excessive MP was observed in these cavities possibly due to HOM couplers, as discussed in Chapter 9. The average number of tests needed to qualify a high-β cavity was 1.4. The yield of vertical test results on $\beta=0.81$ cavities at various Q values is shown in Fig. 11.46. The yield at the specified gradient and Q is about 50%, hence the need for 2.4 tests to qualify a cavity. Figure 11.47(a) summarizes the final spread in peak electric field for cavities which were declared to pass the vertical test.

Several cryomodules were tested at JLab before shipping to SNS. Figure 11.47(b) compares horizontal and vertical test results showing close performance.

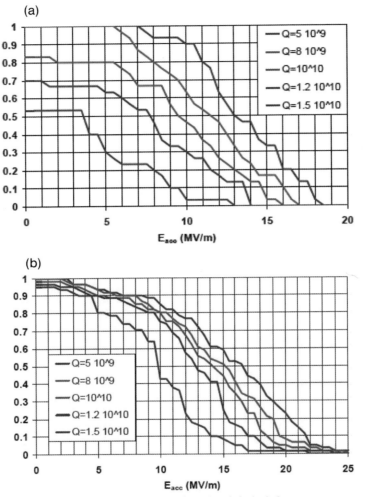

Fig. 11.46 Yield of SNS cavities: (a) medium-β and (b) high-β cavities at various Q values [700, 701] (courtesy of JLab).

After delivery and installation of all but one leaky module, SNS was commissioned in 2006 at 4.4 K operating temperature to reach the design energy of 1 GeV and 300 kW of beam on target. The maximum peak current was 20 mA, pulse length of 880 μs, at 30 Hz rep rate. More than 10^{13} protons were delivered to the target and neutron flux goals exceeded. The 60 Hz mode was demonstrated at 860 MeV with beam to target. Seventy-five cavities were powered at 2.1 K. The overall trip rate with this mode of operation is 0.1/day. While SNS is in the process of ramping up the beam power, it routinely provides neutron beams to several scattering instruments.

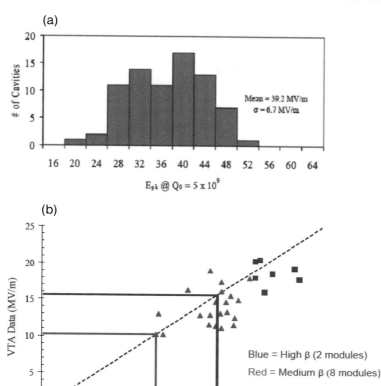

Fig. 11.47 Distribution of peak surface electric fields in vertical tests for the qualified SNS cavities [701]. (b) Cryomodule test versus vertical test results at JLab. $E_{pk}/E_{acc}=2.7$ and 2.2 for medium- and high-β cavities [700, 701] (courtesy of JLab).

On-line performance compares favorably with vertical test and CM test results at JLab before shipping. Figure 11.48 shows the gradient distribution on-line without beam. Medium-β cavities show an average maximum gradient of 17.6 MV/m (to be compared with the design value of 10.2 MV/m). The high-β cavities show an average maximum gradient of 18.0 MV/m, compared to the design value of 15.6 MV/m. Some cavities were powered to lower fields to maintain protection limits, such as need for the LLRF board. Several cavities reached fields of 20–25 MV/m. The cavity that exceeded 25 MV/m has a forward power of 510 kW peak; a record for a pulsed coupler connected to a superconducting cavity in a full cryomodule under real operating conditions.

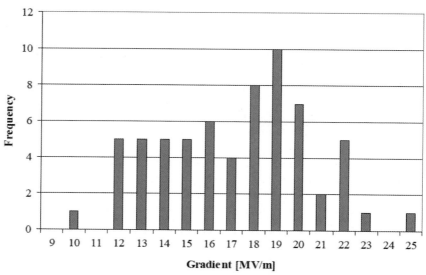

Fig. 11.48 Gradient distribution on-line without beam for 65 out of 81 cavities. The average is 17.8 MV/m [702] (courtesy of SNS).

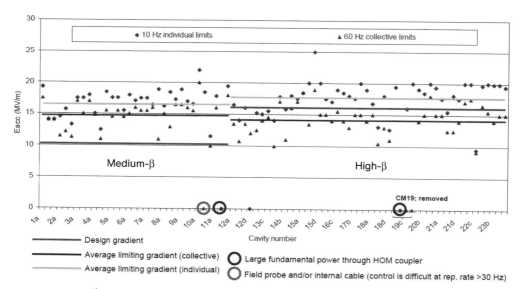

Fig. 11.49 Operational gradients for SNS in 2007. The heavy lines show the design gradients [696].

Operational gradients were maintained at 80% of maximum fields. Figure 11.49 shows the operating gradient limits set during 2007. In a number of cases the HOM power transmitted by the filter at the fundamental mode exceeded values safe for the rf feedthroughs so that cavity fields were prevented from reaching their natural limit by the excessive power going through those ports. During testing at JLab two cryomodules were vented due to failures of HOM feedthroughs subjected to excessive power, so at SNS the power is monitored and a limit imposed on the transmitted power. Improved feedthroughs will eventually lead to higher power and field levels. The Lorentz-force detuning was successfully handled even though it is quite significant in the medium-β ones. The LLRF system routinely gives better than 1%/1° amplitude/phase stability.

One of the main benefits of the superconducting approach realized was the operational flexibility available due to individually powered superconducting cavities. It is possible to "tune around" cavities with reduced gradients, or malfunctioning tuners. Given the large distribution of gradients and the uncertainty of the actual values for operation, algorithms have been developed to set up the machine for optimum utilization of the superconducting cavities available gradients [696].

SNS plans to upgrade the beam power to 2 MW by increasing the beam current from 26 to 42 mA and the beam energy from 1.0 to 1.3 GeV. This will require nine additional high-β cryomodules, which include 36 high-β cavities with improvements to the HOM and tuners.

12
Future Applications

12.1
Overview for Next Generation Light Sources

The next generation light sources will deliver orders of magnitude higher brightness and optical beam quality through multiparticle coherence, smaller source beam emittance, shorter bunch lengths, and high average currents. Such future sources will open a wide range of scientific applications, some of which were mentioned in Chapter 11 when discussing FELs. The demands for high beam quality can be met using low-emittance injectors and cw SRF-based linear accelerators. For modest average current, but high-bunch charge and low-emittance beams, a single-pass linac configuration can be used. Typical parameters are a few GeV, for nm wavelength x-rays, ~1 nC per bunch, and bunch repetition rates in the 100 kHz to 1 MHz range. The energy required for high-power FELs is in the range of 100 MeV for infrared (IR) to less than 1 GeV for ultraviolet (UV). For high currents, especially at high energy, an energy recovery linac (ERL) configuration is necessary. Average currents up to 100 mA are generally proposed at lower charge per bunch (<0.1 nC). Both single-pass and ERL configurations benefit from the efficient operation and large beam apertures afforded by cw SRF cavities.

The oscillator configuration for FELs is generally limited in wavelength to ~160 nm by the availability of mirrors. UV and x-ray lasers cannot use the optical cavity configuration. Instead, long undulators provide single-pass amplification via SASE. The light spectrum and pulse shape can be made more reproducible with a seed laser than with SASE alone. Since seed lasers do not exist at all desired wavelengths, high-gain-harmonic generation (HGHG) is proposed [703] for "upconversion." With a very short pulse (<100 fs) seed laser (e.g., Ti:Sa), the external seed overlaps the bunch and modulates the energy in a short undulator. A dispersive chicane converts this to a spatial modulation with higher harmonic content. A second undulator, tuned to a harmonic, then generates coherent radiation that seeds the next stage. The cascade repeats until the desired wavelength (down to 1.25 nm) is reached. The modulated part of the bunch radiates coherently at a harmonic of the seed laser (limited to about $n \leq 5$). The seed laser rather than the electron bunch determines the properties of the output photon pulses, yielding reproducible "clean" pulses both temporally and spectrally. In addition, several beam lines can be separately seeded providing high flexibility for a range of users. Seeding demands precise beam timing (<100 fs) and position control with constant beam quality along the bunch. Hence, the FEL output depends critically on the performance of the SRF linac and the success of the timing distribution system.

Both the single-pass and ERL approaches to future light sources require low-emittance injectors (1–10 μm normalized rms), third harmonic cavities to linearize the rf waveform (see Section 11.1.3), and very precise rf controls of the cavities, such as 0.02° in phase, and 0.02% in amplitude control. The technology of digital rf control systems has advanced to meet such needs. The Cornell digital cavity control system was tested at the JLab ERL at a current of 5 mA and external Q of 1.2×10^8, achieving an amplitude stability of about 10^{-4} and phase stability of 0.02° [487].

12.1.1
Single-Pass FELs

There are several studies in progress for future one-pass FELs to provide VUV to x-rays as reviewed in [704]. Many concepts are based on the high gradient SRF technology developed at TTF for the XFEL and the ILC; a few examples are presented.

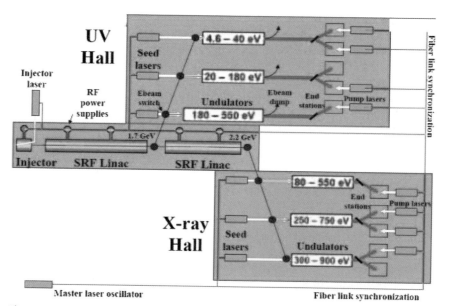

Fig. 12.1 Wisconsin and MIT x-ray laser facility [705].

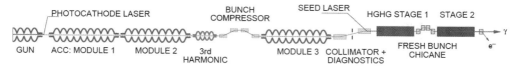

Fig. 12.2 HGHG demonstrator for the BESSY FEL [706].

A collaboration between University of Wisconsin and MIT [705] is studying an x-ray laser facility (Fig. 12.1) to span a broad range of wavelengths from the ultraviolet to the hard x-ray region using fundamental and third harmonic generations. The accelerator configuration would include a staged superconducting electron linac from 1.7 to 2.2 GeV, multiple undulators with seed lasers, and multiple beam lines. The average current is 1 mA with 1 µm emittance at 0.2 pC. All undulators operate simultaneously at repetition rate up to 1 MHz each. The advantages are the modest beam power levels and the challenges are the seeding.

At BESSY in Berlin, an initial project (STARS) is proposed to demonstrate a two-stage HGHG-cascaded FEL [706] to lase between 40 and 70 nm at a beam energy of 325 MeV (Fig. 12.2). There will be three TESLA-style superconducting modules with 20 cavities in all. The cavities and cryomodules will be modified to operate cw. Many of the components will serve as prototypes for the future BESSY-FEL. At the HoBiCaT cavity test facility, a $Q > 2 \times 10^{10}$ has been demonstrated at 1.8 K for a TTF 9-cell cavity at 17 MV/m [707]. High-Q demonstrations of this nature are essential to lower the operating cost of future cw FELs and cw ERLs. Measurements at HoBiCaT also show that the rms microphonic levels are of the order of 3 Hz or less with peak excursions around 15 Hz. At 17.2 MV/m accelerating field, a peak rf power of 5.9 kW (3.1 kW average) is necessary to deal with microphonic detuning values of 5 Hz rms and 25 Hz peak. The optimal cavity bandwidth of 50 Hz corresponds to an external coupling of 2.6×10^7, which can be readily handled by TTF-III couplers (see Chapter 8). Ultimately, BESSY proposes a single-pass 2.3 GeV FEL for wavelengths between 63 and 1.5 nm with 75 µA average current and 1.5 µm emittance at 2.5 nC bunch charge.

In France, the ARC-EN-CIEL (Accelerator-Radiation Complex for Enhanced Coherent Intense Extended Light) project (Fig. 12.3) aims at coherent, femtosecond light pulses from UV to soft x-ray [708]. A cw superconducting linear accelerator will deliver high charge, sub-ps, low-emittance electron bunches with a high repetition rate (1 kHz). This FEL also incorporates HGHG. The first phase accelerator scheme follows the TTF2 layout (see Chapter 11) with a rf gun followed by modules with two cavities each. When the electron energy reaches 100 MeV after the first module, a third harmonic cavity compensates the nonlinearity of the longitudinal phase space before bunch compression. The second module increases the electron energy to 220 MeV. The phase II plans six additional modules to reach 1 GeV. Recirculation is planned for higher energy. Beam break up issues in multipass linacs need to be addressed for recirculation.

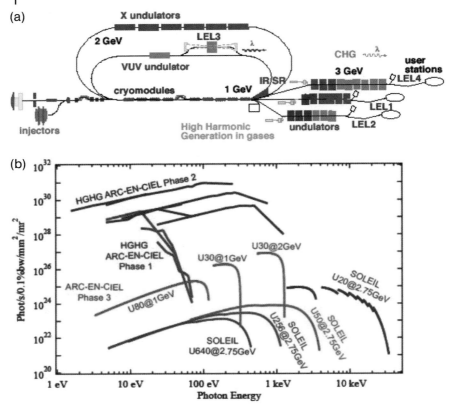

Fig. 12.3 (a) Layout of the ARC-EN-CIEL project in France.
(b) A comparison of the peak brilliance expected with that from the storage ring light source SOLEIL [708] (courtesy of Saclay and Orsay).

12.1.2
ERL-Based Light Sources

Most major synchrotron radiation (SR) sources are based on storage rings. The characteristics of x-rays depend on the qualities of electron beams available to produce the SR in the insertion devices. For a storage ring, the beam emittance, bunch profile, and energy spread are determined by an equilibrium between radiation damping and quantum fluctuations with the emission of SR. Thus, the beam emittance in electron storage rings is dominated by the optics of the ring, and not by the quality of the injected beam. The stochastic nature of synchrotron emission randomizes the motion of electrons over many turns and makes the equilibrium beam distribution poorer than the injected distribution. Linacs do not suffer such emittance dilution and are inherently capable of preserving beams with the small emittance generated by a good injector. Photo-injectors provide bunches with emittances, shapes, and length which are superior to bunches in storage rings. The resulting x-ray beams from SRF-based linacs therefore promise significantly higher spectral brightness [709].

In third-generation storage rings, such as ESRF and APS, beam characteristics are near limits. Although some improvement is possible, storage ring technology is at the point that future improvements will come with larger rings. For the next generation of high-flux light sources, it is desirable to have (1) high average current with low electron beam emittance in order to increase the flux, brilliance and coherence of SR; (2) very short electron bunches to enable fast time-resolved experiments; and (3) a SR output which does not decay over time.

An SRF-based linac provides an approach to producing high quality, high current beams. SRF linacs offer the following clear advantages. High-gradient, low-impedance structures help to preserve the beam quality produced by the injector. Linacs can ensure amplitude and phase stability of the rf fields at the 10^{-5} level. Linacs also allow operational flexibility via changes in beam energy, bunch length, and pulse patterns. After acceleration by the linac, the low emittance, high-energy bunches pass through undulators to produce SR beams with high-peak brilliance. Figure 12.4 compares the average flux and the peak brilliance of SR from ERL and storage ring light sources.

The key issue for a linac-based source, however, is the enormous power associated with beam currents required to provide radiation flux comparable to storage rings. For example, a 5 GeV, 100 mA electron beam carries 500 MW of

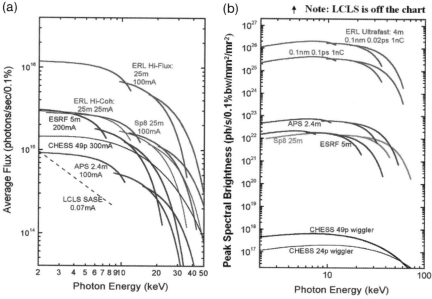

Fig. 12.4 Comparison between ERL and storage ring light sources for (a) average flux and (b) peak brilliance [711].

beam power! It is economically unfeasible to simply dump the electrons at the end of the linac. It is necessary to recover the beam energy, hence the ERL. After producing SR in the insertion devices, the electrons reenter the linac, but 180° out of accelerating phase (Fig. 12.5 a). The bunches decelerate to yield their energy back to the electromagnetic field in the linac. For the best stability, cw

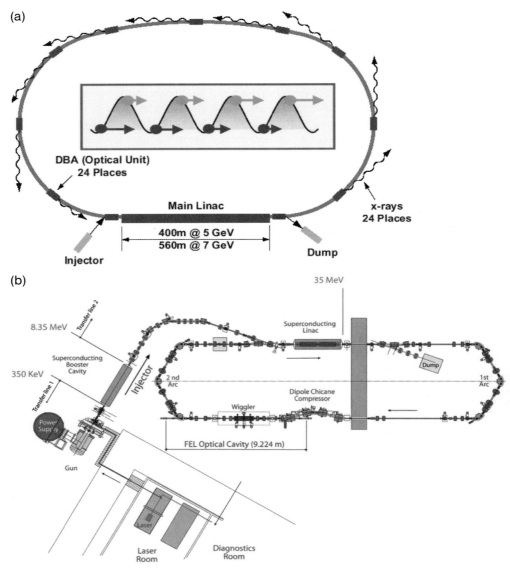

Fig. 12.5 (a) The concept of the energy-recovery-linac-based light source [713]. (b) A prototype under commissioning at Daresbury [714].

operation is a must. When bunches emerge from the linac with the low injector energy (minus SR losses), a weak bending magnet deflects them into a beam dump. The energy recovered by the linac accelerates a fresh bunch for every turn making the ERL free from bunch thermalization typical for a storage ring. In contrast to conventional linacs, where the rf power required for acceleration limits the beam current, the required rf power for an ERL is very modest and virtually independent of the beam current [710].

A storage ring keeps the power costs down by reusing the energetic electrons. Instead of recycling the beam, ERLs recycle the beam energy. Ultimately, ERLs promise efficiencies approaching those of storage rings, while maintaining beam quality characteristics of linacs: superior emittance and energy spread, and sub-ps bunches. The high current ERL enlarges the wide range of applications of third generation light sources by producing beams similar to the cw beams from modern electron storage ring facilities, and with higher brilliance due to the much smaller horizontal emittance and smaller energy spread. At the same time, it can serve more specialized experiments that require ultrasmall emittances for high spatial resolution or ultrashort bunches for high temporal resolution.

Other important issues for the high current ERL include the generation of high average current, high brightness electron beams; acceleration and transport of these beams while preserving their brightness; adequate damping of higher order modes (HOMs) to assure beam stability; removal of large amounts of HOM power from the cryogenic environment; stable rf control of cavities operating at very high external Q; reduction of beam losses to very low levels; and the development of precision diagnostics for beam setup, control, and characterization. Many laboratories are engaged in programs to address these challenges [662, 663].

As discussed in Chapter 11, Jefferson Lab and Japan Atomic Energy Agency used superconducting cavities to demonstrate energy recovery for low energy (50 and 17 MeV) and low current (5 mA) beams. For the UV upgrade, more than 1.3-MW beam power was energy recovered with a beam of 9.1 mA at 150 MeV. These laboratories showed that they could save most of the energy required for beam acceleration up to a beam power of about one MW. A 1-GeV experiment in CEBAF [667, 712] also demonstrated recovery at high beam energy.

Table 12.1 gives the main parameters of several ERLs, some planned, and some operating.

Table 12.1 ERL characteristic examples.

Accelerator	ERLP	4 GLS	Cornell ERL	TJNAF ERL	JAERI ERL	BNL ERL
Avg current (mA)	0.013	100	100	9.1	5.2	500
Frequency (GHz)	1.3	1.3	1.3	1.5	0.5	0.703
E_{acc} (MV/m)	13.5	15.2	16	20	7.5	13
Power/Cavity (kW)	4	10	8	2.5	50	50

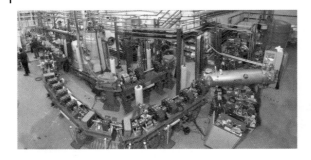

Fig. 12.6 Layout of the ERLP which is a prototype for a future ERL-based light source [716] (courtesy of Daresbury).

Daresbury (UK) is constructing a prototype for a next generation light source project [715]. The prototype aims to demonstrate ERL technologies including gun, superconducting injector, linac, FEL, laser, and timing aspects. Figure 12.5(b) shows the layout of ERLP, and Fig. 12.6 shows the facility status. Two superconducting modules provided by ACCEL are used, one for the injector (8.35 MeV) and the other for the linac (35 MeV). The module uses two TESLA-style 9-cell cavities and is identical to the Rossendorf FEL module (see Chapter 11). A 350-kV dc gun provides the starting beam for the first module, which then injects electrons into the recovery loop. The gun provides a pulse train of 80 pC × 81.25 MHz × 100 ms × 20 Hz.

Cornell is investigating a light source driven by an energy recovering SRF linac operating at 5 GeV with an average current of 100 mA [709, 717]. KEK is considering a 2.5–5 GeV, 100 mA ERL light source along the design of the Cornell ERL [672].

The Cornell ERL is planned as an extension to the CHESS (CESR) ring (Fig. 12.7). The ERL design has two 2.5 GeV linacs. An injector sends 100 mA of

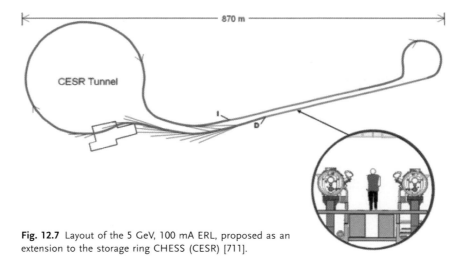

Fig. 12.7 Layout of the 5 GeV, 100 mA ERL, proposed as an extension to the storage ring CHESS (CESR) [711].

1–2 μm emittance, 2 ps long bunches into the first linac which accelerates to 2.5 GeV, and a return loop leads the beam back to the second linac that accelerates further to 5 GeV. The high-energy beam runs through an arc with several x-ray beamlines, before it enters the CESR ring with several additional x-ray beamlines. Subsequently the beam reenters the first linac, decelerates to 2.5 GeV, turns around once again in the return loop, and finally decelerates to 10 MeV to end at the beam dump. Because of energy recovery, the rf power demands are very modest for the 100 mA beam. To establish the required field (15–20 MV/m), the input couplers need to deliver about 5 kW peak power (2 kW average), depending on the chosen operating field level and the microphonics detuning level as discussed in Chapter 8. CW operation means a high dynamic heat load and the helium refrigerator becomes a dominant cost issue, demanding operation at Q_o in 2×10^{10} range. The optimization of a cw machine in terms of capital and operating costs pushes the optimal gradient to a low end of the range [718].

Since neither an electron source, nor an injector system has ever been built for the required large beam current and small transverse and longitudinal emittances, Cornell is building a prototype ERL injector [719–722]. A dc electron gun will send bunches with 77 pC and emittances less than 1.5 μm into a 5-cavity injector linac which will accelerate bunches from 0.5 MeV to 5 MeV with minimal emittance growth. There will be one cryomodule with five 1300 MHz, 2-cell cavities, each providing 1 MV of acceleration, corresponding to an accelerating field of about 5 MV/m in cw operation (Fig. 12.8) [720–722]. To explore the full capabilities of the injector, energy gains up to 3 MV per cavity will be tested at lower (e.g., 33 mA) beam currents. Consequently, the total beam energy gain will be 15 MeV. Besides standard features such as an integrated helium vessel and mechanical tuner, each cavity has two input couplers, symme-

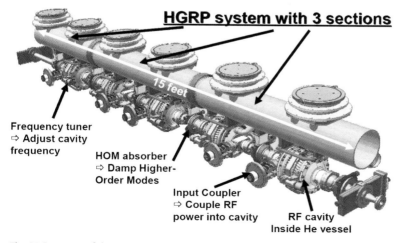

Fig. 12.8 Layout of the 5-cavity injector module for the ERL at CESR [721].

trically placed on the beam pipe to cancel kicks due to coupler fields (see Fig. 8.2). For a 100 mA maximum injected beam current, each coupler must deliver 50 kW of beam power leading to a Q_{ext} of 3×10^4 for matched beam loading conditions. The input coupler is adjustable by a factor of 10. The beam pipe on one side of the cavity is enlarged so that all monopole and dipole modes propagate out of the cell. Ferrite beam pipe loads cooled to liquid nitrogen temperature are positioned on the beam line at a safe distance past the helium vessel. Such loads are described in Chapter 9. More than a 100 W of beam-induced power must be removed through HOM absorbers.

12.2
ERL for Nuclear Physics

Over the past two decades, nuclear science has made important strides in mapping the hadronic structure and nuclear spin structure. Investigations continue to answer crucial questions such as: What is the quark–gluon structure of the proton and neutron? How do quarks and gluons evolve into hadrons, as during the later stages of the Big Bang? What is the quark–gluon origin of nuclear binding? High-luminosity electron–ion colliders beyond HERA at DESY are under exploration as powerful new microscopes to continue to probe nuclear structure. The potential benefits of high-current ERLs have spurred explorations worldwide for electron cooling applications to increase the luminosity of heavy-ion colliders, and for electron–ion colliders.

The relativistic heavy-ion collider (RHIC) at BNL is a planned luminosity upgrade (RHIC-II) via a high-power electron beam to cool the ion beam (Fig. 12.9 a). Electrons with a high bunch charge (5 nC) and a relatively low emittance (3 μm) are introduced into the ion beam. Energy exchange between ions and electrons results in a decrease in the longitudinal and transverse emittance of the ion beam. Such a high current (several hundred milliampere) and high-energy electron beam calls for an energy recovery superconducting linac to minimize power consumption. A module with two 5-cell 704 MHz cavities is proposed to accelerate the electron beam from the gun at about 5 MeV to 54 MeV in a two-pass ERL [723–725].

BNL's future plans also include a high-energy, high-luminosity electron–ion collider (Fig. 12.9 b), eRHIC, to provide 3–20 GeV polarized electrons to collide with high-energy polarized protons, gold ions, and polarized ^{3}He ions. A 4-GeV electron accelerator will be used in a five-pass ERL to provide 20 GeV electrons. An important advantage of the ERL over a storage ring is the high polarization (>80%) available from the linac. The electron current required will be 0.25 A at 10–20 nC bunch charge. The linac will be based on a 703.75 MHz, 5-cell cavity under development (Fig. 9.10). The ERL design assumes 20 MeV energy gain per cavity with 200 cavities in a 600-m linac, consisting of four 150-m-long sections that fit in the RHIC straight sections. The cavity has a low longitudinal loss factor (1.1 V/pC) due to the low frequency and large apertures of 19 and

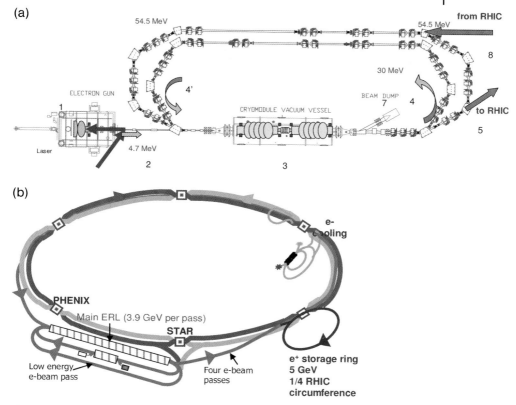

Fig. 12.9 (a) ERL for electron cooling the RHIC beam [723]; (b) ERL for an electron–ion collider [724] (courtesy of BNL).

24 cm diameters. All HOM modes are damped to Q_s of lower than 10^4 (most down to 10^2) with a Cornell type beam-line ferrite damper. Simulations show about 2-A threshold current for the eRHIC ERL in multipass mode.

At JLab, the electron-light-ion collider (eLIC) concept (Fig. 12.10) is a hybrid between ring–ring and linac–ring [726]. It stores the electron beam for ∼100 turns in a circulator ring (CR). The potential advantages are: less electron beam disruption than for a ring–ring option and 100× less average linac current, thus reducing demands on the electron source.

12.3
Nuclear Astrophysics

The next generation radio isotope beam (RIB) facility aims to provide record intensities of beams to derive quantitative information necessary to evaluate theories of stellar evolution and the formation of elements in the cosmos, such as

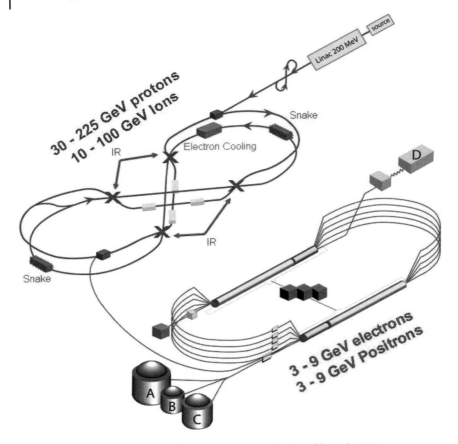

Fig. 12.10 The electron-light-ion collider (eLIC) concept developed by JLab [726].

the origin of heavy elements in supernovae. Nuclear science for predicting the properties of nuclei with unusual neutron-to-proton ratios will be advanced. In addition, the broad range and quantity of radioisotopes produced will provide rich venues in radio-medical research and materials science. US-based studies are described under the Rare Isotope Accelerator [690, 727, 728] (RIA) and European-based studies under EURISOL [729–731]. Compared to existing and near-term facilities such as GSI, SPIRAL2 at GANIL, and RIKEN RIBF, the isotope reach and intensities will be much higher.

12.3.1
Rare Isotope Accelerator (RIA)

A driver linac will accelerate stable isotope from protons through uranium to energies of 400 MeV/u and beam intensities in the 100-kW range. RIA's performance will allow the study of a large number of exotic isotopes. The highly flexible driver linac will provide a variety of beams to utilize combinations of projectile fragmentation, target fragmentation, fission, and spallation to produce a broad assortment of short-lived isotopes. Lighter elements will be used to produce radioactive ion beams by the Isotope Separation On-Line (ISOL) method.

To reach intensity goals, there will be two major sections to RIA. The first stage will be a superconducting multi-ion, multicharge-state driver linac, spanning nearly the entire range of masses from protons to uranium and particle velocities, $0.02 < \beta < 0.84$. For the second stage, a superconducting postaccelerator will provide efficient acceleration and transmission of multicharge-state rare isotopes. The 1.4-GeV driver linac will deliver 100-kW beams on to production targets at energies of 400 MeV/nucleon for uranium, and more than 900 MeV for protons.

At least two groups have been actively pursuing designs for RIA. One group [83] is based at Michigan State University (MSU) and the other is at Argonne National Laboratory (ANL) [67, 68, 727]. Both designs realize the basic requirements. Structure types for both versions have been discussed in Chapter 2. Both designs have a driver linac with initial acceleration provided by normal conducting technology. Some critical issues in the final choice of RIA superconducting cavities will be beam dynamics, microphonics, and rf phase control.

12.3.2
EURISOL Driver

Studied by a large community involving many European nuclear physics laboratories, EURISOL [730] is a project for a European radioactive beam facility based on a 1 GeV, 5 mA cw proton linac. Figure 12.11 shows a tentative layout. The multi-MW beam on a uranium target produces a high flux of radioactive ions by means of ISOL techniques; the radioactive ion beam is then reaccelerated up to about 100 MeV/u by a superconducting linac. The baseline design for the driver beam is superconducting above 85 MeV beam energy. Several schemes are under study for the low and intermediate-β sections, based on a combination of HWR, spoke and elliptical cavities. Structure development work is being carried out at INFN and IPN Orsay. Ladder resonators (see Section 2.3.2) are also under consideration. The front-end of a pulsed EURISOL driver is now under construction at INFN Legnaro (Italy) under the name of SPES [95].

12.4
High-Intensity Proton Linacs

High-intensity superconducting proton linacs will likely fulfill future needs in a variety of arenas: Upgrading the injector chains of proton colliders and accelerators, heavy-ion radioactive beams for nuclear physics, medical therapy, industrial applications, high-intensity spallation neutron sources, transmutation applications for treatment of radioactive nuclear waste, nuclear energy production

402 | 12 Future Applications

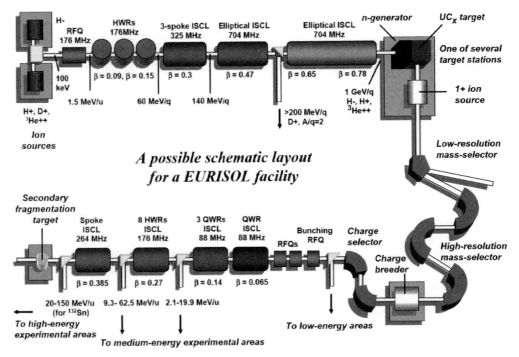

Fig. 12.11 A tentative layout for EURISOL (courtesy of CERN).

using thorium fuel, high-intensity neutrino beam lines, neutrino factories, and muon colliders [730–732].

In a generic definition, a high-intensity proton accelerator delivers proton beams of several milliampere, at an energy between 50 and 2000 MeV. Developmental topics of such machines are the presence of high beam power, space-charge dominated nonrelativistic beams with possible halo formation. Accelerating structures need development for nearly every β value [731]. Typical requirements are low beam losses (<1 W/m) to avoid excess activation, high reliability and, in some cases, the absence of long (>0.3 s) beam interruptions. Although cyclotrons could reach the desired performance with a few milliampere, only linacs are competitive in the higher current range. High average currents (e.g., 100 mA) and cw operation call for all superconducting linacs.

The generic scheme for a high-intensity proton linac is similar to the SNS layout (see Fig. 11.43). This includes a normal conducting injector and an RFQ ($\beta \leq 0.1$), a low- and intermediate-energy section ($0.1 \leq \beta \leq 0.5$) containing the normal to superconducting transition, and a superconducting high-energy section. At the low-energy end after the RFQ, normal conducting structures remain competitive, especially for pulsed operation. The optimum transition energy to superconducting depends on the linac parameters (beam current, time structure, and reliability requirements). Short, low-β superconducting cavities offer advantages of wide velocity acceptance, acceleration of different mass/charge beams (e.g., protons and deuterons), and a more cavity-fault tolerant linac with small loss in total voltage in the case of individual cavity failures. A large number of accelerating gaps in a single structure (typical for normal conducting cases) cannot be completely utilized at low velocity due to rf defocusing. This favors short structures, the strong feature of the superconducting choice. Quarter-wave resonators and two-gap spoke cavities are typical below $\beta=0.5$. Elliptical multicell cavities are the candidates for $\beta \geq 0.5$. Efficiency, gradient, and aperture are larger in superconducting cavities than in normal conducting ones. In addition, the superconducting approach allows a flexible, modular design with excellent vacuum.

12.4.1
Japan Proton Accelerator Research Complex (JPARC)

The Japan Proton Accelerator Research Complex (JPARC), operating in 2007, has a final 180-MeV linac section consisting of normal conducting accelerating structures [733]. In a future second phase, a superconducting linac from 400 to 600 MeV is envisioned [734]. The upgrade linac is based on 972 MHz, 9-cell elliptical shape resonators with $\beta=0.72$ developed in collaboration with JAERI and KEK. A 30 mA H$^-$ beam, pulsed at 25 Hz with 1.5% duty cycle, will be accelerated by 11 cryomodules containing two cavities each. RF couplers have been tested up to the pulsed power of 2.2 MW. A prototype cryomodule with $\beta=0.725$ (424 MeV) has been successfully tested at 2 K. Accelerating gradients of 12 MV/m exceeding the specification were stably achieved in pulsed operation at 2.1 K. Lorentz-force detuning in pulsed operation was systematically investigated, and mechanical vibration modes were measured. Compensation of Lorentz-force detuning by a piezotuner was successfully demonstrated.

12.4.2
Proton Driver

Fermilab has been exploring a future 8 GeV Proton Driver based on a superconducting linac [735] which could supply protons to the Main Injector to produce super beams for neutrinos and intense beams for antiproton production. The idea has evolved into Fermilab's Project X. The main goal is a fivefold increase in the proton intensity with an average beam power of about 2 MW. The linac could also directly produce muons for a future muon collider.

The Proton Driver needs many different types of superconducting accelerating cavities spanning the full range of velocities to accelerate protons from 15 MeV to 8 GeV. The driver is based on about 400 independently phased resonators. The front end would use both room temperature and short superconducting cavities alternating with focusing elements. The warm section includes the ion source, RFQ, and 5-cell cross-bar H-type cavities. The cold section consists of single, double, and triple-spoke resonators accelerating an H ion beam up to 410 MeV. The front-end frequency (325 MHz) is at the fourth subharmonic of 1300 MHz, the high-energy rf frequency chosen to be identical with XFEL and ILC. Multispoke cavities have been developed at 345 MHz in collaboration with Argonne, and demonstrated excellent performance. These will be modified to 325 MHz (same rf frequency as JPARC). At the high end of the medium-β range, from approximately 400 MeV to 1.2 GeV, there will be about 60 squeezed ILC-style 8-cell cavities designed for $\beta=0.81$. Near $\beta=1$, a full three quarters of the proton driver comprises 1.3-GHz ILC-type elliptical cavities.

For an H$^-$ linac with β fixed in coarse increments, it is necessary to phase the rf separately for each cavity. For the 1-GeV linac, SNS obtains maximum phase control by feeding each cavity with a separate klystron. This approach would become quite expensive for an 8-GeV machine. A cost-effective rf distribution system, with one 10-MW klystron driving many cavities, requires electronically controlled ferrite phase shifters [736] to provide phase and amplitude control on individual cavities. The phase and power adjustment is made with a so-called E–H tuner, which consists of a magic T with two of its arms loaded with biased ferrite. Prototypes are under development.

12.4.3
SRF Proton Linac (SPL)

CERN is studying a staged multipurpose SRF proton linac (SPL) to upgrade the injector chain for reaching the full luminosity potential of LHC, to produce heavy-ion radioactive beams for nuclear physics, to provide a superneutrino beam, and the proton source for a possible neutrino factory [732]. The next stage will be a 4 GeV, 200 kW beam power superconducting linac, with a repetition rate of 2 Hz. Two families of superconducting cavities ($\beta=0.65$ and $\beta=1.0$) will accelerate the beam to its top energy. A final extension to 5 GeV will increase the rep rate to 50 Hz and the beam power to 4 MW. In this stage, the beam can be used for the production of neutrinos, and to drive a pulsed RIB facility.

12.4.4
Transmutation Applications

Over the last 50 years, nuclear power and nuclear materials production have generated a large quantity of radioactive wastes presenting a serious problem for safe disposal. To store these in a geological repository is fraught with grave uncertainties because of the 10 000-year lifetime of some of the radioactive waste products. The danger may be reduced if the long-lived species can be transmuted to isotopes with short life. In accelerator-based transmutation of waste (ATW), spallation neutrons transmute long-lived actinide isotopes and fission products to stable isotopes, or to isotopes that decay to stable products over 100 years. This approach can lessen the technical problems of storing long-lived high-level radioactive waste. In an optimistic design, a single accelerator can burn the waste from ten 1-GW reactors, while providing enough power to run itself.

For a 5–10-MW proton beam [737], a superconducting linac significantly lowers the total linac ac power requirement. For example, when Los Alamos studied the Acceleration Demonstration Test Facility consisting of a 600-MeV linac at 13 mA, it concluded that a superconducting linac would need 23 MW of ac power as compared to 80 MW needed for a warm linac. The SRF cavities also have a large bore radius that relaxes alignment and steering tolerances, as well as reducing beam loss.

For dedicated transmutation systems, a subcritical system using external neutrons is attractive: It allows maximum transmutation rate with safe operation. An accelerator-driven system (ADS), coupling a proton accelerator, a spallation target, and a subcritical core [731], could be used as such a "reactor." The high reliability (no beam interruptions longer than 0.3 s) was judged to be an important requirement. The trip-rate issue could be a potential "show-stopper" for ADS. More than a few beam trips per year can significantly damage reactor structures, target, and fuel. High reliability and fault tolerance approaches are therefore some of the important development subjects [731].

A superconducting linac has highly modular systems where the short individual components can be operated well below their performance limits. Independently controlled rf modules with redundancy allow close adjustment of rf phases and amplitudes to compensate for faults of individual cavities, klystrons, or focusing magnets.

A European consortium is studying a reference accelerator for XADS [730], a 10-mA cw, 600 MeV (tunable) proton linac based on superconducting linac technology. The linac is superconducting above 100 MeV, with about a hundred 5- and 6-cell cavities. Bulk niobium elliptical cavities ($\beta=0.65$), operating at 704 MHz and cooled by superfluid helium at 2 K, are used as accelerating structures in the high-energy section of this linac. The input coupler design needs to be optimized to achieve the needed performance and required reliability.

In South Korea, KAERI is studying KOMAK, a multipurpose 1 GeV, 20 mA cw accelerator complex for basic research, waste transmutation, medical therapy, and industrial applications [738]. The project has been launched to 100 MeV.

With abundant thorium resources, India is considering development of ADS systems for nuclear energy production using thorium fuel. A multipurpose facility, including an ADS and a neutron spallation source, is envisioned at CAT Indore [739] to develop eventually into a 1 GeV, 10 mA SC linac starting first with the technologies for a 100-MeV linac. CAT is studying a superconducting design based on 352-MHz reentrant cavities [740].

12.5
International Linear Collider

Understanding the basic features of energy, matter, and space–time remains the focus of elementary particle physics. Both hadron and electron machines continue to play complementary roles. Answering some of the dominant questions in particle physics will require accelerators with energies and luminosities beyond current capabilities. The driving questions are: what is the origin of mass; what is the mechanism of electro-weak symmetry breaking; is supersymmetry a feature of our universe; what is the nature of dark matter and dark energy; can four dimensions describe the universe; can the fundamental interactions be completely unified?

To complement the LHC, the center-of-mass (CM) energy for the next lepton collider will be in the range of 1–5 TeV, a factor of at least 5 over LEP-II. In the past, however, the energy steps for lepton colliders have been smaller, because the needed technologies are only mastered in incremental stages. According to the above arguments, the widespread consensus is that the next lepton collider should have an initial CM energy of 500 GeV, and luminosity in excess of 10^{34} (cgs units). It should eventually be capable of reaching 1 TeV. The initial collider would have the potential for discovery of the much-sought-after Higgs particle with its connection to the origin of mass. It will also provide well-controlled experimental conditions for precise measurements to elucidate why the electromagnetic and weak forces are so different in nature and strength.

At the TeV energy scale, a storage ring larger than LEP-II becomes unaffordable because energy losses from synchrotron radiation increase with the fourth power of the beam energy. On the other hand, for a linear collider, the length and rf power scale linearly with energy for a fixed gradient.

In August 2004, the International Technology Recommendation Panel (ITRP) selected superconducting technology for the International Linear Collider [741]. One attractive feature of superconducting approach is that the peak rf power/m is low compared to the normal conducting option. Since rf dissipation in the structure walls is miniscule, mainly the beam power needs to be supplied with rf sources. With respect to the goal of high luminosity, an important advantage that stems from superconductivity is the affordability of low-rf-frequency (1.3 GHz) structures. For the same length of accelerating structure, and the same accelerating voltage, the rf energy stored in a structure increases as the square of the wavelength; but the large amount of stored energy in a superconducting structure becomes affordable because rf dissipation is low. The large aperture at low frequency yields the pleasant consequence of low short- and long-range wakefields, avoiding the main enemies of high luminosity. Low-wall losses also permit a long rf pulse length, so there can be a long separation between bunches. This offers the possibility of measuring individual bunch position variations, and to make corrections to subsequent bunches. Such corrections can be made a few times along the linac.

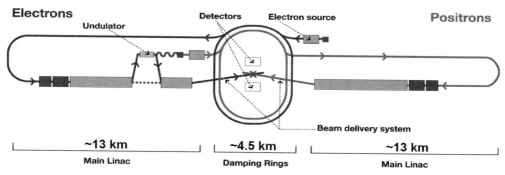

Fig. 12.12 Schematic layout of the 500 GeV ILC [743].

The superconducting linear collider collaboration has made significant advances during the past decade, as discussed for the development of TESLA-style cavities, couplers, and modules. More than hundred 9-cell structures have been produced by industry. The use of new techniques, electropolishing and mild baking (100–120 °C), have provided proof-of-principle gradients between 35 and 40 MV/m (see Chapter 7).

The ILC layout (Fig. 12.12) shows the major accelerator systems. Electron and positron sources produce the beam. The large beam power demands a copious source of positrons. Damping rings reduce the beam emittances by synchrotron radiation. Bunch compressors reduce the bunch length. The main linacs accelerate electrons and positrons to the desired energy without significantly degrading the emittances from the damping rings. The linac gradient choice is 31.5 MV/m. Figure 12.13 shows the two-tunnel layout, one for the accelerator and the other for rf power sources. The final focus system demagnifies the beam to several nanometers in size for final collision [742].

The peak design luminosity goal is 2×10^{34} cgs units, with an availability of 75%. The average beam current during the rf pulse (800 ms) is 9 mA, and the

Fig. 12.13 Two-tunnel layout for the ILC. Each main linac tunnel houses 840 cryomodules and the parallel rf power tunnel houses 280 klystrons. The tunnel diameter is 4.5 m [742].

overall ac power consumption is 230 MW. Several parameter sets have been selected. The range of beam parameters defined span 1000–6000 bunches, bunch charge in the range 1×10^{10} to 2×10^{10} per bunch, and the beam power range is from 5 to 11 MW. With bunch length between 200 and 500 μm at the interaction point, the final vertical spot size range is 3.5–9 nm.

The ILC design is based on a cryomodule average gradient of 31.5 MV/m. Each linac is roughly 11 km long with 280 rf units to accelerate electron and positron beams from 13 GeV to 250 GeV. Each rf unit is planned to power three cryomodules, two with nine 9-cell cavities each and one with eight 9-cell cavities plus the focusing quadrupole and beam position monitor package. All modules are 12.65 m long. The overall effective filling factor is ~67%. The rf distribution system will provide ~310 kW per cavity. The ILC design is based on two 4.5 m tunnels with active components in service tunnel for access. The tunnels are sized to allow for passage during installation with personnel cross-over every 500 m.

As proof-of-principle, several 1-m-long cavities have been operated individually in cryomodules at the DESY cryomodule test bed between 32 and 35 MV/m [294, 744]. Several tens of meters of superconducting cavities in FLASH accelerator modules have been operated at the 25 MV/m level. These results are discussed in more detail in Chapter 11.

ILC R&D is focused on reaching a high yield (more than 80% in the first test) at a cavity gradient of 35 MV/m, and a cryomodule average gradient of 31.5 MV/m [745]. For a good production batch of about 30 cavities each the yield for 35 MV/m and $Q = 10^{10}$ is about 50%, and needs to be increased substantially [745]. Both quench and field emission are responsible for a wide spread in gradient. R&D is in progress at several laboratories to improve the yield of cavity gradients by addressing the sources of quench and field limitations. Ethanol rinsing and ultrasonic degreasing after final chemistry is promising to lower the field emission, as discussed in Chapter 7. The production and preparation of 1000 cavities and 100 cryomodules for XFEL will be an important contribution toward ILC goals [746].

12.6
Muon Collider and Neutrino Factory

Electron–positron colliders beyond 1–2 TeV CM energy are likely to be limited by a high background from beamstrahlung [747]. Proton colliders beyond 14 TeV CM are likely to be limited by the sheer size of the multihundred kilometer circumference. Being 200 times more massive than electrons, muons do not suffer from beamstrahlung limits [748]. Muons can therefore be accelerated in recirculating linacs with several turns [749]. A 3-TeV muon collider is likely to fit on a site such as Fermilab or CERN due to the multifold layout of recirculators. However, the number of components such as cavities and magnets will still be very large. Interest in the muon colliders is growing. The path to a multi-TeV muon collider is filled with many challenges. Muons are unstable. Problems from muon decay are the heat load, large detector background, and neutrino radiation hazards. The last adversity can be turned into the fortune of an intense neutrino source.

As a first step, the muon collaboration [750] is interested in developing a muon-storage-ring-based neutrino factory. Atmospheric neutrino, solar neutrino, and short baseline accelerator experiments have accumulated evidence to show that neutrinos have a small but finite mass. According to theory, neutrinos with mass should oscillate in flavor, which opens up exciting fields of neutrino flavor physics, such as the search for CP violation in neutrino interactions. The goal of the neutrino factory would be to provide 3×10^{20} muon decays per year.

Figure 12.14 shows a generic layout for 20-GeV neutrino factory. Newer ideas for cooling are flourishing [748]. The muon acceleration system starts with a linac from about 200 MeV to 2.5 GeV followed by a 4-pass recirculating linac to the final energy of 20 GeV. Some designs call for a final energy of 50 GeV with a second recirculating linac. A preaccelerator linac is necessary in the first stage because the beam is not relativistic and phase slippage in a recirculating linac would reduce acceleration efficiency.

Acceleration of a muon beam is challenging because of the large phase space and short muon lifetime. The need for very large beam acceptances drives the rf design to a low frequency of 200 MHz. To minimize muon loss from decay, the highest possible gradient is necessary. At gradients of 15 MV/m SRF reduces the peak rf power by virtue of long fill times made affordable by superconductivity. SRF cavities also provide a large aperture that helps preserve beam quality and beam stability.

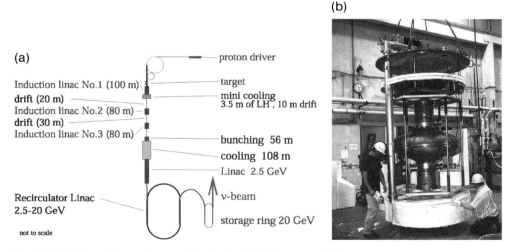

Fig. 12.14 (a) Generic layout for a neutrino factory [750] (courtesy of BNL). (b) A CERN-Cornell 200 MHz Nb–Cu cavity for the muon collider or neutrino factory [751].

The neutrino factory calls for several hundred meters of 200 MHz cavities to provide gradients of 15 MV/m. Nb–Cu is the technology of choice driven by the cost factor. The success of 400 MHz LHC cavities in reaching comparable gradients is encouraging.

A CERN–Cornell collaboration designed, fabricated, and tested a 200-MHz single-cell Nb–Cu cavity (Fig. 12.14). In the first unit, an accelerating field of 10 MV/m was reached at a Q near 10^9 [751]. The strong Q-slope typical for Nb films has limited performance.

The role of rf superconductivity for a muon collider was described in [1] and is only briefly covered here. A muon collider concept has been put forward for 4 TeV CM at a luminosity of 3×10^{34} cgs units. Many aspects of the system are still in need of substantial exploration, such as cooling the muons. After cooling and preacceleration, the beam will be accelerated to full energy using a cascade of superconducting recirculators that would accelerate the beam in stages. The rf frequency would increase as the bunch length decreases. An early stage could be based on 200 MHz superconducting cavities under development for the neutrino factory. The final stage would be based on several kilometers of 1.3-GHz linear collider style superconducting cavities operating in the pulsed mode with 1% duty factor and a gradient of 20 MV/m.

12.7
Concluding Remarks

Superconducting cavities have been operating routinely in a variety of accelerators with a range of demanding applications. With the success of completed projects, niobium cavities have become an enabling technology offering upgrade paths for existing facilities, and pushing frontier accelerators for nuclear physics, high-energy physics, materials, and life sciences. With continued progress in basic understanding of superconductivity, the performance of cavities has steadily improved to approach theoretical capabilities. Besides offering exciting options for traditional applications, superconducting cavities are branching out into new applications to light sources, neutron sources, and high intensity proton sources to fulfill a variety of needs. The most ambitious application for niobium cavities is the superconducting linear collider for high-energy physics.

We can remain confident that the rf superconductivity community has both the creativity and determination to face the upcoming challenges and successfully realize these exciting prospects.

References

1. H. Padamsee, J. Knobloch and T. Hays: *RF Superconductivity for Accelerators* John Wiley & Sons, New York (1998), ISBN 0-471-15432-6.
2. P. Schmueser: Basic Principles of RF Superconductivity and Superconducting Cavities, *Proceedings of 11th Workshop on RF Superconductivity*, Travemünde (2003), paper MoTo1.
3. H. Padamsee: The Science and Technology of Superconducting Cavities for Accelerators, *Supercond. Sci. Technol.* **14**, R28 (2001).
4. H. Padamsee: The Quest for High Gradient Superconducting Cavities. In: *Frontiers of Accelerator Technology*, S. I. Kurokawa et al. (eds), World Scientific, Singapore (1996), p. 383.
5. D. Proch: Superconducting Cavities for $V_p=c$ Linacs, Storage Rings, and Synchrotrons. *Handbook of Accelerator Physics and Engineering*, World Scientific, Singapore (1999), p. 530.
6. K. Shepard: Superconducting Cavities for $V_p<c$, Linacs, *Handbook of Accelerator Physics and Engineering*, World Scientific, Singapore (1999), p. 537.
7. H. Padamsee: Superconducting RF Cavities. In: *Encyclopedia of Electrical and Electronics Engineering*, J. G. Webster (ed), John Wiley & Sons, New York (1999).
8. D. Proch: Superconducting Cavities for Accelerators, *Rep. Prog. Phys.* **61**, 431 (1998).
9. H. Padamsee: Accelerating Applications of RF Superconductivity – Success Stories, *Proceedings of 2004 Applied Superconductivity Conference*, Jacksonville (2004), p. 2432.
10. W. Weingarten: Superconducting Cavities – Basics. In: *Proceedings of the Joint US-CERN-Japan International School, Frontiers of Accelerator Technology*, S. I. Kurokawa et al. (eds), World Scientific, Singapore (1994), p. 311.
11. H. Padamsee and J. Knobloch: Issues in Superconducting RF Technology, *Proceedings of the Joint US-CERN-Japan International School, Frontiers of Accelerator Technology*, S. I. Kurokawa et al. (eds), World Scientific, Singapore (1994), p. 101.
12. H. Padamsee et al.: Physics and Accelerator Applications of RF Superconductivity, *Ann. Rev. Nucl. Sci.*, **43B**, 635 (1993).
13. W. Buckel: *Supraleitung*, VCH Verlagsgesellschaft, Weinheim (1990).
14. D. R. Tilley and J. Tilley: *Superfluidity and Superconductivity*, Institute of Physics Publishing, Bristol (1990).
15. N. W. Ashcroft and N. D. Mermin: *Solid State Physics*, W. B. Saunders, Philadelphia (1976).
16. J. Bardeen, L. N. Cooper and J. R. Schrieffer: Theory of Superconductivity, *Phys. Rev.* **108**, 1175 (1957).
17. S. Belomestnykh and V. Shemelin: High-Cavity Design, *Proceedings of 12th Workshop on RF Superconductivity*, Ithaca (2006), p. 2.
18. J. R. Delayen: Low and Intermediate Beta Cavity Design – A Tutorial, *Proceedings of 11th Workshop on RF Superconductivity*, Travemünde (2003), paper TuT01.
19. J. R. Delayen: Medium- Superconducting Accelerating Structures, *Proceedings of 10th Workshop on RF Superconductivity*, Tsukuba (2001), p. 152.

20. A. Facco: Tutorial on Low-Beta Cavity Design, *Proceedings of 12th Workshop on RF Superconductivity*, Ithaca (2005), p. 21.
21. A. Facco: Low and Medium Beta Superconducting Cavities, *Proceedings of EPAC 2004, European Particle Accelerator Conference*, Lucerne (2004), p. 142.
22. M. Kelly: Overview of TEM-Class Superconducting Cavities For Proton and Ion Acceleration, *Proceedings of LINAC 2006, Linear Accelerator Conference*, Knoxville (2006), p. 23.
23. C. Pagani et al.: Design Criteria for Elliptical Cavities, *Proceedings of 10th Workshop on RF Superconductivity*, Tsukuba (2001), p. 115.
24. E. Haebel et al.: Cavity Shape Optimization for Superconducting Linear Collider, *Proceedings of HEACC 1992, High Energy Accelerator Conference*, Hamburg (1992).
25. B. Aune et al.: Superconducting TESLA Cavities, *Phys. Rev. ST AB*, **3**, 092001 (2000).
26. H. Padamsee et al.: Design Challenges for High Current Storage Rings, *Part. Accel.* **40**, 17 (1992).
27. H. Padamsee et al.: Accelerating Cavity Development for the Cornell B-Factory, CESR-B, *Proceedings of PAC1991, Particle Accelerator Conference*, Vol. 2, San Francisco (1991), p. 786.
28. T. Furuya et al.: *Proceedings of the 7th Workshop on RF Superconductivity*, Gif-sur-Yvette (1995), p. 729.
29. P. Kneisel et al.: First Results on Elliptically Shaped Cavities, *Nucl. Instrum. Methods Phys. Res.* **188**, p. 669 (1981).
30. U. Klein and D. Proch: *Proceedings of Conference of Future Possibilities for Electron Accelerators*, Charlottesville (1979), p. N1 and Wuppertal Report WU B 78-31, November (1978).
31. D. Moffat: Superconducting Niobium RF Cavities Designed to Attain High Surface Electric Fields, *Proceedings of 4th Workshop on RF Superconductivity*, KEK, Tsukuba (1990), p. 445.
32. J. Delayen and K.W. Shepard: Tests of a Superconducting RF Quadrupole Device, *Appl. Phys. Lett.* **57**, 514 (1990).
33. R. L. Geng et al.: High Gradient Studies for ILC with Single-Cell Re-Entrant Shape and Elliptical Shape Cavities Made of Fine-Grain and Large-Grain Niobium, *Proceedings of PAC 2007, Particle Accelerator Conference*, Albuquerque (2007), p. 2337.
34. F. Furuta et al.: Experimental Comparison at KEK of High Gradient Performance of Different Single Cell Superconducting Cavity Designs, *Proceedings of EPAC 2006, European Particle Accelerator Conference*, Edinburgh (2006), p. 750.
35. J. Graber: A World Record Accelerating Gradient in a Niobium Superconducting Accelerator Cavity, *Proceedings of PAC1993 Particle Accelerator Conference*, Washington (1993), p. 892.
36. V. Shemelin and H. Padamsee: The Optimal Shape of Cells of a Superconducting Accelerating Section, SRF Internal Report SRF 020128-01, Cornell University, Ithaca, January (2001).
37. V. Shemelin and H. Padamsee: The Optimal Shape of Cells of a Superconducting Accelerating Section, TESLA Report 2002-1 (2002).
38. V. Shemelin et al.: Optimal Cell for TESLA Superconducting Structure, *Nucl. Instrum. Methods Phys. Res.*, A **496**, 1 (2003).
39. J. Sekutowicz et al.: Cavities for JLab's 12 GeV Upgrade, *Proceedings of PAC 2003 Particle Accelerator Conference*, Portland (2003), p. 1396.
40. J. Sekutowicz et al.: Low Loss Cavity for the 12 GeV CEBAF Upgrade, *JLab TN-02-023*, June (2002).
41. J. Sekutowicz: Design of a Low Loss SRF Cavity for the ILC, *Proceedings of PAC 2005, Particle Accelerator Conference*, Knoxville (2005), p. 3342.
42. K. Saito: Strategy for 50 MV/m, *Presentation at TESLA Collaboration (TTC) Meeting*, DESY, Hamburg (2004).
43. T. Saeki: Initial Studies of 9-Cell High-Gradient Superconducting Cavities at KEK, *Proceedings of LINAC 2006, Linear Accelerator Conference*, Knoxville (2006), p. 794.
44. D. Proch: The TESLA Cavity, Design Considerations and RF Properties, *Proceedings of 6th Workshop on RF Superconductivity*, Newport News (1993), p. 382.
45. R. L. Geng: Review of New Shapes for Higher Gradient, *Proceedings of 12th*

Workshop on *RF Superconductivity*, Ithaca (2005), p. 175.
45 R. L. Geng: Review of New Shapes for Higher Gradients, *Physica C*, **441**, 145 (2006).
47 J. Sekutowicz: New Geometries: Elliptical Cavities, *Proceedings of Workshop on Pushing the Limits of RF Superconductivity*, Argonne National Laboratory, Report, ANL-05/10, Argonne (2005), p. 3.
48 I. Zagorodnov and N. Solyak: Wakefield Effects of New ILC Cavity Shapes, *Proceedings of EPAC 2006, European Particle Accelerator Conference*, Edinburgh (2006), p. 2862.
49 R. L. Geng et al.: World Record Accelerating Gradient Achieved in a Superconducting Niobium RF Cavity, *Proceedings of PAC 2005, Particle Accelerator Conference*, Knoxville (2005), p. 653.
50 E. Kako et al.: Characteristics of the Results of Measurement on 1.3 GHz High Gradient Superconducting Cavities, *Proceedings of 7th Workshop on RF Superconductivity*, Gif-sur-Yvette (1995), p. 425.
51 M. Ono et al.: Achievement of 40 MV/M Accelerating Field in L-Band SCC at KEK, *Proceedings of 8th Workshop on RF Superconductivity*, Padova (1998), p. 472.
52 K. Saito et al.: High Gradient Performance by Electropolishing with 1300 MHz Single and Multi-Cell Niobium Superconducting Cavities, *Proceedings of 9th Workshop on RF Superconductivity*, Santa Fe (1999), p. 283.
53 E. Kako et al.: Improvement of Cavity Performance in the Saclay/Cornell/DESY's SC Cavities, *Proceedings of 9th Workshop on RF Superconductivity*, Santa Fe (1999), p. 179.
54 L. Lilje et al.: Electropolishing and In-Situ Baking of 1.3 GHz Niobium Cavities, *Proceedings of 9th Workshop on RF Superconductivity*, Santa Fe (2000), p. 74.
55 L. Lilje: High Accelerating Gradients in 1.3 GHz Niobium Cavities, *Proceedings of 10th Workshop on RF Superconductivity*, Tsukuba (2001), p. 287.
56 L. Lilje et al.: Improved Surface Treatment of the Superconducting TESLA Cavities, *Nucl. Instrum. Methods Phys. Res. A* **516**, 213 (2004).

57 J. Ozelis: Performance of Medium-Beta and High-Beta Elliptical Cavities, *Proceedings of 12th Workshop on RF Superconductivity*, Ithaca (2005), Talk MoP07.
58 J. Delayen et al.: Performance Overview of the Production Superconducting RF Cavities for the Spallation Neutron Source Linac, *Proceedings of PAC 2005, Particle Accelerator Conference*, Knoxville (2005), p. 4048.
59 C. C. Compton et al.: Prototyping of a Multi-Cell Superconducting Cavity for Acceleration of Medium Velocity Beams, *Phys. Rev. ST AB* **8**(4) (2005).
60 G. Ciovati et al.: Superconducting Prototype Cavities for the Spallation Neutron Source (SNS) Project, *Proceedings of PAC 2001, Particle Accelerator Conference*, Chicago (2001), p. 484.
61 W. Hartung et al.: Status Report on Multi-Cell Superconducting Cavity Development for Medium-Velocity Beams, *Proceedings of PAC 2003, Particle Accelerator Conference*, Portland (2003), p. 1362.
62 A. Facco et al.: A Superconductive Low-Beta Single Gap Cavity for a High Intensity Proton Linac, *Proceedings of LINAC 2000, Linear Accelerator Conference*, Monterey (2001), p. 929.
63 P. H. Ceperley et al.: Superconducting Reentrant Cavities for Heavy Ion Linacs, *Proceedings of PAC 1975, Particle Accelerator Conference*, Washington (1975), p. 1153.
64 D. Barni et al.: SC Cavity Design for the 700 MHz TRASCO Linac, *Proceedings of EPAC 2000, European Particle Accelerator Conference*, Vienna (2000), p. 2019.
65 J. R. Delayen: Medium-Superconducting Accelerating Structures, USPAS Baton Rouge (2003); Low and Medium-β Cavities and Accelerators, *Proceedings of 13th Workshop on RF Superconductivity*, Beijing Tutorial 4a (2007).
66 K. W. Shepard et al.: Prototype 350 MHz, Niobium Spoke-Loaded Cavities, *Proceedings of PAC 1999, Particle Accelerator Conference*, New York (1999), p. 955.
67 M. P. Kelly: Status of SC Spoke Cavity Development, *Proceedings of 13th Workshop on RF Superconductivity*, Beijing (2007), WE302.

68 K. W. Shepard et al.: Development of Spoke Cavities for RIA, *Proceedings of 12th Workshop on RF Superconductivity*, Ithaca (2005), p. 334.

69 K. W. Shepard et al.: Development of Spoke Cavities for RIA, *Physica C*, **441**, 205 (2006).

70 V. Andreev: Comparison of Elliptical and Triple-Spoke Cavities for the Rare Isotope Accelerator, *NSCL-RIA-2004-001* (2004).

71 K. W. Shepard: Triple-Spoke Compared with Elliptical-Cell Cavities, *12th Workshop on RF Superconductivity*, Ithaca (2005), Talk WeHT3.

72 K. W. Shepard: Status of Low and Intermediate Velocity Superconducting Accelerating Structures, *Proceedings of PAC 2003, Particle Accelerator Conference*, Portland (2003), p. 81.

73 E. Zaplatin: Electrodynamics and Mechanical Features of H-Type Superconducting Structures for Low Energy Part of ESS Linac, *ESS 116-01-L*, July (2001).

74 R. Eichhorn and U. Ratzinger: Superconducting H-Mode Structures for Medium Energy Beams, *Proceedings of LINAC 2000, Linear Accelerator Conference*, Monterey (2000), p. 926.

75 First Tests of the Superconducting CH-Structure, *Proceedings of PAC 2005, Particle Accelerator Conference*, Knoxville (2005), p. 3414.

76 H. Podlech: Status of the Superconducting CH-Structure, *Proceedings of EPAC 2004, European Particle Accelerator Conference*, Lucerne (2004), p. 991.

77 G. Devanz: SPIRAL2 Resonators, *Proceedings of 12th Workshop on RF Superconductivity*, Ithaca (2005), p. 108.

78 G. Devanz: SPIRAL2 Resonators, *Physica C*, **441**, 173 (2006).

79 G. Devanz et al.: Quarter-Wave Cavities for the SPIRAL2 Project, *Proceedings of 11th Workshop on RF Superconductivity*, Travemünde (2003), paper TuP06.

80 K. W. Shepard and G. P. Zinkann: A Superconducting Accelerating Structure for Particle Velocities from 0.12 to 0.23c, *Proceedings of PAC1983, Particle Accelerator Conference, IEEE Transact. Nucl. Sci. NS-30* **4**(8), 3339 (1983).

81 K. Shepard: Superconducting RF Technology at Argonne, *DOE and NSF Workshop on RF Superconductvity*, July (2003).

82 P. N. Ostroumov and K. W. Shepard.: *Phys. Rev. ST. Accel. Beams* **11**, 030101 (2001).

83 T. L. Grimm et al.: Superconducting RF Activities for the Rare Isotope Accelerator at Michigan State University, *Proceedings of 11th Workshop on RF Superconductivity*, Travemünde, September (2003), paper MoP03.

84 V. Zvyagintsev et al.: Mechanical Design of a 161 MHz, $\beta=0.16$ Superconducting Quarter Wave Resonator with Steering Correction for RIA, *Proceedings of 11th Workshop on RF Superconductivity*, Travemünde (2003), paper TuP08.

85 K. W. Shepard: Development of SC Intermediate Velocity Structures for the U. S. RIA Project, *Proceedings of 11th Workshop on RF Superconductivity*, Travemünde (2003), paper WeO08.

86 N. Added et al.: Upgraded Phase Control System for Superconducting Low Velocity Accelerating Structure, *Proceedings of LINAC 1992, Linear Accelerator Conference*, Ottawa (1992), p. 181.

87 A. Facco et al.: On-Line Performance of the LNL Mechanically Damped Superconducting Low-Beta Resonators, *Proceedings of EPAC 1998, European Particle Accelerator Conference*, Stockholm (1998), p. 1846.

88 A. Facco: Mechanical Mode Damping in Superconducting Low- Resonators, *Part. Accel.*, **61**, 265 (1998).

89 K. W. Shepard: Superconducting Low-Velocity Linac for the Argonne Positive-Ion Injector, *Proceedings of PAC 1989, Particle Accelerator Conference* (1989), p. 1974.

90 A. Roy: SRF Activities at IUAC, New Delhi and Other Laboratories in India, *Proceedings of 13th Workshop on RF Superconductivity*, Beijing (2007), paper MO303.

91 P. N. Prakash et al.: Quarter Wave Coaxial Line Cavity for New Delhi Linac Booster, *Proceedings of 8th Workshop on RF Superconductivity*, Abano Terme (1996), p. 663.

92 R. E. Laxdal et al.: Recent Progress in the Superconducting RF Program at TRIUMF/ISAC, *Proceedings of 12th Work-*

shop on RF Superconductivity, Ithaca (2005), p. 128.

93. M. Pekeler et al.: Performance of a Prototype 176 MHz $\beta=0.09$ Half-Wave Resonator for the SARAF Linac, Proceedings of 12th Workshop on RF Superconductivity, Ithaca (2005), p. 331.

94. T. L. Grimm et al.: Superconducting RF Activities for the Rare Isotope Accelerator at Michigan State University, Proceedings of 11th Workshop on RF Superconductivity, Travemünde (2003), paper MoPO3.

95. A. Facco et al.: Construction and Testing of the $\beta=0.31$, 352 MHz Superconducting Half-Wave Resonator for the SPES project, Proceedings of EPAC 2004, European Particle Accelerator Conference, Lucerne (2004), p. 1012.

96. G. Bisoffi: Superconducting RFQ's Ready for Ion Beam Operation at INFN-LNL, Proceedings of EPAC 2002, European Particle Accelerator Conference, Paris (2002), p. 266.

97. I. M. Kapchinskii and V. A. Teplyakov: Linear Ion Accelerator with Spatially Homogeneous Strong Focusing, Pub. Tekh, Eksp., **2**, 19 (1970).

98. R. H. Stokes and T. P. Wangler: Radiofrequency Quadrupole Accelerators and their Applications, Ann. Rev. Nucl. Part. Sci., **38**, 97 (1988).

99. G. Bisoffi: First Results with the Full Niobium Superconducting RFQ Resonator at INFN-LNL, Proceedings of EPAC 2000, European Particle Accelerator Conference, Vienna (2000), p. 324.

100. G. Bisoffi: Superconducting RFQs Ready for Ion Beam Operation at INFN-LNL, Proceedings of EPAC 2002, European Particle Accelerator Conference, Paris (2002), p. 266.

101. G. Bisoffi: Superconducting RFQs, Proceedings of the 12th Workshop on RF Superconductivity, Ithaca (2005), p. 123.

102. G. Bisoffi: Superconducting RFQs, Physica C, **441**, 185 (2006).

103. G. Bisoffi: Completion of the Commissioning of the Superconducting Heavy Ion Injector Piave at INFN-LNL, Proceedings of EPAC 2006, European Particle Accelerator Conference, Edinburgh (2006), p. 1597.

104. J. R. Delayen: Ponderomotive Instabilities and Microphonics, a Tutorial, Proceedings of 12th Workshop on RF Superconductivity, Ithaca (2005), p. 35.

105. J. R. Delayen: Ponderomotive Instabilities and Microphonics: A Tutorial, Physica C, **441**, 1 (2006).

106. P. H. Ceperley: Ponderomotive Oscillations in a Superconducting Helical Resonator, IEEE Trans. Nucl. Sci., **2**, 217 (1972).

107. G. H. Luo: NSRRC, B-cell mechanical calculations, reported in [17].

108. M. C. Lin: Coupled-Field Analysis on a 500 MHz Superconducting Radio Frequency Niobium Cavity, Proceedings of EPAC 2002, European Particle Accelerator Conference, Paris (2002), p. 2259.

109. A. Roy et al.: RF Design of a Single-Cell Superconducting Elliptical Cavity with Input Coupler, Proceedings of 12th Workshop on RF Superconductivity, Ithaca (2005), p. 272.

110. A. Roy et al.: Structural Analysis of Single-Cell Superconducting Elliptical Cavity with Static Lorentz Force, Proceedings of 12th Workshop on RF Superconductivity, Ithaca (2005), p. 275.

111. J. Mondal et al.: Optimization of Wall Thickness of Superconducting 700 MHz Bulk Niobium and Niobium Coated OFHC Copper Cavities by Thermal/Structural Analysis, Proceedings of 12th Workshop on RF Superconductivity, Ithaca (2005), p. 278.

112. H. Gassot et al.: Mechanical Studies of the Multi-Gap Spoke Cavity for European Project Hippi, Proceedings of 12th Workshop on RF Superconductivity, Ithaca (2005), p. 312.

113. G. Apollinari: Design of 325 MHz Single Spoke Resonator at FNAL, Proceedings of 12th Workshop on RF Superconductivity, Ithaca (2005), p. 314.

114. M. Liepe: ERL, raising the frequency of mechanical vibrations, reported in [17].

115. R. Mitchell et al.: Lorentz Force Detuning Analysis of the Spallation Neutron Source (SNS) Accelerating Cavities, Proceedings of 10th Workshop on RF Superconductivity, Tsukuba (2001), p. 236.

116. J. R. Delayen: Tutorials: LLRF Control Systems and Low and Medium-β

Cavities and Accelerators, *Proceedings of 13th Workshop on RF Superconductivity*, Beijing (2007).

117 P. Brown et al.: Operating Experience with the LEP2 Superconducting RF System, *Proceedings of 10th Workshop on RF Superconductivity*, Tsukuba (2001), p. 185.

118 I. Gonin: in J.Sekutowicz (ed.) Tutorial 2a, *Proceedings of 13th Workshop on RF Superconductivity*, Beijing (2007).

119 M. Liepe et al.: Dynamic Lorentz Force Compensation with a Fast Piezoelectric Tuner, *Proceedings of PAC 2001, Particle Accelerator Conference*, Chicago (2001), p. 1074.

120 S. Simrock: Advances in RF Control for High Gradients, *Proceedings of 9th Workshop on RF Superconductivity*, Santa Fe (1999), p. 92.

121 S. Simrock: Achieving Phase and Amplitude Stability in Pulsed Superconducting Cavities, *Proceedings of 10th Workshop on RF Superconductivity*, Tsukuba (2001), p. 231.

122 A. Brandt et al.: General Automation of LLRF Control for Superconducting Accelerators, *Proceedings of 12th Workshop on RF Superconductivity*, Ithaca (2005), p. 194.

123 A. Brandt et al.: General Automation of LLRF Control for Superconducting Accelerators, *Physica C*, **441**, 263 (2006).

124 G. Ciovati: Superconducting Prototype Cavities for the Spallation Neutron Source (SNS) Project, *Proceedings of PAC 2001, Particle Accelerator Conference*, Chicago (2001), p. 484.

125 A. Bosotti: RF Tests of the $\beta=0.5$ Five-Cell TRASCO Cavities, *Proceedings of EPAC 2004, European Particle Accelerator Conference*, Lucerne (2004), p. 1024.

126 G. Olry et al.: Development of Spoke Cavities for the EURISOL and EUROTRANS Projects, *Proceedings of 12th Workshop on RF Superconductivity*, Ithaca (2005), p. 328.

127 Poisson/Superfish: *http://laacg.lanl.gov/laacg/services/download_sf.phtml*.

128 D.G. Myakishev and V.P. Yakovlev: The New Possibilities of SuperLANS Code for Evaluation of Axisymmetric Cavities, *Proceedings of PAC* (1995), p. 2348.

129 W. Bruns: GdfidL: A Finite Difference Program with Reduced Memory and CPU ge, *PAC97, Proceedings of PAC 1997, Particle Accelerator Conference*, Vancouver (1997), p. 2651.

130 MAFIA: CST GMbH, Buedinger Str. 2a, 64289 Darmstadt, Germany, *http://www.cst.com/Content/Products/MAFIA/Overview.aspx*.

131 CST Microwave Studio: CST GMbH, Buedinger Str. 2a, 64289, Darmstadt, Germany, *http://www.cst.com/Content/Products/MWS/Overview.aspx*.

132 HFSS, Ansoft Corp.: Ansoft Corporate Headquarters, 225 West Station Square Drive Suite 200, Pittsburgh, PA 15219, USA, *http://www.ansoft.com/products/hf/hfss/*.

133 Vector Fields 3D SOPRANO: Vector Fields Inc, 1700 N Farnsworth Ave, Aurora, IL 60505, USA.

134 K. Ko et al.: Advances in Electromagnetic Modeling through High Performance Computing, *Proceedings of 12th Workshop on RF Superconductivity*, Ithaca (2005), p. 428.

135 K. Ko et al.: Advances in Electromagnetic Modeling through High Performance Computing, *Physica C*, **441**, 258 (2006).

136 ANSYS: ANSYS Inc., Southpointe, 275 Technology Drive, Canonsburg, PA 15317, USA. *http://www.ansys.com/*.

137 COSMOS: Structural Research and Analysis Corp., 3000 Ocean Park Boulevard Suite 2001, Santa Monica, CA 90405-3030, USA.

138 F. London and H. London: The Electromagnetic Equations of the Supraconductor, *Proceedings of R. Soc. London, Ser. A* **149**, 71 (1935).

139 D.C. Mattis and J. Bardeen, Theory of the Anomalous Skin Effect in Normal and Superconducting Metals, *Phys. Rev.* **111**, 412 (1958).

140 J. Halbritter: On Surface Resistance of Superconductors, *Z. Physik* **266**, 209 (1974).

141 J. Halbritter: Z. Physik, FORTRAN Program for the Computation of the Surface Impedance of Superconductors, FZK 3/70-6, Karlsruhe, **238**, 466 (1970).

142 P. Kneisel: R&D Paths Towards Achieving Ultimate Capabilities, *Proceedings of Workshop on Pushing the Limits of RF Superconductivity*, Argonne National Laboratory, Report ANL-05/10, Argonne (2004), p. 106.

143 G. Ciovati: Investigation of the Superconducting Properties of Niobium Radio-Frequency Cavities, PhD Thesis, Old Dominion University, Norfolk (2005).

144 G. Ciovati et al.: Effect of Low Temperature Baking on Nb Cavities, *Proceedings of 11th Workshop on RF Superconductivity*, Travemünde (2003), paper WeO14.

145 J. P. Turneaure et al.: The Surface Impedance of Superconductors and Normal Conductors: The Mattis–Bardeen Theory, *J. Superconductivity*, **4**, 341 (1991).

146 A. B. Pippard: An Experimental and Theoretical Study of the Relation between Magnetic Field and Current in a Superconductor, *Proceedings of R. Soc. London, Ser. A* **216**, 547 (1953).

147 A. Gurevich: Thermal RF Breakdown of Superconducting Cavities, *Proceedings of Workshop on Pushing the Limits of RF Superconductivity*, Argonne National Laboratory, Report ANL-05/10, Argonne (2004), p. 17.

148 P. Kneisel: Preliminary Experience with In-Situ Baking of Niobium Cavities, *Proceedings of 9th Workshop on RF Superconductivity*, Santa Fe (1999), p. 328.

149 C. Benvenuti and S. Calatroni: Diffusion of Oxygen in Niobium During Bake-Out, *Proceedings of 10th Workshop on RF Superconductivity*, Tsukuba (2001), p. 441.

150 A. Dacca et al.: XPS Analysis of the Surface Composition of Nb for SRF Cavities, *Appl. Surf. Sci.* **126**, 219, 1998.

151 B. Visentin: Low, Medium, High Field Q-Slopes Change with Surface Treatments, *Proceedings of Workshop on Pushing the Limits of RF Superconductivity*, Argonne National Laboratory, Report ANL-05/10, Argonne (2004), p. 94.

152 A. Gurevich: Multiscale Mechanisms of SRF Breakdown, *Proceedings of the 12th Workshop on RF Superconductivity*, Ithaca (2005), p. 156.

153 A. Gurevich: Multiscale Mechanisms of SRF Breakdown, *Physica C* **441**, 38 (2006).

154 G. Ciovati: High Q at Low and Medium Field, *Proceedings of Workshop on Pushing the Limits of RF Superconductivity*, Argonne National Laboratory, Report ANL-05/10, Argonne (2004), p. 52.

155 G. Ciovati: Analysis of the Medium Field Q-slope in Superconducting Cavities Made of Bulk Niobium, *Proceedings of 12th Workshop on RF Superconductivity*, Ithaca (2005), p. 230.

156 J. Halbritter: *Proceedings of of the 38th Eloisitron Workshop*, Erice (1999), p. 59.

157 P. Bauer: Evidence for Non-Linear BCS Resistance in SRF Cavities, *Proceedings of 12th Workshop on RF Superconductivity*, Ithaca, p. 223.

158 J. Vines: Systematic Trends for the Medium Field Q-Slope, *Proceedings of 13th Workshop on RF Superconductivity*, Beijing (2007), TUP27.

159 J. Halbritter: Granular Superconductors and Their Intrinsic and Extrinsic Surface Impedance, *J. Supercond.* **8**, 691–703 (1995).

160 J. Halbritter: Transport in Superconducting Niobium Films for Radio Frequency Applications, *J. Appl. Phys.* **97**, 083904 (2005).

161 J. Halbritter: Material Science of Nb RF Accelerator Cavities: Where Do We Stand 2001? *Proceedings of 10th Workshop on RF Superconductivity*, Tsukuba (2001), p. 292.

162 V. Palmieri: The Problem of Q-Drop in Superconducting Resonators Revised by the Analysis of Fundamental Concepts from RF-Superconductivity Theory, *Proceedings of 12th Workshop on RF Superconductivity*, Ithaca (2005), p. 162.

163 G. Eremeev: High Field Q-Slope's Studies Using Thermometry, *Proceedings of 12th Workshop on RF Superconductivity*, Ithaca (2005), p. 189.

164 K. Mittag: Kapitza Conductance and Thermal Conductivity of Copper, Niobium and Aluminium in the Range from 1.3 To 2.1 K, *Cryogenics* **13**, 94 (1973).

165 C. Johannes: *Proceedings of 3rd International Cryogenics Engineering. Conference*, ICEC3 (1970), p. 97.

166 F. Koechlin and B. Bonin: Parametrization of the Niobium Thermal Conduc-

tivity in the Superconducting State, *Supercond. Sci. Technol.* **9**, 453 (1996).

167 D. Reschke: Limits in Cavity Performance, *Proceedings of 13th Workshop on RF Superconductivity*, Beijing (2007), Tutorial 3c.

168 J. Graber: High Power RF Processing Studies of 3 GHz Niobium Superconducting Accelerator Cavities, PhD Thesis, Cornell University, Ithaca (1993).

169 P. Bauer: Evidence for Non-Linear BCS Resistance in SRF Cavities, *Physica C* **441**(1–2), 51 (2006).

170 T. Khabiboulline: 3.9 GHz Superconducting Accelerating 9-Cell Cavity Vertical Test Results, *Proceedings of PAC 2007, Particle Accelerator Conference*, Albuquerque (2007), p. 2295.

171 H. Padamsee: Status Report on TESLA Activities, *Proceedings of 5th Workshop on RF Superconductivity*, DESY, Hamburg (1991), p. 963.

172 Y. Xie and H. Padamsee: Medium-Field Q-slope Calculated from Nonlinear BCS Surface Resistance and Comparisons with Experimental Data, SRF Note 080108-05, Cornell University.

173 H. Safa: High-Field Behavior of SCRF cavities, *Proceedings of 10th Workshop on RF Superconductivity*, Tsukuba (2001), p. 279.

174 J. Knobloch and H. Padamsee: Reduction of the Resistance in Superconducting Cavities Due to Gas Discharge, *Proceedings of 8th Workshop on RF Superconductivity*, Abano Terme (1998), p. 345.

175 K. Kowalski: *In Situ* XPS Investigation of the Baking Effect on the Surface Oxide Structure Formed on Niobium Sheets Used for Superconducting RF Cavity Production, *Proceedings of 11th Workshop on RF Superconductivity*, Travemünde (2003), paper ThP09.

176 B. Bonin and R. Roth: Q Degradation of Niobium Cavities Due to Hydrogen Contamination, *Proceedings of 5th Workshop on RF Superconductivity*, Hamburg (1991), p. 210.

177 J. F. Smith: *Bull. Alloy Phase Diagrams*, **4**, 39 (1983).

178 TESLA Technology Collaboration Cavity Database: RF Tests. *http://teslanew. desy.de/cavity_ database/rf_tests/index_ eng.html*

179 B. Bonin and R. Roth: Q-Degradation for Accelerating Cavities Due to Hydrogen Contamination, *Proceedings of 5th Workshop on RF Superconductivity*, DESY, Hamburg (1991), p. 210.

180 J. K. Norskov and F. Besenbacher: Theory of Hydrogen Interaction with Metals, *J. Less-Common Metals*, **130**, 475 (1987).

181 J. Knobloch and H. Padamsee: Enhanced Susceptibility of Nb Cavity Equator Welds to the Hydrogen Related Q-Virus, *Proceedings of 8th Workshop on RF Superconductivity*, Abano Terme (1998).

182 W. Singer: SC Cavities; Material, Fabrication and QA, *Proceedings of 13th Workshop on RF Superconductivity*, Beijing (2007), tutorial 4a2.

183 C. Vallet et al.: Flux trapping in superconducting cavities, *Proceedings of EPAC 1992, European Particle Accelerator Conference*, Berlin (1992), p. 1295.

184 C. Benvenuti et al.: RF Superconductivity at CERN, *Proceedings of 2nd Workshop on RF Superconductivity*, CERN, Geneva (1984), p. 25.

185 J. Knobloch and H. Padamsee: Flux Trapping in Niobium Cavities During Breakdown Events, *Proceedings of 8th Workshop on RF Superconductivity*, Abano Terme (1998), p. 337.

186 A. A. Abrikosov: On the Magnetic Properties of Superconductors of the Second Group, *Zh. Eksp. Teor. Fiz.* **32**, 1442 (1957), or *Soviet Phys. JETP* **5**, 1174 (1957).

187 A. Gurevich: General Aspects of Superconductivity, Tutorial 1a, *Proceedings of 13th Workshop on RF Superconductivity*, Beijing (2007).

188 K. Saito: Theoretical Critical Field in RF Application, *Proceedings of 11th Workshop on RF Superconductivity*, Travemünde (2003), paper MoO02.

189 D. K. Finnemore, T. F. Stromberg and C. A. Swenson: Superconducting Properties of High-Purity Nb, *Phys. Rev.* **149**, 231 (1966).

190 R. A. French: Intrinsic Type-2 Superconductivity in Pure Niobium, *Cryogenics* **8**, 301 (1968).

191 S. Casalbuoni et al.: Superconductivity Above the Upper Critical Field as a Probe for Niobium RF-Cavity Surfaces, *Proceedings of 11th Workshop on RF Superconductivity*, Travemünde (2003), paper WeO13.

192 M. Bahte et al.: Magnetization and Susceptibility Measurements on Niobium Samples for Cavity Production, *Proceedings of the 8th Workshop on RF Superconductivity*, Abano Terme (1997), p. 881.

193 G. Meuller: Superconducting Niobium in High RF Magnetic Fields, *Proceedings of 3rd Workshop on RF Superconductivity*, Argonne (1987), p. 331.

194 H. Padamsee et al.: Critical Field Deviations in Superconducting Lead–Indium Alloys, *Phys. Lett.* **41A**, 427 (1973).

195 D. Saint-James and P. G. de Gennes: Onset of Superconductivity in Decreasing Fields, *Phys. Lett.* **7**, 306 (1963).

196 B. Steffen et al.: Susceptibility Measurements on Surface Treated Niobium Samples, *Proceedings of 11th Workshop on RF Superconductivity*, Travemünde (2003), paper MoP49.

197 S. Casalbuoni et al.: Surface Superconductivity in Niobium for Superconducting Cavities, *Nucl. Instrum. Methods Phys. Res. A*, **538**, 45 (2005).

198 L. von Sawilski et al.: Surface Superconducting of Niobium: Onset on Long-Range Coherence, *Proceedings of 11th Workshop on RF Superconductivity*, Travemünde (2003), paper MoP12.

199 R. B. Flippen: The Radial Velocity of Magnetic Field Penetration in Type II Superconductors, *Phys. Lett. A* **17**, 193 (1965).

200 J. Matricon and D. Saint-James: Superheating Fields in Superconductors, *Phys. Lett. A* **24**, 241 (1967).

201 V. Guritanu et al.: Specific Heat of Nb_3Sn: The Case for a Second Energy Gap, *Phys. Rev.* **70B**, 184526 (2004).

202 T. Yogi et al.: Critical RF Magnetic Fields for Some Type-I and Type-II Superconductors, *Phys. Rev. Lett.* **39**, 826 (1977).

203 T. Yogi: Radio Frequency Studies of Surface Resistance and Critical Magnetic Field of Type I and Type II Superconductors, PhD Thesis, California Institute of Technology, Pasadena (1997).

204 J. Sethna: Outstanding Issues in RF Superconductivity: What Can Theory Tell Us? *Proceedings of 13th Workshop on RF Superconductivity*, Beijing (2007), Talk TU101.

205 T. Hays and H. Padamsee: Measuring the RF Critical Field of Pb, Nb, and Nb_3Sn, *Proceedings of 8th Workshop on RF Superconductivity*, Abano Terme (1997), p. 789.

206 T. Hays and H. Padamsee: Response of Superconducting Cavities to High Peak Power, *Proceedings of PAC 1995, Particle Accelerator Conference*, Dallas (1995).

207 I. E. Campisi: High-Field RF Superconductivity, to Pulse or Not to Pulse? *IEEE Trans. Magn.*, **25**, 134 (1985).

208 R. L. Geng: Multipacting Simulations for Superconducting Cavities and RF Coupler Waveguide, *Proceedings of PAC 2003, Particle Accelerator Conference*, Portland (2003), p. 264.

209 G. Devanz: Multipactor Simulations in Superconducting Cavities and Power Couplers, *Phys. Rev. Spec. Top. – Accel. Beams*, **4**, 012001 (2001).

210 F. Krawczyk: Status of Multipacting Simulation Capabilities for SCRF Applications, *Proceedings of 10th Workshop on RF Superconductivity*, Tsukuba (2001), p. 108.

211 E. Somersalo et al.: Analysis of Multipacting in Coaxial lines, *Proceedings of PAC1995, Particle Accelerator Conference*, Dallas (1995), p. 1500.

212 E. Somersalo et al.: Computational Methods for Analyzing Electron Multipacting in RF Structures, *Part. Accel.*, **59**, 107 (1998).

213 R. Yla-Oijala: Multipacting in TESLA Cavities and Input Couplers, *Part. Accel.*, **63**, 105 (1999).

214 P. Ylä-Oijala and D. Proch: MultiPac – Multipacting Simulation Package with 2D FEM Field Solver, *Proceedings of of 10th Workshop on RF Superconductivity*, Tsukuba (2001), p. 105.

215 K. Saito et al.: Superiority of Electropolishing over Chemical Polishing on High Gradients, *Proceedings of 8th Work-*

shop on RF Superconductivity, Abano Terme (1998), p. 810.

216 V. Shemelin: Multipacting in Crossed RF Fields Near Cavity Equator, *Proceedings of EPAC 2004, European Particle Accelerator Conference*, Lucerne (2004), p. 1075.

217 J. Knobloch et al.: Multipacting in 1.5-GHz Superconducting Niobium Cavities of the CEBAF Shape, *Proceedings of 8th Workshop on RF Superconductivity*, Abano Terme (1998), p. 1017.

218 K. Twarowski et al.: Multipacting in 9-Cell Tesla Cavities, DESY, 22603 Hamburg, Germany.

219 P. Fabbricatore et al.: Experimental Evidence of MP Discharges in Spherical Cavities at 3 GHz, *Proceedings of 7th Workshop on RF Superconductivity*, Gif sur Yvette (1995), p. 385.

220 S. Belomestnykh and V. Shemelin: Multipacting-Free Transitions between Cavities and Beam-Pipes, *TESLA Report 2008-4* (2008).

221 T. Saeki et al.: Initial Studies of 9-Cell High-Gradient Superconducting Cavities at KEK, *Proceedings of LINAC 2006, Linear Accelerator Conference*, Knoxville (2006), p. 794.

222 C. Ng: State-of-the-Art in EM Field Computation, *Proceedings of of EPAC 2006, European Particle Accelerator Conference*, Edinburgh (2006), p. 2763.

223 R. L. Geng: Fabrication and Performance of Superconducting RF Cavities for the Cornell ERL Injector, *Proceedings of PAC07*, Albuquerque (2007), p. 2340.

224 E. Donoghue et al.: Studies of Electron Activities in SNS Cavities Using FishPact, *Proceedings of 12th Workshop on RF Superconductivity*, Ithaca (2005), p. 402.

225 H. Padamsee: Overview of Advances in Basic Understanding of Dark Current and Breakdown in RF Cavities, *PAC 1997 Particle Accelerator Conference*, Vancouver (1997), p. 2884.

226 H. Padamsee and J. Knobloch: The Nature of Field Emission from Microparticles and the Ensuing Voltage Breakdown. In: *High Energy Density Microwaves*, R. M. Phillips (ed), *AIP Conference Proc.* **474** (1998).

227 R. H. Fowler and L. Nordheim: Electron Emission in Intense Electric Fields, *Proceedings of of the Royal Society London*, **A119**, 173 (1928).

228 E. L. Murphy and R. H. Good, Jr.: Thermionic Emission, Field Emission and the Transition Region. *Phys. Rev.* **102**(6), 1464 (1956).

229 P. Niedermann: Experiments on Enhanced Field Emission, PhD Thesis, University of Geneva, Geneva, Dissertation Number 2197 (1986).

230 E. Mahner: Understanding and Supressing Field Emission Using DC, *Part. Accel.*, **46**, 67 (1994). Also published in *Proceedings of 6th Workshop on RF Superconductivity*, Newport News (1993), p. 252.

231 P. Kneisel et al.: Experience with High Pressure Ultrapure Water Rinsing of Niobium Cavities, *Proceedings of 6th Workshop on RF Superconductivity*, Newport News (1994), p. 628.

232 P. Kneisel and B.Lewis: Advanced Surface Cleaning Methods, Three Years of Experience with High Pressure Ultrapure Water Rinsing of Superconducting Cavities, *Proceedings of 7th Workshop on RF Superconductivity*, Gif sur Yvette (1995), p. 311.

233 A. Dangwal: DC Field Emission Scanning Measurements on Electropolished Niobium Samples, *Proceedings of 12th Workshop on RF Superconductivity*, Ithaca (2005), p. 255.

234 D. Moffat et al.: Studies on the Nature of Field Emission Sites, *Part. Accel.* **40**, 85 (1992).

235 J. Graber et al.: Microscopic Investigation of High Gradient Superconducting Cavities after Reduction of Field Emission, *Nucl. Instrum. Methods Phys. Res.* **A 350**, 582 (1994).

236 J. Graber: High Power Processing Studies of 3 GHz Niobium Superconducting Accelerator Cavities, PhD Thesis, Cornell University, Laboratory of Nuclear Studies Report CLNS 93-1 (1993).

237 J. Graber et al.: Reduction of Field Emission in Superconducting Cavities with High Power Pulsed RF, *Nucl.*

Instrum. Methods Phys. Res. **A 350**, 572 (1994).
238 C. Crawford et al.: High Gradients in Linear Collider Superconducting Accelerator Cavities by High Pulsed Power to Suppress Field Emission, *Part. Accel.* **49**, 1 (1995).
239 H. A. Schwettman et al.: Evidence for Surface-State-Enhanced Field Emission in RF Superconducting Cavities, *J. Appl. Phys.*, **45**, 914 (1974).
240 W. Weingarten: Electron Loading, *Proceedings of 2nd Workshop on RF Superconductivity*, CERN, Geneva (1984), p. 551.
241 J. Graber: An Update on High Peak Power (HPP) RF Processing of 3 GHz Nine-Cell Niobium Accelerator Cavities, *PAC 93, Particle Accelerator Conference*, Washington DC (1993), p. 886.
242 A. Hutton: SRF Operating Experience at JLab, *TTC KEK06 meeting*.
243 J. Knobloch: Advanced Thermometry Studies of Superconducting Radio-Frequency Cavities, PhD Thesis, Cornell University, Ithaca (1997).
244 S. Musser: X-Ray Imaging of Superconducting RF Cavities, *Proceedings of 12th Workshop on RF Superconductivity*, Ithaca (2005), p. 295.
245 K. Saito et al.: Study of Ultra-Clean Surface for Niobium SC Cavities, *Proceedings of 6th Workshop on RF Superconductivity*, Newport News (1994), p. 1151.
246 T. Habermann: Raster-Microscopic Investigation of Field Emission from Metal and Diamond Cathodes, PhD Thesis, Bergische Universität Wuppertal (1999).
247 G. Müller: DC Field Emission of Nb Samples Prepared in TESLA Cavities, *Proceedings of EPAC 1998, European Particle Accelerator Conference*, Stockholm (1998), p. 1867.
248 H. Padamsee: A Statistical Model for Field Emission in Superconducting Cavities, *Proceedings of PAC1993, Particle Accelerator Conference*, Washington DC (1993), p. 998.
249 J. Wiener and H. Padamsee: Improvements in Field Emission: An Updated Statistical Model for Electropolished Baked Cavities, TESLA Report 2008-2.

250 B. van der Horst: Update on Cavity Preparation for High Gradient Superconducting Multicell Cavities at DESY, *Proceedings of 13th Workshop on RF Superconductivity*, Beijing (2007), paper TUP30.
251 R. L. Geng: Latest Results of ILC High-Gradient R&D 9-Cell Cavities at JLab, *Proceedings of 13th Workshop on RF Superconductivity*, Beijing (2007), paper WEP28.
252 D. Reschke: Dry-Ice Cleaning: The Most Effective Cleaning Process for SRF Cavities? *Proceedings of 13th Workshop on RF Superconductivity*, Beijing (2007), paper TUP48.
253 D. Reschke: First Experience with Dry-Ice Cleaning on SRF Cavities, *Proceedings of LINAC 2004, Linear Accelerator Conference*, Lübeck (2004), p. 776.
254 M. Jimenez et al.: Electron Field Emission from Large-Area Cathodes: Evidence for the Projection Model, *J. Phys. D* **27**, 1038 (1994).
255 G. R. Werner: Investigation of Voltage Breakdown Caused by Microparticles, *Proceedings of the 2001 Particle Accelerator Conference*, Chicago (2001), p. 1071.
256 G. R. Werner: Probing and Modeling Voltage Breakdown in Vacuum, PhD Thesis, Cornell University, Ithaca (2004).
257 P. Brown: Ultimate Performance of the LEP RF System, *Proceedings of PAC 2001, Particle Accelerator Conference*, Chicago (2001), p. 1059.
258 J. Knobloch and H. Padamsee: Microscopic Investigation of Field Emitters Located by Thermometry in 1.5 GHz Superconducting Niobium Cavities, *Part. Accel.* **53**, 53 (1996). Also published in the *Proceedings of 7th Workshop on RF Superconductivity*, Gif-sur-Yvette (1995), p. 95.
259 J. Graber: Microscopic Investigation of RF Surfaces of 3 GHz Niobium Accelerator Cavities Following RF Processing, *Proceedings of PAC93, Particle Accelerator Conference*, Washington DC (1993), p. 889.
260 J. Knobloch and H. Padamsee: Explosive Field Emitter Processing in Superconducting RF Cavities, *Proceedings of*

8th Workshop on RF Superconductivity, Abano Terme (1998), p. 994.

261 T. Hays et al.: Microscopic Examination and Elemental Analysis of Field Emission Sites in 5.8 GHz Superconducting Mushroom Cavities, *Proceedings of 6th Workshop on RF Superconductivity*, Newport News (1993), p. 750.

262 H. Padamsee: The Nature of Field Emission from Microparticles and the Ensuing Voltage Breakdown, in: *High Energy Density Microwaves*, R. M. Phillips (ed.), AIP Conference Proc. **474** (1998).

263 C. Crawford et al.: Achieving the TESLA Gradient of 25 MV/m in Multicell Structures at 1.3 GHz, *Proceedings of EPAC 2004, European Particle Accelerator Conference*, Lucerne (1994), p. 37.

264 A. Boechner et al.: *Proceedings of EPAC 2006, European Particle Accelerator Conference*, Edinburgh (2006), p. 413.

265 D. Reschke: Field Emission Overview: Cleanliness and Processing, *Proceedings of Workshop on Pushing the Limits of RF Superconductivity*, Argonne National Laboratory, Report ANL-05/10, Argonne (2005).

266 C. Z. Antoine et al.: Morphological and Chemical Studies of Nb Samples after Various Surface Treatment, *Proceedings of 9th Workshop on RF Superconductivity*, Santa Fe (1999), p. 295.

267 G. Werner et al.: Voltage Breakdown on Niobium and Copper Surfaces, *Proceedings of 10th Workshop on RF Superconductivity*, Tsukuba (2001), p. 268.

268 MASK: developed by Science Applications International Corporation, 10260 Campus Point Dr., San Diego, CA 92121, USA.

269 OOPICPRO: maintained by Tech-X Corporation, 5621 Arapahoe Ave, Boulder, CO 80303, USA.

270 C. Antoine: Summary Talk, *Fermilab Materials Workshop* (2007).

271 D. K. Davies and M. A. Biondi: Detection of Electrode Vapor Between Plane Parallel Copper Electrodes Prior to Current Amplification and Breakdown in Vacuum, *J. Appl. Phys.*, **41**, 88 (1970).

272 L. Laurent et al.: RF Breakdown Experiments at SLAC, in: *High Energy Density Microwaves*, R. M. Phillips (ed.), AIP Conference Proc. **474** (1998), p. 261.

273 C. B. Duke and M. E. Alferief: Field Emission through Atoms Adsorbed on a Metal, *Surf. J. Chem. Phys.* **46**, 923 (1967).

274 Q. S. Shu et al.: Influence of Condensed Gases on Field Emission and the Performance of Superconducting RF Cavities, *IEEE Transactions on Magnetics: Proceedings of 14th Applied Superconductivity Conference MAG-25* (1989), p. 1868.

275 K. Saito: Q-Drop at High Gradient, Prospect of Higher Q for CW, Critical RF SC Field, *Proceedings of 11th Workshop on RF Superconductivity*, Travemünde (2003), paper THWG1.

276 B. Visentin: Improvements of Superconducting Cavity Performance at High Accelerating Gradients, *Proceedings of EPAC 1998, European Particle Accelerator Conference*, Vol. III, Stockholm (1998), p. 1885.

277 H. Padamsee et al.: Cornell Status Report, *Proceedings of 10th Workshop on RF Superconductivity*, Tsukuba (2001), p. 15.

278 L. Lilje et al.: Electropolishing and In-Situ Baking of 1.3 GHz Niobium Cavities, *Proceedings of 9th Workshop on RF superconductivity*, Santa Fe (1999), paper TuA001.

279 G. Ciovati and P. Kneisel: Measurements of the High-Field Q-Drop in TE_{011}/TM_{010} Mode in a Single-Cell Cavity, *Proceedings of Workshop on Pushing the Limits of RF Superconductivity*, Argonne National Laboratory, Report ANL-05/10, Argonne (2005).

280 G. Ciovati: Review of Frontier Workshop and Q-Slope Results, *Proceedings of 12th Workshop on RF Superconductivity*, Ithaca (2005), p. 167.

281 J. Halbritter: Degradation of Superconducting RF Cavity Performances by Extrinsic Properties, *Proceedings of 11th Workshop on RF Superconductivity*, Travemünde (2003), paper MoP44.

282 G. Eremeev: A Comparison of Large Grain and Fine Grain Cavities Using Thermometry, *Proceedings of EPAC*

2006, European Particle Accelerator Conference, Edinburgh (2006), p. 475.
283. B. Visentin et al.: High Gradient Q-Slope: Non In-Situ Baking, Surface Treatment by Plasma and Similarities between BCP and EP Cavities, Proceedings of 11th Workshop on RF Superconductivity, Travemünde (2003), MoP19.
284. G. Eremeev: High Field Q-slope's Studies Using Thermometry, PhD Thesis, Cornell University, Ithaca, (2008).
285. J. Hao: Low Temperature Heat Treatment Effect on High-Field EP Cavities, Proceedings of 11th Workshop on RF Superconductivity, Travemünde (2003), paper MoP16.
286. L. Lilje et al.: Improved Surface Treatment of the Superconducting TESLA Cavities, Nucl. Instrum. Methods Phys. Res. A, **516**, 213 (2004).
287. J. Wiener and H. Padamsee: High Field Q-Slope Onset in EP and BCP Cavities Before Bake, SRF Note 080326-06, Cornell University.
288. B. Visentin et al.: Cavity Baking: A Cure for the High Accelerator Field Q_0 Drop, Proceedings of 9th Workshop on RF Superconductivity, Santa Fe (2000), p. 198.
289. L. Lilje et al.: Electropolishing and In-Situ Baking of 1.3 GHz Niobium Cavities, Proceedings of 9th Workshop on RF Superconductivity, Santa Fe (2000), p. 74.
290. B. Visentin: Q-Slope at High Gradients: Review of Experiments and Theories, Proceedings of 11th Workshop on RF Superconductivity, Travemünde (2003), paper TuO01.
291. H. Padamsee et al.: Recent Q-Slope and Related Surface Studies at Cornell, Proceedings of Workshop on Pushing the Limits of RF Superconductivity, September, 2004, Argonne National Laboratory, Report ANL-05/10 (2004), p. 291.
292. B. Visentin: A Non-Electropolished Niobium Cavity Reached 40 MV/m at Saclay, Proceedings of EPAC 2002, European Particle Accelerator Conference, Paris (2002), p. 2292.
293. R. Losito et al.: Report on Superconducting RF Activities at CERN, Proceedings of 10th Workshop on RF Superconductivity, Tsukuba (2001), paper TL004.
294. L. Lilje: R&D in RF Superconductivity to Support the International Linear Collider, Proceedings of PAC 2007, Particle Accelerator Conference, Albuquerque (2007), p. 2559.
295. G. Eremeev: Effect of Mild Baking on High Field Q-Drop of BCP Cavity, Proceedings of 11th Workshop on RF Superconductivity, Travemünde (2003), paper MoP18.
296. L. Lilje: Experimental Investigations on Superconducting Niobium Cavities at Highest Radio-Frequency Fields, PhD Dissertation, University of Hamburg (2001).
297. K. Saito: Surface Smoothness for High Gradient Niobium SCRF Cavities, Proceedings of 11th Workshop on RF Superconductivity, Travemünde (2003), paper THP15.
298. C. Boffo: Electropolishing on Small Samples at Fermilab, Proceedings of 12th Workshop on RF Superconductivity, Ithaca (2005), p. 447.
299. H. Tian: Recent XPS Studies of the Effect of Processing on Nb SRF Surfaces, Fermilab Materials Workshop (2007).
300. J. Knobloch et al.: High-Field Q Slope in Superconducting Cavities Due to Magnetic Field Enhancement, Proceedings of 9th Workshop on RF Superconductivity, Santa Fe (1999), p. 77.
301. J. Shipman: A Comparison of Grain Boundary Morphology of RRR 300 and RRR 500 Niobium after Various Polishing Processes, and Possible Implications for Q-Slope, Master Thesis, Cornell University, Ithaca, (2004).
302. R.L. Geng et al.: Microstructures of RF Surfaces in the Electron Beam Weld Region of Niobium, Proceedings of 9th Workshop on RF Superconductivity, Santa Fe (1999), p. 238.
303. J. Kaufman: Cornell University, Ithaca, private communication.
304. T. Hays: Cornell University, Ithaca, private communication.
305. C.C. Koch et al.: Effects of Interstitial Oxygen on the Superconductivity of Niobium, Phys. Rev. B **9** 888 (1974).
306. G. Horz et al. (eds.): Physics Data: Gases and Carbon in Metals, Part VIII:

Nb, Fachinformationszentrum Energie, Physik, Mathematik, Karlsruhe (1981).
307 W. Weingarten: On the Dependence of the Q-Value on the Accelerating Gradient for Superconducting Cavities, *Proceedings of 13th Workshop on RF Superconductivity*, Beijing (2007), paper TuP16.
308 B. Visentin: First Results on Fast Baking, *Proceedings of 12th Workshop on RF Superconductivity*, Ithaca (2005), p. 241.
309 H. Tian: Recent XPS Studies of the Effect of Processing on Nb SRF Surfaces, *Proceedings of 13th Workshop on RF Superconductivity*, Beijing (2007), paper TuP18.
310 A. T. Fromhold and E. L. Cook: Kinetics of Oxide Film Growth on Metal Crystals: Electron Tunneling and Ionic Diffusion, *Phys. Rev.* **158**, 158 (1967).
311 G. Eremeev and H. Padamsee: Change in High Field Q-slope By Baking and Anodizing, p. 236, and *Physica C* **441**(1–2), 62 (2006).
312 G. Ciovati: Measurement of the High-Field Q Drop in a High-Purity Large-Grain Niobium Cavity for Different Oxidation Processes, *Phys. Rev. Special Topics – Accel. Beams* **10**, 062002 (2007).
313 C. Antoine: Overview of Surface Measurements: What Do Surface Studies Tell Us about Q-Slope? *Proceedings of Workshop on Pushing the Limits of RF Superconductivity*, Argonne National Laboratory, Argonne, Report ANL-05/10 (2005), p. 65; and C. Antoine et al.: Surface Studies: Method of Analysis and Results, *Proceedings of 10th Workshop on RF Superconductivity*, Tsukuba (2001), p. 272.
314 M. Rabinowitz: Analysis of Critical Power Loss in a Superconductor, *J. Appl. Phys.* **42**, 88 (1971).
315 P. J. Lee et al.: Grain Boundary Flux Penetration and Resistivity in Large Grain Niobium Sheet, *Proceedings of 12th Workshop on RF Superconductivity*, Ithaca (2005), p. 372.
316 P. J. Lee et al.: Flux Penetration into Grain Boundaries Large Grain Niobium Sheet for SRF Cavities, *Single Crystal-Large Grain Niobium Technology Workshop*, CBMM Arax, Braz 2006, *AIP Conf. Proc.* **927** (2006), p. 113.
317 M. Rabinowitz: Frequency Dependence of Superconducting Cavity Q and Magnetic Breakdown Field, *Appl. Phys. Lett.* **19**, 73 (1971).
318 C. P. Bean and J. D. Livingston: Surface Barrier in Type-II Superconductors, *Phys. Rev. Lett.* **12**, 14 (1964); D. Blois and W. de Sorbo: Surface Barrier in Type-II Superconductors, *Phys. Rev. Lett.* **12**, 499 (1964).
319 A. Joseph and W. J. Tomasch: Experimental Evidence for Delayed Entry of Flux into a Type-II Superconductor, *Phys. Rev. Lett.* **12**, 219 (1964).
320 P. Kneisel: Progress on Large Grain and Single Grain Niobium – Ingots and Sheet, *Proceedings of 13th Workshop on RF Superconductivity*, Beijing (2007), paper TH102.
321 W. Singer: Experiences with Multi-Cell Large Grain Superconducting RF Cavities, *TTC Meeting at FNAL*, Batavia (2007).
322 P. Kneisel: private communication.
323 P. Kneisel et al.: Performance of Large Grain and Single Crystal Niobium Cavities, *Proceedings of 12th Workshop on RF Superconductivity*, Ithaca (2005), p. 134.
324 P. Kneisel: Surface Characterization of Bulk Nb: What Has Been Done? What Has Been Learnt? *Proceedings of 11th Workshop on RF Superconductivity*, Travemünde (2003), paper TuO02.
325 M. J. Kelly: Overview of Characterization Methodology, *SRF Materials Workshop*, Fermilab (2007).
326 H. Tian et al.: Near-Surface Composition of Electropolished Niobium by Variable Photon Energy XPS, *Proceedings of 11th Workshop on RF Superconductivity*, Travemünde (2003), paper MoP15.
327 H. Tian et al.: Surface Studies of Niobium Chemically Polished under Conditions for SRF Cavity Production, *Proceedings of 12th Workshop on RF Superconductivity*, Ithaca (2005), paper TuA10.
328 A. Dacca: PhD Thesis, INFN and Universita di Genova (2000).

329 R. Ballantini et al.: Improvement of the Maximum Field of Accelerating Cavities by Dry Oxidation, *Proceedings of 9th Workshop on RF Superconductivity*, Santa Fe (1999), p. 211.

330 T. B. Massalski et al.: Binary Alloy Phase Diagrams, ASM International, Materials Park (1990).

331 F. Palmer: Influence of Oxide Layers on the Microwave Surface Resistance of Niobium, *IEEE Trans. Magn.* **23**, 1617 (1987).

332 F. Palmer et al.: Oxide Overlayers and the Superconducting RF Properties of Yttrium-Processed High Purity Nb, *Nucl. Instrum. Methods Phys. Res. A* **297**, 321 (1990).

333 A. Romanenko: Review of High Field Q-slope, Surface Measurement, *Proceedings of 13th Workshop on RF Superconductivity*, Beijing (2007), TU103.

334 B. Visentin: Review on the Q-Drop Mechanism, *Thin Films Workshop*, Legnaro (2006).

335 A. D. Batchelor et al.: TEM and SIMS Analysis of (100), (110), and 222 Single Crystal Niobium, *Single Crystal-Large Grain Niobium Technology Workshop*, CBMM Arax, Braz October 2006, *AIP Conf. Proc.* **927**, 72 (2006).

336 J. Sebastian et al.: Atom-Probe Tomography Analyses of Niobium Superconducting RF Cavity Materials, *Proceedings of 12th Workshop on RF Superconductivity*, Ithaca (2005), p. 244.

337 H. Padamsee et al.: Calculation for Breakdown Induced by 'Large Defects' in Superconducting Niobium Cavities, *IEEE Trans. Magn.*, **19**, 1322 (1983).

338 H. Padamsee: The Technology of Nb Production and Purification, *Proceedings of 2nd Workshop on RF Superconductivity*, CERN, Geneva (1984), p. 339.

339 H. Piel: Superconducting Cavities, *CERN Accelerator School*, May–June 1988, CERN 89-04, S. Turner (ed) (1989), p. 149.

340 K. Krafft: Thermal Transport and Thermal-Magnetic Breakdown in Superconducting Cavities Made of High Thermal Conductivity Niobium, PhD Thesis, Cornell University, Ithaca (1983).

341 L. Lilje: DESY, private communication.

342 J. Preble: CEBAF Energy Upgrade Program Including Re-Work of CEBAF Cavities, *Proceedings of 13th Workshop on RF Superconductivity*, Beijing (2007), paper FR104.

343 J. Wiener and H. Padamsee: Thermal and Statistical Models for Quench in Superconducting Cavities, TTC-Report 2008-08 (2008).

344 XFEL TDR, Chapter 4, p. 58 (2008).

345 G. R. Myneni: Physical and Mechanical Properties of Nb for SRF Science and Technology, *Single Crystal-Large Grain Niobium Technology Workshop*, CBMM Arax, Braz 2006, *AIP Conf. Proc.* **927**, 41 (2006).

346 G. R. Myneni: Physical and Mechanical Properties of Single and Large Crystal High-RRR Niobium, *Proceedings of 12th Workshop on RF Superconductivity*, Ithaca (2005), paper MoP08.

347 ATI Wah Chang: Teledyne Wah Chang, Albany, 97321-0136 USA.

348 W. C. Heraeus: 63450 Hanau, Germany.

349 Tokyo Denkai Co.: Ltd., Higashisuna, Japan.

350 Plansee Vertriebsniederlassung Deutschland: Schützenstraße 29, 72574 Bad Urach, Germany.

351 Ningxia: 109 Metallurgy Rod, Dawukou Borough, Shizuishan, Ningxia, China.

352 Cabot: Cabot Performance Materials, Boyertown, PA 19512-1607, USA.

353 CBMM: Companhia Brasileira de Metalurgia e Mineracão, Córrego da Mata S/N, Caixa Postal 8, Araxá, Brazil.

354 H. C. Starck GmbH: Lorenz-Hutschenreuther-Str. 81, 95100 Selb, Germany and 45 Industrial Place, Newton MA, USA.

355 Giredmet: Bolshoy Tolmachevsky Per., 5, 109017 Moscow, Russia.

356 A. Brinkmann: Progress of the Test Cavity Program for the European XFEL, *Proceedings of 13th Workshop on RF Superconductivity*, Beijing (2007), paper TuP74.

357 H. Umezawa: Current Status and Future Plan for RRR-Grade Niobium Production in Tokyo Denkai, *Single Crystal-Large Grain Niobium Technology Workshop*, CBMM Arax, BrazOctober 2006, *AIP Conf. Proc.* **927**, 186 (2006).

358 R. A. Graham: RRR Niobium Manufacturing Experience, *Single Crystal-Large Grain Niobium Technology Workshop*, CBMM Arax, BrazOctober 2006, *AIP Conf. Proc.* **927**, 191 (2006).

359 H. Padamsee et al.: Advances in Production of High Purity Nb for RF Superconductivity, *IEEE Trans. Magn.* **23**, 1607 (1987).

360 P. Jepsen, H. C. Starck: *Single Crystal-Large Grain Niobium Technology Workshop*, CBMM Arax, October (2006).

361 W. Singer et al.: Diagnostic of Defects in High Purity Niobium, *Proceedings of 8th Workshop on RF Superconductivity*, Abano Terme (1998), p. 850.

362 W. Singer et al.: A. Farr et al.: SQUID-Based Scanning System for Detecting Defects in Nb Sheets for RF Cavities, *Proceedings of 12th Workshop on RF Superconductivity*, Ithaca (2005), paper TuP49.

363 A. Farr et al.: Testing of Nb Sheets on a SQUID NDE System with Large Scale x/y Table for Use in Industrial Environment, *Proceedings of 11th Workshop on RF Superconductivity*, Travemünde (2003), paper ThP40.

364 B. Spaniol: W. C. Heraeus GmbH and Its Activities Regarding Large Grain Niobium Discs, *Single Crystal-Large Grain Niobium Technology Workshop*, CBMM Arax, Braz 2006, *AIP Conf. Proc.* **927**, 179 and Talk (2006).

365 D. Proch: TESLA Cavity, Cryomodule Production Model, *DESY Main Linac Meeting*, May 11–14 (2006).

366 W. Singer et al.: Large Grain/Single Crystal RF Cavities, *Single Crystal-Large Grain Niobium Technology Workshop*, CBMM Arax, Braz 2006, *AIP Conf. Proc.* **927**, 123 and 133 (2006).

367 M. Pekeler et al.: Experiences in Large Grain/Single Crystal Cavity Fabrication, *Single Crystal-Large Grain Niobium Technology Workshop*, CBMM Arax, Braz 2006, *AIP Conf. Proc.* **927**, 141 (2006).

368 P. Kneisel et al.: Development of Large Grain/Single Crystal Niobium Cavity Technology at Jefferson Lab, *Single Crystal-Large Grain Niobium Technology Workshop*, CBMM Arax, BrazOctober 2006, *AIP Conf. Proc.* **927**, 84 (2006).

369 K. Saito: Gradient Yield Improvement Efforts for Single and Multi-Cells and Progress for Very High Gradient Cavities, *Proceedings of 13th Workshop on RF Superconductivity*, Beijing (2007), paper TU202.

370 T. Saeki et al.: Series Tests of High-Gradient Single-Cell Superconducting Cavity for the Establishment of KEK Recipe, *Proceedings of EPAC 2006, European Particle Accelerator Conference*, Edinburgh (2006), p. 756.

371 W. Singer: Large-Grain Superconducting RF Cavities at DESY, *Proceedings of LINAC 2006, Linear Accelerator Conference*, Knoxville (2006), p. 327.

372 W. Singer: Advances in Large Grain/Single Crystal SC Resonators at DESY, *Proceedings of PAC 2007, Particle Accelerator Conference*, Albuquerque (2007), p. 2569.

373 P. Bauer et al.: Recent RRR Measurements on Niobium at Fermilab, *Proceedings of 12th Workshop on RF Superconductivity*, Ithaca (2005), p. 352.

374 J. Sears: Developments in Electron Beam Welding of Niobium Cavities, *Proceedings of 12th Workshop on RF Superconductivity*, Ithaca (2005), p. 481.

375 M. P. Kelly et al.: Superconducting $\beta=0.15$ Quarter-Wave Cavity for RIA, *Proceedings of LINAC 2004, Linear Accelerator Conference* (2004), p. 605.

376 K. W. Shepard: Development of Spoke Cavities for RIA, *Proceedings of 12th Workshop on RF Superconductivity*, Ithaca (2005), p. 334.

377 W. Singer: Seamless/Bonded Niobium Cavities, *Proceedings of 12th Workshop on RF Superconductivity*, Ithaca (2005), p. 143.

378 W. Singer: Progress in Seamless Cavities, *Proceedings of 13th Workshop on RF Superconductivity*, Beijing (2007), paper WE301.

379 K. Ueno: Seamless Cavity R&D in KEK, *TTC Meeting at KEK*, September 26 (2006).

380 I. Itoh et al.: Hot Roll Bonding Method for Nb/Cu Clad Seamless SC Cavity, *Proceedings of 11th Workshop on RF Superconductivity*, Travemünde (2003).

381 K. Enami et al.: Development of Nb/Cu Clad Seamless Cavity, *Proceedings of*

12th Workshop on RF Superconductivity, Ithaca (2005), paper TuP59.

382 P. Kneisel: Development of a Superconducting Connection for Niobium Cavities, *Proceedings of PAC 2007, Particle Accelerator Conference*, Albuquerque (2007), p. 2484.

383 W. Singer: Hydroforming of Superconducting TESLA Cavities, *Proceedings of 10th Workshop on RF Superconductivity*, Tsukuba (2001), p. 170.

384 V. Palmieri: Advancements on Spinning of Seamless Multicell Reentrant Cavities, *Proceedings of 11th Workshop on RF Superconductivity*, Travemünde (2003), paper TuP26.

385 S. Calatroni: 20 Years of Experience with the Nb/Cu Technology for Superconducting Cavities and Perspectives for Future Developments, *Proceedings of 12th Workshop on RF Superconductivity*, Ithaca (2005), p. 149.

386 E. Chiaveri: The CERN Nb/Cu Programme for the LHC and Reduced-Superconducting Cavities, *Proceedings of 9th Workshop on RF Superconductivity*, Santa Fe (1999), p. 352.

387 A. Mosnier et al.: Design of a Heavily Damped Superconducting Cavity for SOLE *Proceedings of PAC 97, Particle Accelerator Conference*, Vancouver (1997), p. 1709.

388 P. Bosland: Completion of the Soleil Cryomodule, *Proceedings of 9th Workshop on RF Superconductivity*, Santa Fe (1999), paper WEP009.

389 P. Bosland et al.: Upgrade of the Cryomodule Prototype before its Implementation in SOLEIL, *Proceedings of EPAC2004, European Particle Accelerator Conference*, Lucerne (2004), p. 2329.

390 C. Benvenuti et al.: CERN Studies on Niobium-Coated 1.5 GHz Copper Cavities, *Proceedings of 10th Workshop on RF Superconductivity*, Tsukuba (2001), p. 252.

391 S. Calatroni: 20 Years of Experience with the Nb/Cu Technology for Superconducting Cavities and Perspectives for Future Developments, *Physica C* **441**, 95 (2006).

392 M. Svandrlik et al.: The SUPER-3HC Project: An Idle Superconducting Harmonic Cavity for Bunch Length Manipulation, *Proceedings of EPAC'00*, Vienna (2000), p. 2052.

393 M. Pedrozzi et al.: First Operational Results of the 3rd Harmonic Superconducting Cavities in SLS and ELETTRA, *Proceedings of PAC 2003, Particle Accelerator Conference*, Portland (2003), p. 878.

394 O. Aberle: Technical Developments on Reduced- Superconducting Cavities at CERN, *Proceedings of the 1999 Particle Accelerator Conference*, New York (1999), p. 878.

395 A. M. Porcellato et al.: Niobium Sputtered QWRs, *Proceedings of 12th Workshop on RF Superconductivity*, Ithaca (2005), p. 118.

396 Chen Jiaer: The Growth of SRF in *Proceedings of 13th Workshop on RF Superconductivity*, Beijing (2007), paper MO101.

397 A. Gurevich: Dynamics of Vortex Penetration, Jumpwise Instabilities and High-Field Surface Resistance, *Proceedings of 13th Workshop on RF Superconductivity*, Beijing (2007), paper Tu104.

398 D. Tonini et al.: Morphology of Niobium Films Sputtered at Different Target-Substrate Angle, *Proceedings of 11th Workshop on RF Superconductivity*, Travemünde (2003), paper ThP11.

399 R. Russo et al.: Cathodic Arc Grown Niobium Films for RF Superconducting Cavity Applications, *Proceedings of 12th Workshop on RF Superconductivity*, Ithaca, p. 385 (2005) and High Quality Superconducting Niobium Films Produced by an Ultrahigh Vacuum Cathodic Arc, *Supercond. Sci. Technol.* **18** (2005).

400 A. Romanenko and H. Padamsee: RF Properties at 6 GHz of Cathodic Arc Films up to 300 Oe, *Proceedings of 12th Workshop on RF Superconductivity*, Ithaca (2005), p. 388.

401 G. Wu et al.: A Prototype of 500 MHz Cavity Coating System by ECR Plasma, *Proceedings of 12th Workshop on RF Superconductivity*, Ithaca (2005), paper TuP64.

402 P. Kneisel et al.: Performance of 1300 MHz KEK-Type Single-Cell Nio-

bium Cavities, *Proceedings of 8th Workshop on RF Superconductivity*, Abano Terme (1997), p. 463.

403 P. Kneisel and B. Lewis: Advanced Surface Cleaning Methods – Three Years of Experience with High Pressure Ultra-Pure Water Rinsing of Superconducting Cavities, *Part. Accel.* **53**, 97 (1996).

404 Q. S. Shu et al.: Reducing Field Emission in Superconducting RF Cavities for the Next Generation of Particle Accelerators, *Proceedings of the 1990 Applied Superconductivity Conference, IEEE Trans. Magn.*, **27**, 1935 (1991).

405 H. Safa: High Gradients in SCRF Cavities, *Proceedings of 8th Workshop on RF Superconductivity*, Abano Terme (1998), p. 814.

406 E. Kako et al.: Improvement of Cavity Performance in the Saclay/Cornell/DESY SC Cavities, *Proceedings of 9th Workshop on RF Superconductivity*, Santa Fe (1999), p. 179.

407 A. Labanc: Electrical Axes of TESLA-Type Cavities, TESLA Report 2008-01 (2008).

408 Y. Iwashita: Development of High Resolution Camera and Observations of Superconducting Cavities, *Proceedings of EPAC 2008*, European Particle Accelerator Conference, Genoa (2008), p. 1956.

409 A. Matheisen: Cavity Preparation, *Proceedings of 13th Workshop on RF Superconductivity*, Beijing (2007), tutorial 3b.

410 K. T. Higuchi et al.: Centrifugal Barrel Polishing of L-band Niobium Cavities, *Proceedings of 10th Workshop on RF Superconductivity*, Tsukuba (2001), p. 431.

411 G. Issarovitch et al.: Development of Centrifugal Barrel Polishing for Treatment of Superconducting Cavities, *Proceedings of 11th Workshop on RF Superconductivity*, Travemünde (2003), paper TuP56.

412 R. L. Geng et al.: Continuous Current Oscillation Electropolishing and Application to Half-Cells, *Proceedings of 11th Workshop on RF Superconductivity*, Travemünde (2003), paper TuP13.

413 R. L. Geng et al.: Vertical Electropolishing Niobium Cavities, *Proceedings of 12th Workshop on RF Superconductivity*, Ithaca (2005), p. 459.

414 H. Diepers et al.: A New Method of Electropolishing Niobium, *Phys. Lett.*, **37A**, 139, (1971).

415 N. Steinhau-Kuehl et al.: Electropolishing at DESY, *Proceedings of 11th Workshop on RF Superconductivity*, Travemünde (2003), paper TuP46.

416 N. Steinhau-Kuehl et al.: Update on the Experiences of Electropolishing of Multi-Cell Resonators at DESY, *Proceedings of 12th Workshop on RF Superconductivity*, Ithaca (2005), p. 464.

417 A. Matheisen et al.: Preparation Sequences for Electropolished High Gradient Multi-Cell Cavities at DESY, *Proceedings of 12th Workshop on RF Superconductivity*, Ithaca (2005), p. 470.

418 K. Saito et al.: R&D of Superconducting Cavities at KEK, *Proceedings of 4th Workshop on RF Superconductivity*, KEK, Tsukuba (1989), p. 18.

419 K. Saito et al.: Electropolishing of L-band cavities, *Proceedings of 4th Workshop on RF Superconductivity*, KEK, Tsukuba (1989), p. 635.

420 B. Visentin: SRF Activities in Europe, *TTC Meeting, Fermi National Accelerator Laboratory*, 23–26 April (2007).

421 V. Palmieri: Fundamentals of Electrochemistry – The Electrolytic Polishing of Metals: Application to Copper and Niobium, *Proceedings of 11th Workshop on RF Superconductivity*, Travemünde (2003), paper WeT02.

422 C. Bonavolonta et al.: Application of Flux Gate Magnetometry to Electropolishing, *Proceedings of 11th Workshop on RF Superconductivity*, Travemünde (2003), paper TuP25.

423 P. A. Jacquet: On the Anodic Behavior of Copper in Aqueous Solutions of Orthophosphoric Acid, *Trans. Electrochem. Soc*, **69**, 629 (1936).

424 W. C. Elmore: Electrolytic Polishing, *J. Appl. Phys.*, **10**, 724 (1939).

425 F. Eozenou et al.: Efficiency of Electropolishing Versus Bath Composition and Aging: First Results, *Proceedings of 12th Workshop on RF Superconductivity*, Ithaca (2005), p. 451.

426 A. Aspart et al.: Aluminum and Sulphur Impurities in Electropolishing Baths, *Proceedings of 12th Workshop on*

RF Superconductivity, Ithaca (2005), paper 455.
427 C. Hartmann: Analysis Technique of the HF-H_2SO_4-Electrolyte, *TTC-Meeting at KEK 2006*, September (2006).
428 C. Antoine: Some Current Issues About EP, *TTC Meeting at KEK*, September (2006).
429 K. Saito: EP Control, *TTC-Meeting at KEK*, September (2006).
430 M. Kelly: EP review at Argonne, private communication.
431 T. Tajima: Design of a New Electropolishing System for SRF Cavities, *Proceedings of EPAC 2006, European Particle Accelerator Conference*, Edinburgh (2006), p. 485.
432 J. Mammosser et al.: ILC Cavity Qualifications – Americas, *TTC meeting at FNAL*, April (2007).
433 B. van der Horst: Update on Cavity Preparation for High Gradient Superconducting Multicell Cavities at DESY, *Proceedings of 13th Workshop on RF Superconductivity*, Beijing (2007), paper TUP30.
434 H. Padamsee: Results on 9-Cell ILC and 9-Cell Re-entrant Cavities, *Proceedings of PAC07, Particle Accelerator Conference*, Albuquerque (2007), p. 2443.
435 T. Higuchi and K. Saito: Development of Hydrogen-Free EP and Hydrogen Absorption Phenomena, *Proceedings of 11th Workshop on RF Superconductivity*, Travemünde (2003), paper WeO15.
436 M. Grunder: Dissertation, University of Karlsruhe (1977).
437 A. Morgan: Surface Studies of Contaminants Generated During Electropolishing, *Proceedings of PAC07, Particle Accelerator Conference*, Albuquerque (2007), p. 2346.
438 H. Martens et al.: High Critical Magnetic Flux Densities in a Single Piece TM_{010}-X-Band Cavities of Niobium, *Phys. Lett.* 44A, p. 213 (1973).
439 H. Padamsee et al.: Report on 9-Cell ILC and 9-Cell Re-entrant Cavities, *Proceedings of PAC 2007, Particle Accelerator Conference*, Albuquerque (2007), p. 2343.
440 W. J. Ashmanskas et al.: Status of ILC Cavity Processing and Testing at Cornell, *Proceedings of 13th Workshop on RF Superconductivity*, Beijing (2007).
441 G. Wu, private communication, FNAL.
442 C. Pagani, Overview of ILC, Single Crystal Niobium Technology Workshop, Avaxá (2006).
443 P. Bernard et al.: *Proceedings of EPAC92, European Particle Accelerator Conference*, Berlin (1992), p. 1269.
444 K. Saito et al.: *Proceedings of 6th Workshop on RF Superconductivity*, Newport News (1993), p. 1151.
445 P. Kneisel et al.: *Proceedings of the 6th Workshop on RF Superconductivity*, Newport News (1993), p. 628.
446 K. Escherich et al.: Clean-Room Facilities for High Gradient Resonator Preparation, *Proceedings of 12th Workshop on RF Superconductivity*, Ithaca (2005), p. 486.
447 N. Krupka et al.: Quality Control at the TTF-Cleanroom Infrastructure for Cavity-Processing, *Proceedings of 11th Workshop on RF Superconductivity*, Travemünde (2003), paper TuP18.
448 N. Krupka et al.: Update on Quality Control of the Clean-Room for Superconducting Multi-Cell Cavities at DESY, *Proceedings of 12th Workshop on RF Superconductivity*, Ithaca (2005), p. 483.
449 D. Reschke: Cleanliness Techniques, *Proceedings of 12th Workshop on RF Superconductivity*, Ithaca (2005), p. 71.
450 T. Ebeling et al.: Processing of TTF Cavities at DESY, *Proceedings of 11th Workshop on RF Superconductivity*, Travemünde (2003), paper TuP07.
451 E. Cavaliere: High Pressure Rinsing Parameters Measurements, *Proceedings of 12th Workshop on RF Superconductivity*, Ithaca (2005), paper THP10.
452 M. Luong et al.: Understanding and Processing of the Field Emission Enhanced by Conducting Protrusions, *Proceedings of 8th Workshop on RF Superconductivity*, Abano Terme (1998), p. 981.
453 J. Knobloch and R. Freyman: Effect of High-Pressure Rinsing on Niobium, Cornell SRF Note: 980223-01 (1998).
454 C. Reece: Investigation of HPR-Induced Damage, TTC08, Hamburg (2008).
455 K. Saito et al.: Water Rinsing of the Contaminated Superconducting RF Cav-

ities, *Proceedings of 7th Workshop on RF Superconductivity*, Gif sur Yvette (1995), p. 379.

456 D. L. Tolliver (ed.): Handbook of Contamination Control in Microelectronics, Noyes Publications, Berkshire (1988), ISBN 0-8155-1151-5.

457 W. Kern (ed.): Handbook of Semiconductor Cleaning Technology, Noyes Publications, Berkshire (1993), ISBN 0-8155-1331-3.

458 M. S. Champion: RF Input Couplers and Windows: Performances, Limitations, and Recent Developments, *Proceedings of 7th Workshop on RF Superconductivity*, Gif-sur-Yvette (1995), p. 195.

459 D. Proch: Techniques in High-Power Components for SRF Cavities, a Look to the Future, *Proceedings of LINAC2002, Linear Accelerator Conference*, Gyeongju (2002), p. 529.

460 B. Rusnak: RF Power and HOM Coupler Tutorial, *Proceedings of 11th Workshop on RF Superconductivity*, Travemünde (2003).

461 I. Campisi: Fundamental Power Couplers for Superconducting Cavities, *Proceedings of 10th Workshop on RF Superconductivity*, Tsukuba (2001), p. 132.

462 I. Campisi: State-of-the-Art Power Couplers for Superconducting RF Cavities, *Proceedings of EPAC 2002, European Particle Accelerator Conference*, Paris (2002), p. 144.

463 S. Belomestnykh: Review of High Power CW Couplers for Superconducting Cavities, *Proceedings of Workshop on High-Power Couplers for Superconducting Accelerators*, Newport News (2002).

464 S. Belomestnykh: Overview of Input Power Coupler Developments, Pulsed and CW, *Proceedings of 13th Workshop on RF Superconductivity*, Beijing (2007), paper WE305.

465 T. Garvey: The Design and Performance of CW and Pulsed Power Couplers – A Review, *Proceedings of 12th Workshop on RF Superconductivity*, Ithaca (2005), p. 423.

466 T. Garvey: Design and Performance of CW and Pulsed High Power Couplers – A Review, *Physica C*, **441**, 209 (2006).

467 A. Variola: High Power Couplers for Linear Accelerators, *Proceedings of LINAC 2006, Linear Accelerator Conference*, Knoxville (2006), p. 531.

468 S. Noguchi et al.: Recent Status of the TRISTAN Superconducting RF System, *Proceedings of EPAC 1994, European Particle Accelerator Conference*, London (1994), p. 1891.

469 E. Kako et al.: Long Term Performance of the TRISTAN Superconducting RF Cavities, *Proceedings of 1991 Particle Accelerator Conference*, San Francisco (1991), p. 2408.

470 G. Cavallari et al.: Acceptance Tests of Superconducting Cavities and Modules for LEP from Industry, *Proceedings of EPAC 1994, European Particle Accelerator Conference*, London (1994), p. 2042.

471 J. Tückmantel et al.: Improvements to Power Couplers for the LEP2 Superconducting Cavities, *Proceedings of PAC 1995, Particle Accelerator Conference*, Dallas (1995), p. 1642.

472 B. Dwersteg et al.: Operating Experience with Superconducting Cavities in HERA, *Proceedings of EPAC 1994, European Particle Accelerator Conference*, London (1994), p. 2039.

473 T. Tajima et al.: The Superconducting Cavity System for KEKB, *Proceedings of PAC 1999, Particle Accelerator Conference*, New York (1999), p. 440.

474 S. Mitsunobu et al.: Status and Development of Superconducting Cavity for KEKB, *Proceedings of the 1997 Particle Accelerator Conference*, Vancouver (1997), p. 2908.

475 Y. Kijima et al.: Input Coupler of Superconducting Cavity for KEKB, *Proceedings of EPAC 2000, European Particle Accelerator Conference*, Vienna (2000), p. 2040.

476 E. Chojnacki et al.: RF Power Coupler Performance at CESR and Study of a Multipactor Inhibited Coupler, *Proceedings of 9th Workshop on RF Superconductivity*, Santa Fe (1999), p. 560.

477 E. Chojnacki et al.: Tests and Designs of High-Power Waveguide Vacuum Windows at Cornell, *Part. Accel.*, **61**, 309 (1998).

478 H. P. Kindermann et al.: The Variable Power Coupler for the LHC Supercon-

ducting Cavity, *Proceedings of 9th Workshop on RF Superconductivity*, Santa Fe (1999), p. 566.
479 W. B. Haynes et al.: Testing Status of the Superconducting RF Power Coupler for the APT Accelerator, *Proceedings of 9th Workshop on RF Superconductivity*, Santa Fe (1999), p. 570.
480 W. D. Moeller: High Power Coupler for the TESLA Test Facility, *Proceedings of 9th Workshop on RF Superconductivity*, Santa Fe (1999), p. 577.
481 M. Stirbet et al.: High Power RF Tests on Fundamental Power Couplers for the SNS Project, *Proceedings of EPAC 2002, European Particle Accelerator Conference*, Paris (2002), p. 2283.
482 E. Kako et al.: High Power Input Coupler for the STF Baseline Cavity System, *Proceedings of 13th Workshop on RF Superconductivity*, Beijing (2007), paper TUP60.
483 H. Matsumoto et al.: A New Design for a Superconducting Cavity Input Coupler, *Proceedings of PAC 2005, Particle Accelerator Conference*, Knoxville (2005), p. 4141.
484 T. Saeki et al.: The First Processing of Capacitive coupling Coupler at Room Temperature in a Cryomodule at STF, *Proceedings of 13th Workshop on RF Superconductivity*, Beijing (2007), paper WEP45.
485 V. Shemelin et al.: Dipole-Mode-Free and Kick-Free 2-Cell Cavity for the SC ERL Injector, *Proceedings of PAC 2003, Particle Accelerator Conference*, Portland (2003), p. 2059.
486 M. Liepe: Microphonics Detuning in the 500 MHz Superconducting CESR Cavities, *Proceedings of PAC 2003, Particle Accelerator Conference*, Portland (2003), p. 1326.
487 M. Liepe et al.: Pushing the Limits: RF Field Control at High Loaded Q, *Proceedings of PAC 2005, Particle Accelerator Conference*, Knoxville (2005).
488 M. Liepe: RF Parameter and Field Stability Requirements for the Cornell ERL Prototype, *Proceedings of PAC 2003, Particle Accelerator Conference*, Portland (2003), p. 1329.

489 S. Belomestnykh and H. Padamsee: Performance of the CESR Superconducting RF System and Future Plans, *Proceedings of 10th Workshop on RF Superconductivity*, Tsukuba (2001), p. 197.
490 S. Belomestnykh et al.: Superconducting RF System Upgrade for Short Bunch Operation of CESR, *Proceedings of PAC 2001, Particle Accelerator Conference*, Chicago (2001).
491 J. R. Delayen et al.: An RF Input Coupler System for the CEBAF Energy Upgrade Cryomodule, *Proceedings of PAC 1999, Particle Accelerator Conference*, New York (1999), p. 1462.
492 V. Nguyen et al.: Development of a 50 kW CW L-Band Rectangular Window for Jefferson Lab FEL Cryomodule, *Proceedings of PAC 1999, Particle Accelerator Conference*, New York (1999), p. 1459.
493 E. Somersalo et al.: Analysis of Multipacting in Coaxial Lines, *Proceedings of PAC 1995, Particle Accelerator Conference*, Dallas (1995), p. 1500.
494 P. Ylä-Oijala: Suppressing Electron Multipacting in Coaxial Lines by DC Voltage, TESLA Report 97-21 (1997).
495 J. Tuckmantel: Technical Report 94-25, CERN (1994), CERN LEP-2 Notes.
496 J. Tuckmantel et al.: Improvements to Power Couplers for the LEP2 Superconducting Cavities, *Proceedings of PAC 1995, Particle Accelerator Conference*, Dallas (1995), p. 1642.
497 R. L. Geng and H. Padamsee: Exploring Multipacting Characteristics do a Rectangular Waveguide, *Proceedings of PAC 1999, Particle Accelerator Conference*, New York (1999), p. 429.
498 R. L. Geng et al.: Multipacting in a Rectangular Waveguide, *Proceedings of PAC 2001, Particle Accelerator Conference*, Chicago (2001).
499 B. Yunn and R. M. Sundelin: Field Emitted Electron Trajectories for the CEBAF Cavity, *Proceedings of PAC1993, Particle Accelerator Conference*, Washington DC (1993), p. 1092.
500 L. Phillips et al.: New Window Design Options for CEBAF Energy Upgrade, *Proceedings of PAC 1997, Particle Accelerator Conference*, Vancouver (1997), p. 3102.

501 W. D. Moeller et al.: Development and Testing of RF Double Window Input Power Couplers for TESLA, *Proceedings of 12th Workshop on RF Superconductivity*, Ithaca (2005), p. 571.

502 B. Dwersteg: High Power Windows at DESY, *Proceedings of 8th Workshop on RF Superconductivity*, Abano Terme (1998), p. 740.

503 B. Dwersteg et al.: TESLA RF Power Couplers Development at DESY, *Proceedings of 10th Workshop on RF Superconductivity*, Tsukuba (2001), p. 443.

504 M. Pisharody et al.: High Power Window Tests on a 500 MHz Planar Waveguide Window for the CESR Upgrade, *Proceedings of PAC 1995, Particle Accelerator Conference*, Dallas (1995), p. 1720.

505 N. Jacobsen et al.: Infrared Propagation through Various Waveguide Inner Surface Geometries, Cornell University, Ithaca, SRF 990301-01 (1999).

506 B. Dwersteg: SC-Cavity Operation via WG Transformer, *Proceedings of 4th Workshop on RF Superconductivity*, KEK, Tsukuba (1989), p. 593.

507 V. Veshcherevich and S. Belomestnykh: Correction of the Coupling of CESR RF Cavities to Klystrons Using Three-Post Waveguide Transformers, Report SRF020220-02, Laboratory for Elementary-Particle Physics, Cornell University, Ithaca (2002).

508 W.-D. Moeller, Input Couplers for Superconducting Cavities – Design and Test, *Proceedings of 12th Workshop on RF Superconductivity*, Ithaca (2005), paper SuA03.

509 J. Knobloch et al.: CW Operation of the TTF-III Input Coupler, *Proceedings of PAC 2005, Particle Accelerator Conference*, Knoxville (2005), p. 3292.

510 J. Knobloch et al.: CW Operation of Superconducting TESLA Cavities, *Proceedings of 13th Workshop on RF Superconductivity*, Beijing (2007), paper TUP40.

511 V. Veshcherevich et al.: Design of High Power Input Coupler for Cornell ERL Injector Cavities, *Proceedings of 12th Workshop on RF Superconductivity*, Ithaca (2005), p. 590.

512 V. Veshcherevich et al.: High Power Tests of Input Couplers for Cornell ERL Injector, *Proceedings of 13th Workshop on RF Superconductivity*, Beijing (2007), paper WEP26.

513 M. Kelly: Status of Superconducting Spoke Cavity Development, *Proceedings of 13th Workshop on RF Superconductivity*, Beijing (2007), paper WE302.

514 R. E. Laxdal: Recent Progress in the Superconducting RF Program at TRIUMF/ISAC, *Proceedings of 13th Workshop on RF Superconductivity*, Beijing (2007), paper MO402.

515 S. Belomestnykh and H. Padamsee: Performance of the CESR Superconducting RF System and Future Plans, *Proceedings of the 10th Workshop on RF Superconductivity*, Tsukuba (2001).

516 S. Prat: Industrialization Process for XFEL Power Couplers and Volume Manufacturing, *TTC Meeting at Fermilab*, April (2007).

517 S. Prat et al.: Industrialization Process for XFEL Power Couplers and Volume Manufacturing, *Proceedings of 13th Workshop on RF Superconductivity*, Beijing (2007), paper TH202.

518 T. Garvey et al.: The TESLA High Power Coupler Program at Orsay, *Proceedings of 11th Workshop on RF Superconductivity*, Travemünde (2003), paper MOP02.

519 M. Fouaidy: RRR of Copper Coating and Low Temperature Electrical Resistivity of Material for TTF Couplers, *Proceedings of 12th Workshop on RF Superconductivity*, Ithaca (2005), p. 392.

520 X. Singer: Properties and Structure of Electrodeposited Copper Layers in Parts of the TTF Main Coupler, *Proceedings of 11th Workshop on RF Superconductivity*, Travemünde (2003), paper ThP18.

521 J. Lokiewicz: Characteristics of Tin Anti-Multipactor Layers Reached by Titanium Vapor Deposition on Alumina Coupler Windows, *Proceedings of 11th Workshop on RF Superconductivity*, Travemünde (2003), paper ThP31.

522 T. M. Huang: Some Fabrication Issues on the Spare High Power Input Coupler for BEPCII Superconductor Cavities, *Proceedings of 13th Workshop on RF*

Superconductivity, Beijing (2007), paper WEP51.
523 P. Oijala: Analysis of Electron Multipacting in Coaxial Lines with Traveling and Mixed Waves, TESLA Report, TESLA 97-20 (1997).
524 P. Oijala and M. Ukkola: Multipacting Simulations on the Coaxial SNS Coupler, University of Helsinki Report, November (2000).
525 L. Ge: Multipacting Simulations of TTF-III Power Coupler Components, *Proceedings of PAC 2007, Particle Accelerator Conference*, Albuquerque (2007), p. 2436.
526 R. L. Geng and H. Padamsee: Exploring Multipacting Characteristics of a Rectangular Waveguide, *Proceedings of PAC 1999, Particle Accelerator Conference*, New York (1999), p. 429.
527 R. L. Geng et al.: Multipacting in a Rectangular Waveguide, *Proceedings of PAC 2001, Particle Accelerator Conference*, Chicago (2001), p. 1228.
528 R. L. Geng: Experimental Studies of Electron Multipacting in CESR Type Rectangular Waveguide Couplers, *Proceedings of EPAC 2002, European Particle Accelerator Conference*, Paris (2002), p. 2238.
529 R. L. Geng et al.: Suppression of Multipacting in Rectangular Coupler Waveguides, *Nucl. Instrum. Methods Phys. Res.* A508, 227 (2003).
530 P. Goudket: Studies of Electron Multipacting in CESR Type Rectangular Waveguide Couplers, *Proceedings of EPAC 2004, European Particle Accelerator Conference*, Lucerne (2004), p. 1057.
531 P. Goudket et al.: Multipactor Studies in Rectangular Waveguides, *Proceedings of 11th Workshop on RF Superconductivity*, Travemünde (2003), paper ThP25.
532 R. L. Geng et al.: Dynamics of Multipacting in Rectangular Coupler Waveguides and Suppression Methods, *Proceedings of 11th Workshop on RF Superconductivity*, Travemünde (2003), paper ThP24.
533 MAGIC: ATK Mission Research, 8560 Cinderbed Rd., Suite 700, Newington VA 22122, http://www.mrcwdc.com/magic/.

534 D. Kostin et al.: Testing the FLASH Superconducting Accelerating Modules, *This Workshop, Proceedings of 13th Workshop on RF Superconductivity*, Beijing (2007), WEP05.
535 M. Liepe: Operational Aspects of SC RF Cavities with Beam, *Proceedings of 13th Workshop on RF Superconductivity*, Beijing (2007), Tutorial 5b; M. Dohlus, Short Range Longitudinal Wakefields and Cryogenic Load, Snowmass 2001, WG M3 (2001).
536 K. Ko: Advances in Electromagnetic Modelling through High Performance Computing, *Proceedings of 12th Workshop on RF Superconductivity*, Ithaca (2005), p. 428.
537 R. Wanzenberg: Monopole, Dipole and Quadrupole Passbands of the TESLA 9-Cell Cavity, TESLA Note, 2001-33 (2001).
538 J. Sekutowicz: Higher Order Mode Coupler for TESLA, TESLA note 1994-07 (1994).
539 N. Mildner: A Beam Line HOM Absorber for the European XFEL Linac, *Proceedings of 12th Workshop on RF Superconductivity*, Ithaca (2005), p. 593.
540 C. J. Glasman: Simulations of Transverse Higher Order Deflecting Modes in the Main Linacs of ILC, *Proceedings of 13th Workshop on RF Superconductivity*, Beijing (2007), paper WEP80.
541 C. Magne: Measurement with Beam of the Deflecting Higher Order Modes in the TTF Superconducting Cavities, *Proceedings of PAC 2001, Particle Accelerator Conference*, Chicago (2001), p. 3771.
542 S. Fartoukh: Evidence for a Strongly Coupled Dipole Mode with Insufficient Damping in TTF First Accelerating Module, *Proceedings of PAC 1999, Particle Accelerator Conference*, New York (1999), p. 922.
543 M. Wendt: Beam Based HOM Analysis of Accelerating Structures at the Tesla Test Facility Linac, *Proceedings DIPAC 2003*, Mainz (2003), paper CT11.
544 S. Molloy et al.: Using Higher Order Modes in Superconducting Accelerating Cavities for Beam Monitoring, *Proceedings of LINAC 2006, Linear Accelerator Conference*, Knoxville (2006), p. 271.

545 J. Frisch et al.: High Precision SC Cavity Diagnostics with HOM Measurements, *Proceedings of EPAC 2006, European Particle Accelerator Conference*, Edinburgh (2006), p. 920.

546 L. K. Ko: Comparison of Q_{ext} between Measurements and Calculations using the 3D Code, OMEGA3P, from J. Sekutowicz, ILC Workshop, KEK (2004).

547 L. Xiao: Modeling Imperfection Effects on Dipole Modes in the Tesla Cavity, *Proceedings of PAC 2007, Particle Accelerator Conference*, Albuquerque (2007), p. 2454.

548 J. Sekutowicz: HOM Damping and Power Extraction from Superconducting Cavities, *Proceedings of LINAC 2006, Linear Accelerator Conference*, Knoxville (2006), p. 506.

549 I. Ben-Zvi et al.: Electron Cooling of RHIC, *Proceedings of PAC 2005, Particle Accelerator Conference*, Knoxville (2005); R. Calaga et al.: Study of Higher Order Modes in 5-Cell SRF Cavity for e-Cooling at RHIC, *Proceedings of 11th Workshop on RF Superconductivity*, Travemünde (2003), paper TuP05.

550 S. Belomestnykh et al.: Comparison of the Predicted and Measured Loss Factor of the Superconducting Cavity Assembly for the CESR Upgrade, *Proceedings of PAC 1995, Particle Accelerator Conference*, Dallas (1996), p. 3394.

551 J. Sekutowicz: Higher Order Mode Coupler for TESLA, *Proceedings of 6th Workshop on RF Superconductivity*, Jefferson Lab, Newport News (1993), p. 426; J. Campisi: HOM Filter Issues at SNS, *TTC Meeting at KEK* (2006).

552 P. Kneisel et al.: Coaxial HOM Coupler Designs Tested on a Single-Cell Niobium Cavity, *Proceedings of LINAC 2006, Linear Accelerator Conference*, Knoxville (2006), p. 716.

553 S. Tariq and T. Khabiboulline: FNAL 3.9 GHz HOM Coupler and Coaxial Cable Thermal FEA, *Proceedings of 12th Workshop on RF Superconductivity*, Ithaca (2005), p. 604.

554 E. Harms: Status of 3.9-GHz Superconducting RF Cavity Technology at Fermilab, *Proceedings of LINAC 2006, Linear Accelerator Conference*, Knoxville (2006), p. 695.

555 K. Watanabe: HOM Coupler Study for the KEK(STF) TESLA-Like Cavities, KEK-TTC Meeting WG5, September 26 (2006).

556 G. Wu et al.: Electromagnetic Simulations of Coaxial Type HOM Coupler, *Proceedings of 12th Workshop on RF Superconductivity*, Ithaca (2005), p. 600.

557 N. Solyak: High-Field Test Results of Superconducting 3.9 GHz Accelerating Cavities at FNAL, *Proceedings of LINAC 2006, Linear Accelerator Conference*, Knoxville (2006), p. 722.

558 N. Solyak: New Design of the 3.9 GHz HOM coupler, *TTC Meeting, KEK*, September 25–28 (2006).

559 C. E. Reece et al.: High Thermal Conductivity Cryogenic RF Feedthroughs for Higher Order Mode Couplers, *Proceedings of PAC 2005, Particle Accelerator Conference*, Knoxville (2005).

560 Z. Li et al.: Optimization of the Low Loss SRF Cavity for the ILC, *Proceedings of PAC 2007, Particle Accelerator Conference*, Albuquerque (2007), p. 2439.

561 I. Zagorodnov and M. Dohlus: Coupler Kick, ILC LCWS, Workshop, DESY 31 May (2007).

562 I. Campisi: Artificial Dielectric Ceramics for CEBAF's Higher-Order Mode Loads, *Proceedings of 6th Workshop on RF Superconductivity*, Newport News (1993), p. 587.

563 R. Rimmer et al.: Strongly HOM-Damped Multi-Cell RF Cavities for High-Current Applications, *Proceedings of 11th Workshop on RF Superconductivity*, Travemünde (2003), paper TuP36.

564 R. Rimmer et al.: The JLab Ampere-Class Cryomodule, *Proceedings of 12th Workshop on RF Superconductivity*, Ithaca (2005), p. 567.

565 R. Rimmer: Higher-Order Mode Calculations, Predictions and Overview of Damping Schemes for Energy Recovery Linacs, *ERL 2005 Workshop*, Newport News and *Nucl. Instrum. Methods Phys. Res. A*, **557**(1), 259 (2006).

566 T. Furuya et al.: Superconducting Accelerator Cavity for KEK B-Factory, *Proceedings of 7th Workshop on RF Superconductivity*, Gif-sur-Yvette (1995), p. 729.

567 J. Kirchgessner: The Use of Superconducting RF for High Current Applications, *Part. Accel.*, **46**(1), 151 (1994).

568 S. Belomestnykh: Superconducting RF in Storage-Ring-Based Light Sources, *Proceedings of 13th Workshop on RF Superconductivity*, Beijing (2007), paper MO302.

569 H. Hahn: R-Square Impedance of ERL Ferrite HOM Absorber, *Proceedings of 12th Workshop on RF Superconductivity*, Ithaca (2005), p. 596.

570 T. Furuya et al.: Achievements of the Superconducting Damped Cavities in KEKB Accelerator, *Proceedings of 11th Workshop on RF Superconductivity*, Travemünde (2003), paper MoP21.

571 V. Shemelin: Status of HOM Load for the Cornell ERL Injector, *Proceedings of EPAC 2006, European Particle Accelerator Conference*, Edinburgh, Scotland (2006), p. 478.

572 Transtech Corp., 5520 Adamstown, Road, Adamstown, MD 21710, USA.

573 Ceradyne, Inc., 3169 Red Hill Avenue, Costa Mesa, CA 92626, USA.

574 Elconite is a copper-tungsten alloy supplied by Toshiba Materials 1-1, Shibaura 1-Chome, Minato-ku, Tokyo, 105-8001, Japan or by MALLORY Alloys Group, Mallory House, 106 Park Street Lane, St Albans, Hertfordshire, AL2 2JQ, UK.

575 S. Simrock: Review of Slow and Fast Tuners, *Proceedings of 12th Workshop on RF Superconductivity*, Ithaca (2005), paper ThA07.

576 E. F. Daly: Overview of Existing Mechanical Tuners, *ERL Workshop*, 18–23 March (2005).

577 S. Noguchi: Review of New Tuner Designs, *Proceedings of 13th Workshop on RF Superconductivity*, Beijing (2007), paper WE303.

578 G. Cavallari et al.: The Tuner System for the SC 352 MHz LEP 4-Cell Cavities, *Proceedings of 3rd Workshop on RF Superconductivity*, Argonne (1987), p. 625.

579 S. Simrock: Control of Microphones and Lorentz Force Detuning with a Fast Mechanical Tuner, *Proceedings of 11th Workshop on RF Superconductivity*, Travemünde (2003), paper TuO09.

580 A. Mavanur et al.: Magnetostrictive Tuners for SRF Cavities, *Proceedings of PAC 2003, Particle Accelerator Conference*, Portland (2003), p. 1407.

581 R. A. Rimmer: JLab CW Cryomodules for 4th Generation Light Sources, *Proceedings of 13th Workshop on RF Superconductivity*, Beijing (2007), paper TuP65.

582 M. Liepe: Tuners, Lorentz Force Detuning, ILC BCD Document Preparation Activity at Snowmass. *http://lcdev.kek.jp/GDE/BCD/64471.Liepe.pdf*

583 A. Mosnier: MACSE Superconducting Cavity RF Drive System, *Proceedings of EPAC 90, European Particle Accelerator Conference*, Nice (1990), p. 1064.

584 P. Bosland: Tuning Systems for Superconducting Cavities at Saclay, *SOLEIL Workshop* (2007). 2007/ESLS-RF/ESLS-RF-PRESENTATIONS/07-ESLS07-PBosland.pdf

585 G. Devanz: Active Compensation of Lorentz Force Detuning of a TTF 9-Cell Cavity in CRYHOLAB, *Proceedings of LINAC 2006, Linear Accelerator Conference*, Knoxville (2006), p. 598.

586 D. Barni et al.: A New Tuner for TESLA, *Proceedings of EPAC 2002, European Particle Accelerator Conference*, Paris (2002), p. 2205.

587 H. Kaiser: New Approaches to Tuning of TESLA Resonators, *Proceedings of 9th Workshop on RF Superconductivity*, Santa Fe (1999), p. 324.

588 C. Pagani: ILC Coaxial Blade Tuner, *Proceedings of EPAC 2006, European Particle Accelerator Conference*, Edinburgh (2006), p. 466.

589 J. Sekutowicz: Cold and Beam Test of the First Prototypes of the Superstructure for the TESLA Collider, *Proceedings of PAC 2003, Particle Accelerator Conference*, Portland (2003), p. 467.

590 C. Pagani: Comparative Analysis of Blade Tuner Optimization Options for the ILC, *Proceedings of APAC 2007, Asian Particle Accelerator Conference*, Raja Ramanna Centre for Advanced Technology (RRCAT), Indore (2007), p. 494.

591 T. Higo: Wide-Range Frequency Compensation by Coaxial Ball-Screw Tuner,

THP037, *Proceedings of LINAC 2006, Linear Accelerator Conference*, Knoxville (2006), p. 658.

592 Y. Higashi et al.: Coaxial Ball Screw Tuner for ICHRO 9-Cell Cavity, *Proceedings of 12th Workshop on RF Superconductivity*, Ithaca (2005), paper ThP61.

593 K. Sennyu: Design and Fabrication of Superconducting Cavities for STF, *Proceedings of PAC 2007, Particle Accelerator Conference*, Albuquerque (2007), p. 2674.

594 J. Preble: SNS Cryomodule Performance, *Proceedings of PAC 2003, Particle Accelerator Conference*, Portland (2003), p. 457.

595 G. Davis: Development and Testing of a Prototype Tuner for the CEBAF Upgrade Cryomodule, *Proceedings of PAC 2001, Particle Accelerator Conference*, Chicago (2001), p. 1149.

596 J. R. Delayen and G. K. Davis: Piezoelectric Tuner Compensation of Lorentz Detuning in Superconducting Cavities, *Proceedings of 11th Workshop on RF Superconductivity*, Travemünde (2003), paper ThP22.

597 P. Sekalski et al.: Lorentz Force Detuning Compensation System for Accelerating Field Gradients up to 35 MV/m for Superconducting XFEL and TESLA Nine-Cell Cavities, *CARE Conf 04-001-SRF* (2004).

598 S. N. Simrock: Lorentz Force Compensation of Pulsed SRF Cavities, *Proceedings of LINAC 2002, Linear Accelerator Conference*, Gyeongju (2002), p. 554.

599 R. Amirikas et al.: Vibration Stability Studies of a Superconducting Accelerating Module at Room Temperature, *Proceedings of PAC07, Particle Accelerator Conference*, Albuquerque (2007), p. 2062.

600 P. Pierini et al.: The Wire Position Monitor (WPM) as a Sensor for Mechanical Vibration for TTF Cryomodules, *Proceedings of 12th Workshop on RF Superconductivity*, Ithaca (2005), p. 558.

601 A. Neumann: Microphonics in CW TESLA Cavities and their Compensation with Fast Tuners, *Proceedings of 13th Workshop on RF Superconductivity*, Beijing (2007), paper WE201.

602 P. Sekalski and S. Simrock: Piezoelectric Stack-Based System for Lorentz Force Compensation, *Proceedings of 12th Workshop on RF Superconductivity*, Ithaca (2005), p. 218.

603 M. Fouaidy et al.: Electromechanical, Thermal Properties, and Radiation Hardness Tests of Piezoelectric Actuators at Low Temperature, *Proceedings of 12th Workshop on RF Superconductivity*, Ithaca (2005), p. 608.

604 A. Bosotti et al.: PI Piezo Lifetime Test Report, INFN Milan, Internal Report.

605 C. Tai: A Magnetostrictive Tuning System for Particle Accelerators, *Proceedings of PAC 2005, Particle Accelerator Conference*, Knoxville (2005), p. 3762.

606 P. Sekalski: JRA1-SRF Work Package 8 Tuners, *JRA1 SRF – CARE Annual Meeting*, Frascati, November (2006).

607 J. D. Fuerst: Progress on Cavity Fabrication for the ATLAS Energy Upgrade, *Proceedings of 13th Workshop on RF Superconductivity*, Beijing (2007), paper TUP75.

608 R. E. Laxdal et al.: Recent Progress in the Superconducting RF Program at TRIUMF/ISAC, *Proceedings of 12th Workshop on RF Superconductivity*, Ithaca (2005), p. 128.

609 A. Facco: On-Line Mechanical Instabilities Measurements and Tuner Development in SC Low-Beta Resonators, *Proceedings of EPAC 2004, European Particle Accelerator Conference*, Lucerne (2004), p. 2086.

610 A. Facco et al.: On-Line Performance of the LNL Mechanical Damped Superconducting Low-Beta Resonators, *Proceedings of EPAC'98, European Particle Accelerator Conference*, Stockholm (1998), p. 1846.

611 S. Noguchi: Experience on the New Superconducting RF System in TRISTAN, *Proceedings of 7th Workshop on RF Superconductivity*, Gif-sur-Yvette (1995), p. 163.

612 B. Dwersteg et al.: Operating Experience with Superconducting Cavities in the HERA e-Ring, *Proceedings of 7th Workshop on RF Superconductivity*, Gif-sur-Yvette (1995), p. 151.

613 A. Butterworth et al.: The LEP2 Superconducting RF System, *Nucl. Instrum. Methods Phys. Res. A* **587**, 151 (2008).

614 E. Chiaveri: Large-Scale Industrial Production of Superconducting Cavities, *Proceedings of EPAC 1996, European Particle Accelerator Conference*, Sitges, Barcelona (1996), paper WEZ01a.

615 P. Brown et al.: The LEP Superconducting RF System: Overview, ITRP Visit to DESY, 5th/6th April 2004, The LC Cold Option, Poster SRF19 (2004).

616 L. Evans: LHC: Construction and Commissioning Status, *Proceedings of PAC 2007, Particle Accelerator Conference*, Albuquerque (2007), p. 1.

617 R. Losito et al.: Report on Superconducting RF Activities at CERN from 2001 to 2003, *Proceedings of 11th Workshop on RF Superconductivity*, Travemünde (2003), paper MoPO5.

618 E. Chiaveri: Measurements on the First LHC Acceleration Module, *Proceedings of PAC 2001, Particle Accelerator Conference*, Chicago (2001), p. 481.

619 S. Belomestnykh: Commissioning and Operations Results of the Industry Produced CESR-Type SRF Cryomodules, *Proceedings of PAC 2005, Particle Accelerator Conference*, Knoxville (2005), p. 4235.

620 K. Akai et al.: RF Systems for the KEK B-Factory, *Nucl. Instrum. Methods Phys. Res. A* **499**, 45 (2003).

621 D. Rice: CESR-C – A Frontier Machine for QCD and Weak Decay Physics in the Charm Region, *Proceedings of EPAC 2002, European Particle Accelerator Conference*, Paris (2002), p. 428.

622 M. Pekeler: SRF in Storage Rings, *Proceedings of 12th Workshop on RF Superconductivity*, Ithaca (2005), p. 98.

623 K. Akai: Operational Experience of KEKB-SCRF System, *TTC Meeting at KEK*, September (2006).

624 T. Furuya et al.: Achievements of the Superconducting Damped Cavities in the KEKB Accelerator, *Proceedings of 11th Workshop on RF Superconductivity*, Travemünde (2003), paper MoP21.

625 Z. Li et al.: Fabrication and Test of the 500 MHz SC Modules for the BEPC-II, *Proceedings of 13th Workshop on RF Superconductivity*, Beijing (2007), paper WEP23.

626 C. Wang et al.: Operational Experience of the Superconducting RF Module at TLS, *Proceedings of 12th Workshop on RF Superconductivity*, Ithaca (2005), p. 535.

627 R. Tanner et al.: Canadian Light Source Storage Ring RF System, *Proceedings of 12th Workshop on RF Superconductivity*, Ithaca (2005), p. 532.

628 M. Jensen et al.: Operational Experience of the DIAMOND SCRF System, *Proceedings of 13th Workshop on RF Superconductivity*, Beijing (2007), paper WEP73.

629 J. Liu and H. Hou: SSRF Superconducting RF System, *Proceedings of 13th Workshop on RF Superconductivity*, Beijing (2007), paper WEP38.

630 S. Bauer et al.: Latest Results from the Production of 500 MHz SRF Modules for Light Sources and CESR Upgrade, *Proceedings of 10th Workshop on RF Superconductivity*, Tsukuba (2001), p. 602.

631 P. Marchand: Operation of the SOLEIL RF System, *Proceedings of 13th Workshop on RF Superconductivity*, Beijing (2007), paper WEP53 and *Proceedings of PAC 2007, Particle Accelerator Conference*, Albuquerque (2007), p. 2050.

632 Superconducting RF Cavities for Synchrotron Light Sources, *Proc. EPAC 2004, European Particle Accelerator Conference*, Lucerne (2004), p. 21.

633 J. Rose: Conceptual Design of the NSLS-II RF System, *Proceedings of PAC 2007, Particle Accelerator Conference*, Albuquerque (2007), p. 2550.

634 N. Phinney: ILC Reference Design Report – Accelerator Executive Summary, Beam Dynamics Newsletter No. 42 (2007).

635 M. Svandrlik et al.: The SUPER-3HC Project: An Idle Superconducting Harmonic Cavity for Bunch Length Manipulation, *Proceedings of EPAC 2000, European Particle Accelerator Conference*, Vienna (2000), p. 2052.

636 R. B. Palmer: Energy Scaling, Crab Crossing and the Pair Problem, SLAC-PUB-4707 (1988).

637 R. B. Palmer: *Proceedings of of Snowmass DPF Summer Study*, 1988:0613 (1988).

638 K. Hosoyama: Crab Cavity Development, *Proceedings of 12th Workshop on RF Superconductivity*, Ithaca (2005), p. 439.

639 K. Hosoyama: Construction and Commisioning of KEKB Superconducting Crab Cavities, *Proceedings of 13th Workshop on RF Superconductivity*, Beijing (2007), paper MO405.

640 T. Abe et al.: Compensation of the Crossing Angle with Crab Cavities at KEKB, *Proceedings of PAC 2007, Particle Accelerator Conference*, Albuquerque (2007), p. 27.

641 R. Calaga: Small Angle Crab Compensation for LHC IR Upgrade, *Proceedings of PAC 2007, Particle Accelerator Conference*, Albuquerque (2007), p. 1853.

642 G. Burt: Progress Towards Crab Cavity Solutions for the ILC, *Proceedings of EPAC 2006, European Particle Accelerator Conference*, Edinburgh (2006), p. 724.

643 G. Burt: Status of the ILC Crab Cavity Development, *Proceedings of 13th Workshop on RF Superconductivity*, Beijing (2007), paper WEP42.

644 N. Solyak: Test Results of the 3.9 GHz Cavity at Fermilab, *Proceedings of LINAC 2004, Linear Accelerator Conference*, Lübeck (2004), p. 797.

645 L. Bellantoni: Status of 3.9-GHz Deflecting-Mode (Crab) Cavity R&D, *Proceedings of LINAC 2006, Linear Accelerator Conference*, Knoxville (2006), p. 682.

646 I. Gonin et al.: LHC Crab Cavity with Reduced Outer Diameter, LHC CC08 Meeting, BNL, February 25–26 (2008).

647 J. Corlett: Future FELs, *Proceedings of 12th Workshop on RF Superconductivity*, Ithaca (2005), paper FrA05.

648 A. Zholents et al.: Generation of Subpicosecond X-Ray Pulses using RF Orbit Deflection, *Nucl. Instrum. Methods Phys. Res. A* **425**, 385 (1999).

649 K. Harkay et al.: Generation of Short X-Ray Pulses Using Crab Cavities at the Advanced Photon Source, *Proceedings of PAC 2005, Particle Accelerator Conference*, Knoxville (2005), p. 668.

650 M. Borland et al.: Planned Use of Pulsed Crab Cavities for Short X-Ray Pulsed Generation at the Advanced Photon Source, *Proceedings of PAC 2007, Particle Accelerator Conference*, Albuquerque (2007), p. 1127.

651 L. S. Cardman: The JLab 12 GeV Energy Upgrade of CEBAF for QCD and Hadronic Physics, *Proceedings of PAC 2007, Particle Accelerator Conference*, Albuquerque (2007), p. 58.

652 W. Funk: Advancing Superconducting RF in Support of US DOE Office of Science Projects: The Next Step for Jefferson Lab, *ITRP Poster JLab, ITRP Visit to DESY*, The LC Cold Option, 5th/6th April 2004, Poster SRF08 (2004).

653 A. Hutton: SRF Operating Experience at JLab, *TTC Meeting at KEK* (2006).

654 C. Reece: Performance Experience with the CEBAF SRF Cavities, *Proceedings of PAC 1995, Particle Accelerator Conference*, Dallas (1995), p. 1512.

655 R. A. Rimmer: SRF Performance of CEBAF after Thermal Cycle to Ambient Temperature, *Proceedings of PAC 2005, Particle Accelerator Conference*, Knoxville (2005), p. 665.

656 A. Hutton: JLab Hurricane Recovery, *Proceedings of EPAC 2004, European Particle Accelerator Conference*, Lucerne (2004), p. 1102.

657 W. Funk: The Jefferson Lab 12 GeV Upgrade Project, TTC Meeting at DESY, January (2008).

658 J. Sekutowicz et al.: Cavities for JLab's 12 GeV Upgrade, *Proceedings of PAC 2003, Particle Accelerator Conference*, Portland (2003), p. 1395.

659 C. Reece: A 100 MV Cryomodule for CW Operation, *Proceedings of 12th Workshop on RF Superconductivity*, Ithaca (2005), Talk MoA04.

660 G. Neil: FEL Oscillators, *Proceedings of PAC 2003, Particle Accelerator Conference*, Portland (2003), p. 181.

661 J. Knobloch: Review of SRF Linac-Based FELs (Current and Future), *Proceedings of 13th Workshop on RF Superconductivity*, Beijing (2007), Talk MO301.

662 G. R. Neil and L. Merminga: Technical Approaches for High-Average-Power Free-Electron Lasers, *Rev. Modern Phys.*, **74**, 685 (2002).

663 L. Merminga: Technological Challenges of ERLs, *Proceedings of APAC 2007, Asian Particle Accelerator Conference*,

Raja Ramanna Centre for Advanced Technology (RRCAT), Indore (2007), paper MOOPMA03.
664 D. A. G. Deacon et al.: First Operation of a Free-Electron Laser, *Phys. Rev. Lett.* **38**, 892 (1977).
665 H. A. Schwettman et al.: The Stanford Picosecond FEL Center, *Nucl. Instrum. Methods A* **375**, 662 (1996).
666 S. Benson et al.: High Power Operation of the JLab IR FEL Driver Accelerator, *Proceedings of PAC 2007, Particle Accelerator Conference*, Albuquerque (2007), p. 79.
667 A. P. Freyberger et al.: The CEBAF Energy Recovery Experiment: Update and Future Plans, *Proceedings of EPAC 2004, European Particle Accelerator Conference*, Lucerne (2004), p. 524.
668 E. J. Minehara: Highly Efficient and High-Power Industrial FELs Driven by a Compact, Stand-Alone and Zero-Boil-Off Superconducting RF Linac, *Nucl. Instrum. Methods Phys. Res. A* **483**, 8 (2002).
669 N. Nishimori et al.: Sustained Saturation in a Free-Electron Laser Oscillator at Perfect Synchronism of an Optical Cavity, *Phys. Rev. Lett.* **86**, 5707 (2001).
670 M. Brunken: Latest Developments from the S-DALINAC, *Proceedings of EPAC 1998, European Particle Accelerator Conference*, Stockholm (1998), p. 403.
671 U. Lehnert: First Experiences with the FIR-FEL at ELBE, *Proceedings of FEL 2007*, Novosibirsk, Russia (2007), p. 97.
672 R. Hajima: Overview of Energy-Recovery Linacs, *Proceedings of APAC 2007, Asian Particle Accelerator Conference*, Raja Ramanna Centre for Advanced Technology (RRCAT), Indore (2007), p. 11.
673 R. Bonifacio et al.: Collective Instabilities and High-Gain Regime in a Free-Electron Laser, *Opt. Commun.* **50**, 373 (1984).
674 A. M. Kondratenko and E. L. Saldin: Generating of Coherent Radiation by a Relativistic Electron Beam in an Undulator, *Part. Accel.* **10**, 207 (1980).
675 S. Schreiber: Results from the Free-Electron Laser Flash, *Proceedings of APAC 2007, Asian Particle Accelerator Conference*, Raja Ramanna Centre for Advanced Technology (RRCAT), Indore (2007), Talk MOYMA02.
676 J. Rossbach: A VUV Free Electron Laser at the TESLA Test Facility at DESY, *Nucl. Instrum. Methods Phys. Res. A* **375**, 269 (1996).
677 P. Castro: Demonstration of Exponential Growth and Saturation at VUV Wavelengths at the TTF Free-Electron Laser, *Proceedings of EPAC 2002, European Particle Accelerator Conference*, Paris (2002), p. 221.
678 P. Castro: Overview of SASE Experiments, *Proceedings of LINAC 2000, Linear Accelerator Conference*, Monterey (2000), p. 696.
679 S. Schreiber: Status of the TTF FEL, *Proceedings of 11th Workshop on RF Superconductivity*, Travemünde (2003), paper WeO03.
680 E. Vogel: FLASH Progress Report, *Proceedings of 13th Workshop on RF Superconductivity*, Beijing (2007), paper Mo204.
681 H. Weise: Superconducting RF Structures – Test Facilities and Results, *Proceedings of PAC 2003, Particle Accelerator Conference*, Portland (2003), p. 673.
682 J. G. Weisend: The TESLA Test Facility (TTF) Cryomodule: A Summary of Work to Date, *Proceedings of 1999 Cryogenic Engineering Conference*, Montreal (1999), paper CCB1.
683 K. Floettmann: The European XFEL Project, *Proceedings of 12th Workshop on RF Superconductivity*, Ithaca (2005), p. 638; K. Honkavara: Status of FLASH, FEL 2008, Korea (2008).
684 H. Weise: Report from TTF/VUV-FEL (FLASH), *TTC Meeting at KEK*, September (2006).
685 L. Lilje: State-of-the-Art SRF Cavity Performance, *Proceedings of LINAC 2004, Linear Accelerator Conference*, Lübeck (2004), p. 518.
686 L. Lilje: Experience with the TTF, *Proceedings of PAC 2005, Particle Accelerator Conference*, Knoxville (2005), p. 1.
687 J. Galayda: Status of the LCLS, *Proceedings of EPAC 2008, European Particle Accelerator Conference*, Genova (2008).

688 H. Weise: Status of the European XFEL Project, DESY TTC Meeting at KEK, September (2006).

689 G. Bisoffi et al.: ALPI QWR and S-RFQ Operating Experience, *Proceedings of 13th Workshop on RF Superconductivity*, Beijing (2007), paper MO404.

690 R. C. York: Rare Isotope Accelerator (RIA) Project, *Proceedings of 12th Workshop on RF Superconductivity*, Ithaca (2005), p. 626.

691 M. Di Giacomo: The SPIRAL2 Project at GAN *Proceedings of 12th Workshop on RF Superconductivity*, Ithaca (2005), p. 632.

692 T. Junquera: The SPIRAL2 Project: Construction Progress and Recent Developments on the SC Linac Driver, *Proceedings of 13th Workshop on RF Superconductivity*, Beijing (2007), paper MO401.

693 G. Olry: Development of a Beta 0.12, 88 MHz, Quarter-Wave Resonator and its Cryomodule for the SPIRAL2 Project, *Proceedings of 12th Workshop on RF Superconductivity*, Ithaca (2005), p. 323.

694 G. Olry: Status of the Cryomodules and Cavities Development for the SPIRAL2 Superconducting Linac, *Proceedings of 13th Workshop on RF Superconductivity*, Beijing (2007), paper TUP78.

695 X. Wu: MSU Re-Accelerator – The Re-Acceleration of Low Energy RIBs at the NSCL, *Proceedings of 13th Workshop on RF Superconductivity*, Beijing (2007), paper MO304.

696 S. Henderson: Status of the Spallation Neutron Source: Machine and Science, *Proceedings of PAC 2007, Particle Accelerator Conference*, Albuquerque (2007), p. 7.

697 N. Holtkamp: Status of the SNS Project, *Proceedings of PAC 2003, Particle Accelerator Conference*, Portland (2003), p. 11.

698 J. Mammosser: Status of the SNS Superconducting Cavity Testing and Cryomodule Production, *Proceedings of 11th Workshop on RF Superconductivity*, Travemünde (2003), paper We001.

699 D. S. Stout et al.: Installation of the Spallation Neutron Source (SNS) Superconducting Linac, *Proceedings of PAC 2005, Particle Accelerator Conference*, Knoxville (2005), p. 3838.

700 J. R. Delayen et al.: Performance Overview of the Production Superconducting RF Cavities for the Spallation Neutron Source Linac, *Proceedings of PAC 2005, Particle Accelerator Conference*, Knoxville (2005), p. 4048.

701 J. Ozelis: Test Results of $\beta<1$ Superconducting Elliptical Cavities: Experience and Lessons Learned, *Proceedings of 12th Workshop on RF Superconductivity*, Ithaca (2005), Talk MoPO7.

702 I. E. Campisi: Testing of the SNS Superconducting Cavities and Cryomodules, *Proceedings of PAC 2005, Particle Accelerator Conference*, Knoxville (2005), p. 34.

703 A. Doyuran et al.: Characterization of a High-Gain Harmonic-Generation Free-Electron Laser at Saturation, *Phys. Rev. Lett.* **86**, 5902 (2001).

704 R. J. Bakker: Overview of VUV and Soft X-Ray FELs Worldwide, *Proceedings of EPAC 2008, European Particle Accelerator Conference*, Genova (2008).

705 J. Bisognano: The Wisconsin VUV/Soft X-ray Free Election Laser, *Proceedings of PAC 2007, Particle Accelerator Conference*, Albuquerque (2007), p. 1278.

706 M. Abo-Bakr et al.: STARS – An FEL to Demonstrate Cascaded High-Gain Harmonic Generation, *Proceedings of FEL 2007*, Novosibirsk (2007), p. 220.

707 W. Anders: CW Operation of Superconducting TESLA Cavities, *Proceedings of 13th Workshop on RF Superconductivity*, Beijing (2007), paper TUP40.

708 C. Bruni: The ARC-EN-CIEL FEL Proposal, *Proceedings of EPAC 2006, European Particle Accelerator Conference*, Edinburgh (2006), p. 53; M. E. Couprie et al.: The ARC-EN-CIEL Radiation Sources, *Proceedings of EPAC 08, European Particle Accelerator Conference*, Genova (2008), p. 73.

709 G. H. Hoffstaetter: Progress Toward an ERL Extension to CESR, *Proceedings of PAC 2007, Particle Accelerator Conference*, Albuquerque (2007), p. 107.

710 L. Merminga: Physics Challenges for ERL Light Sources, *Proceedings of EPAC 2004, European Particle Accelerator Conference*, Lucerne (2004), p. 16.

711 D. H. Bilderback: Energy Recovery Linac Experimental Challenges, *Proceedings of FLS 2006, ICFA Advanced Beam Dynamics Workshop on Future Light Sources*, Hamburg (2006), p. 1.

712 A. Bogacz et al.: CEBAF Energy Recovery Experiment, *Proceedings of PAC 2003, Particle Accelerator Conference*, Portland (2003), p. 195.

713 D. H. Bilderback: Review of Third and Next Generation Synchrotron Light Sources, *J. Phys. B: At. Mol. Opt. Phys.* **38**, S773 (2005).

714 D. J. Holder et al.: The Status of the Daresbury Energy Recovery Prototype Project, *Proceedings of EPAC 2006, European Particle Accelerator Conference*, Edinburgh (2006), p. 187.

715 S. L. Smith: A Review of ERL Prototype Experience and Light Source Design Challenges, *Proceedings of EPAC 2006, European Particle Accelerator Conference*, Edinburgh (2006), p. 39.

716 S. L. Smith: ERLP Status, *ERL07 Workshop*, Daresbury Laboratory, May (2007).

717 G. H. Hoffstaetter: ERL Upgrade of an Existing X-Ray Facility: CHESS at CESR, *Proceedings of EPAC 2004, European Particle Accelerator Conference*, Lucerne (2004), p. 497.

718 M. Liepe and J. Knobloch: Superconducting RF for Energy-Recovery Linacs, *Nucl. Instrum. Methods Phys. Res. A* **557**, 354 (2006).

719 M. Liepe: The Cornell ERL Superconducting 2-Cell Injector Cavity String and Test Cryomodule, *Proceedings of PAC07*, Albuquerque (2007), p. 2572.

720 S. A. Belomestnykh: First Test of the Cornell Single-Cavity Horizontal Cryomodule, *Proceedings of EPAC 2008, European Particle Accelerator Conference*, Genova (2008), paper MOPP117.

721 V. Medjidzade: Design of the CW Cornell ERL Injector Cryomodule, *Proceedings of PAC 2005, Particle Accelerator Conference*, Knoxville (2005), p. 4290.

722 E. P. Chojnacki: Design and Fabrication of the Cornell ERL Injector Cryomodule, *Proceedings of EPAC 2008, European Particle Accelerator Conference*, Genova (2008), p. 844.

723 I. Ben-Zvi: Electron Cooling and Electron–Ion Colliders at BNL, *Proceedings of 13th Workshop on RF Superconductivity*, Beijing (2007), paper Fr101.

724 I. Ben-Zvi: High Current R&D ERL, *ERL '07 Workshop*, Daresbury Laboratory, May (2007).

725 I. Ben-Zvi: Next Generation Electron–Ion Colliders, APAC 2007, Raja Ramanna Centre for Advanced Technology (RRCAT), Indore (2007), p. 56.

726 A. Bogacz: Design Studies of High-Luminosity Ring-Ring Electron–Ion Collider at CEBAF, *Proceedings of PAC 2007, Particle Accelerator Conference*, Albuquerque (2007), p. 1935.

727 J. A. Nolen: The U.S. Rare Isotope Accelerator Project, *Nucl. Phys. A* **734**, 661 (2004).

728 S. Schriber: Rare Isotope Heavy Ion Accelerators, *Proceedings of 11th Workshop on RF Superconductivity*, Travemünde (2003), paper FrO04.

729 S. Bousson et al.: Spoke Cavity Developments for the Eurisol Driver, *Proceedings of LINAC 2006, Linear Accelerator Conference*, Knoxville (2006), p. 704.

730 J. L. Biarrotte: High Power CW Superconducting Linacs for EURISOL and XADS, *Proceedings of LINAC 2004, Linear Accelerator Conference*, Lübeck (2004), p. 275.

731 A. Facco: High Intensity Proton Sources, *Proceedings of 11th Workshop on RF Superconductivity*, Travemünde (2003), paper Fr003.

732 F. Gerigk: Future High-Intensity Proton Accelerators, *Proceedings of 13th Workshop on RF Superconductivity*, Beijing (2007), paper FR103.

733 Y. Yamazaki: The JAERI-KEK Joint Project for the High Intensity Proton Accelerator, J-PARC, *Proceedings of PAC 2003, Particle Accelerator Conference*, Portland (2003), p. 576.

734 S. Noguchi: Prototype Cryomodule for the ADS Linac, *Proceedings of 11th Workshop on RF Superconductivity*, Travemünde (2003), paper MoP32.

735 G. W. Foster and J. A. MacLachlan: A Multi-Mission 8 GeV Injector Linac

as a Fermilab Booster Replacement, *Proceedings of LINAC2002, Linear Accelerator Conference*, Gyeongju (2002), p. 826.

736 G. W. Foster et al.: High Power Phase Shifter, *Proceedings of PAC 2005, Particle Accelerator Conference*, Knoxville (2005), p. 3123.

737 G. P. Lawrence: Transmutation and Energy Production with High Power Accelerators, *Proceedings of PAC 1995, Particle Accelerator Conference*, Dallas (1995), p. 35.

738 B. H. Choi et al.: High Power Proton Linac Program in Korea, *Proceedings of LINAC 2002, Linear Accelerator Conference*, Gyeongju (2002), p. 26.

739 J. Bhawalkar et al.: Indian Programme to Develop High Current Proton Linac for ADSS, *Proceedings of LINAC2002, Linear Accelerator Conference*, Gyeongju, p. 70 (2002).

740 A. Puntambekar and M. Karmarkar: Approach Towards Development of 350 MHz Bulk Nb SCRF Cavities for 100 MeV LINAC, *Proceedings of 11th Workshop on RF Superconductivity*, Travemünde (2003), paper TuP34.

741 B. C. Barish: The Global Design Effort for an International Linear Collider, *Proceedings of EPAC 2006, European Particle Accelerator Conference*, Edinburgh (2006), p. 1.

742 ILC Reference Design Report: *http://www.linearcollider.org/cms/?pid=1000437*

743 C. Pagani: ILC, Single Crystal-Large Grain Niobium Technology Workshop, CBMM Arax, BrazOctober 2006, *AIP Conf. Proc.* **927** (10), 3 (2006).

744 D. Kostin et al.: Testing the FLASH Superconducting Accelerating Modules, *Proceedings of 13th Workshop on RF Superconductivity*, Beijing (2007), WEP05.

745 H. Padamsee: Progress and Plans for R&D and the Conceptual Design of the ILC High Gradient Structures, *Proceedings of PAC 2005, Particle Accelerator Conference*, Knoxville (2005), p. 461.

746 L. Lilje: XFEL: Plans for 101 Cryomodules, *Proceedings of 13th Workshop on RF Superconductivity*, Beijing (2007), paper MO102.

747 G. Loew (ed.): ILC TRC report, *http://www.slac.stanford.edu/xorg/ilc-trc/2002/2002/report/03rephome.htm*

748 R. B. Palmer: A Complete Scheme for a Muon Collider, *Proceedings of COOL 2007, Workshop on Beam Cooling*, Bad Kreuznach (2007), p. 77.

749 The $\mu^+\mu^-$ Collider Collaboration: $\mu^+\mu^-$ Collier, A Feasibility Study, Technical Report BNL-52503, Brookhaven (1996).

750 The $\mu^+\mu^-$ Collaboration: *Fermi Lab-Conference 96/092*, LBNL-38946 (1996).

751 R. L. Geng: First RF Test at 4.2K of a 200 MHz Superconducting Nb–Cu Cavity, *Proceedings of PAC 2003, Particle Accelerator Conference*, Portland (2003), p. 1309.

Subject Index

a
absorber, higher order mode (HOM) 298, 311
ALICE, *see* energy recovery linac
ALPI, *see* heavy-ion accelerator
ARC-EN-CIEL, *see* energy recovery linac
ATLAS, *see* heavy-ion accelerator
Atomic Probe Tomography 192

b
baking
– final mild baking 238
– H contamination degassing 238
baking benefit, oxygen diffusion 161
baking effect 71, 150, 168, 171, 180
– oxygen pollution model 159, 160
barrel polishing 240, 241, 237, 247
BCP, buffered chemical polishing 242
– surface roughness due to 242
BCS 4, 4144, 46, 47, 51, 55, 57, 75, 80, 161, 238
– non-linear resistance 46–48, 60, 62, 147
– resistance, mean free path dependence, 42
beam pipe HOM coupler 308
– absorber 298, 311
BEPC-II, *see* collider, storage ring
BESSY, *see* storage ring light source

c
Canadian Light Source, *see* storage ring light source
cavities high-β
– crab cavity 349
– elliptical 9, 10
– ICHIRO 14, 16, 17, 89
– low-loss 13, 14
– re-entrant cavity 4, 13, 14, 16
cavities low-β 9, 19, 26, 231
– half-wave resonator (HWR) 29, 30, 92
– ladder resonator 9, 25
– quarter wave resonator (QWR) 26, 28–30, 221
– radio frequency quadrupole (RFQ) 4, 30, 371
– split ring resonator 26, 369
– spoke resonator 9, 22–25, 33, 38, 221
cavities medium-β 9, 18, 22, 26, 33, 35, 38
CEBAF 6, 353
– cavity 356
– cavity pair 353
– electron linac 353
– ERL 395
– operating experience 354, 355
– refurbished cavities 356
– upgrade 356
centrifugal barrel polishing, *see* barrel polishing
CESR, *see also* collider, storage ring
– cavity 338
– operating experience 339
CHESS, *see* storage ring light source
coating, niobium 234
– bias sputtering 234
– cathodic arc 234
– electron cyclotron resonance (ECR) 235
– magnetron sputtering 234
collider, storage ring
– BEPC-II 343
– CESR 6, 11, 338, 339, 341, 342
– LEP 6, 105, 335, 336
– TRISTAN 335
condensed gas
– couplers 293
– field emission 102, 125
– multipacting 87, 287, 339
– residual resistance, 65
Cooper pairs 41, 42, 73, 196

crab cavities 348, 351
crab crossing 348, 349
critical magnetic field,
- H_c 72, 73, 75, 76, 80, 160
- H_{c1} 69, 72, 73, 75, 80, 160, 172, 174, 180
- H_{c2} 72, 75, 78, 230
- H_{c3} 72, 77, 78
- H_{sh} 17, 72, 79, 80, 174
coupler, input 263–265
- coaxial 269
-- examples 272
- conditioning 290, 293
- dc bias 287
- diagnostics 290
- for low-β resonators 280
- materials and fabrication 282
- rf processing 287, 288
- transverse kick 266
- waveguide 268, 269, 280
- window 270, 271
coupler, higher order mode (HOM) 296
- absorber 298, 311
- beam pipe absorber 308
- coaxial 302
- coupler kick 306
- heating 305
- loop coupler 296
- waveguide 296, 306

d

dc bias, see input coaxial couplers
dc magnetization curves, annealing 256
degassing
- couplers 266
- coupler processing 290
- for H 69, 255, 357
DIAMOND, see storage ring light source
dry-ice cleaning 259
dust-free assembly 260

e

EBSD, see surface analysis
eddy current scanning 194, 203, 208, 211
ELBE, see free electron laser
electron beam melting 206, 212, 219
electron cyclotron resonance (ECR) 235
electropolishing 241, 242
- I–V characteristics 246, 248
- contamination issues 253
- continuous EP 243, 244
- current oscillations 243
- horizontal method 249
- intermittent polishing 243
- material removal 250
- surface roughness 242
- theory 248
- ultrasonic cleaning 254
- vertical method 251
- viscous layer 248
ELETTRA, see storage ring light source
eLIC, see energy recovery linac
energy recovery linac (ERL) 7, 300, 309, 360, 389, 390, 392, 394–398
- ALICE, Daresbury 396
- ARC-EN-CIEL 391
- Cornell 396, 397
- Daresbury 396
- eLIC 399
- eRHIC 398
- Japan Atomic Energy Agency (JAEA) 395
- Jefferson Lab 390, 395
- KEK 396
- RHIC 398, 399
e-p collider 335
- HERA 6, 398
eRHIC, see energy recovery linac
ethanol rinsing 251, 254
EURISOL, see heavy ion accelerator

f

ferrites, HOM absorber 309
field emission 5, 13, 85, 92, 94, 95, 96, 100, 102, 103, 127, 129, 256, 258, 408
- coupler 265
- dc 95
- emissive area 100
- emission sites 101
- emitter density 98, 99, 103
- Fowler-Nordheim 94
- helium processing 95, 125, 126, 354, 355, 382
- induced quench 96
- processing 106, 109, 110, 111
- simulation 120, 123
- statistical model 103
- x-rays 96
field enhancement
- Fowler-Nordheim 100
- geometric 104
- grain boundary 157, 198
- magnetic 149, 180
FLASH, see free electron laser
free electron laser (FEL) 7, 357 360, 362, 364, 366, 389–391

Subject Index | 445

- Darmstadt 360
- ELBE FEL 360
- FLASH 6, 10, 361, 362, 364
- high-gain-harmonic generation, HGHG 389, 391
- JAERI 360
- Jefferson Lab (JLab) IR FEL 6, 360
- Rossendorf 396
- S-DALINAC 360
- Tesla Test Facility 362
- WIFEL 390
- XFEL 6, 366

g
GANIL, *see* heavy-ion accelerator
geometric shunt impedance 11, 19, 295
geometry factor 11, 15, 19
Ginsburg-Landau parameter, niobium 172
Ginsburg-Landau theory 75, 80
global thermal instability 55
groove, MP suppression in WG 289

h
H contamination
- degassing 238
- EP 245
half-wave resonator (HWR) 29, 30, 92
heavy-ion accelerator 369, 370
- ADS 405
- ALPI 6, 29, 31, 231, 233, 370
- ATLAS 5, 369
- EURISOL 7, 29, 376, 400, 401
- GANIL 376, 400
- Inter-University Accelerator Centre (IUAC), New Delhi 373
- ISAC 6, 374
- National Superconducting Cyclotron Laboratory (NSCL) 378
- PIAVE, heavy-ion injector 6, 31, 371, 373
- Reaccelerator 378
- RIA 7, 25, 29, 378, 400, 401
- RIB 7, 18, 373, 376, 378, 399
- SARAF 379
- SPIRAL-II 7, 376, 400
HERA, *see* e-p collider
high-field Q-drop 129, 180
- comparison EP/BCP 134
- fluxoids 172
- grain boundary role 174
- Interface Tunneling Exchange 133
- large grain cavity 176
- large grain Nb 142, 150, 176, 180

- magnetic field effect 130
- mild baking benefit 138, 142, 144
- modified pollution model 162, 166
- oxygen pollution model 160
- pollution model 171
- pollution-roughness model 174
- role of hydrogen 171
- roughness effect 147, 150
- single crystal cavity 174
- spatial inhomogeneity 133
- surface studies 181
- thermal feedback effect 145
high-intensity proton linac 7, 401, 402
- JPARC 7, 403
- KAERI 405
- KOMAK 405
- Project X 403
- Proton Driver 403
- SNS 6, 18, 38, 380, 381, 384, 387, 402, 404
- SPL 7, 404
- TRASCO 19
- XADS 7, 405
high-pressure rinsing (HPR) 256
- force calibration 258
- system 257
HoBiCaT 391
HOM couplers 296
- beam pipe absorbers 308
- coaxial HOM couplers 302
- coupler kicks 306
- heating 305
- loop couplers 296
- wavgeuide 306
HOM absorber 298, 311
HOM excitation 295
- mode trapping 301
HOM propagation, asymmetric cavity 302
HPP 95, 96, 110
HWR 29, 30, 92
hydroforming 5, 204, 222, 226
- niobium-copper 226

i
ICHIRO 14, 16, 17, 89
Inspection 239
International Linear Collider (ILC) 6, 7, 205, 365, 406, 407, 408
- crab cavities for 351
Inter-University Accelerator Centre (IUAC), New Delhi 373
ISAC, *see* heavy-ion accelerator

j

JAERI, *see* free electron laser
Japan Atomic Energy Agency, *see* energy recovery linac
Jefferson Lab, *see* also CEBAF
- ERL   390, 395
- FEL   6
- IR FEL   360
Josephson fluxons   48, 51, 62
JPARC, *see* high intensity proton accelerator

k

KAERI, *see* high intensity proton accelerator
KEK, *see* TRISTAN, energy recovery linac
KEK-B   6, 11, 279, 308, 338, 341, 342, 343, 348, 351
- cavity   338
- collider e+e-   338
- crab cavity   349
- operating experience   339
KOMAK, *see* high intensity proton linac

l

ladder resonator   9, 25
large grain cavities
- BCP and EP   215
- cavity fabrication   213
- maximum gradient   214
- medium field Q-slope   56
large grain Nb
- high field Q-drop   142, 150, 176, 180
- niobium material   211, 212, 218
- surface smoothness   211
LEP-II   6, 105, 336
- e+e-collider   335
- operating experience   336
LHC
- collider p-p   338
- crab cavities for   351
loop couplers, HOM   296
Lorentz-force detuning   4, 11, 32, 36, 39, 267, 268, 313, 314, 316, 324, 326, 403
- low-β 403
lower order modes, crab cavity   348
low-field Q-slope   4, 48, 50, 129, 161, 168
low-loss cavity   13, 14
low-β 9, 19, 26, 231
low-β resonator fabrication   221

m

magnetic field bias, suppresssion in waveguides   289
magnetostrictive actuator, fast tuner   314, 329
magnetron sputtering   234
medium-field Q-slope   4, 48, 50, 51, 54–62, 129, 145, 147, 161
medium-β cavities   9, 18, 22, 26, 33, 35, 38
megasonic cleaning   260
microphonics   5, 12, 18, 19, 32, 35, 267, 268, 313, 324, 326, 331
mode trapping, HOM   301
modified pollution model   162, 166
multipacting   5, 11, 85, 89, 92
- rectangular waveguide   287
- scaling law   287
- simulation coupler   285
- simulations waveguide   289
- suppression for couplers   287
multipacting trajectories
- beam pipe   89
- in couplers   5 265, 285, 305
- low-β resonator   92
- scaling laws in coaxial couplers   286
muon collider   408, 410

n

National Superconducting Cyclotron Laboratory (NSCL)   378
niobium material
- production   205
- recrystallization   208
- specifications   204
- vendors   205
Nb-Cu bonding   226
Nb-Cu resonators, low-β 233
Nb-Cu-sputtered cavities   230
neutrino factory   408–410
neutron source   380, *see* also SNS
niobium-hydride, 65

o

optical inspection   239
oxygen diffusion
- baking effect   160
- BCS resistance   63, 161
- low-field Q-slope   50
- weak links   63
oxygen pollution model   166

p

peak surface electric field 11–13, 29
peak surface magnetic field 11, 13, 15, 19, 23
phonon peak 52, 55, 60, 134, 196
PIAVE, see heavy-ion accelerator
– heavy-ion injector 371
piezo elements, fast tuners 314, 326
planar disk window 271
planar rectangular ceramic 271
pollution model 171
postpurification, yield strength 238
power requirement, input coupler 267
Project X, see high intensity proton linac
Proton Driver, see high intensity proton linac

q

Q-disease 65–67, 69, 171, 230, 238
Q-slope, Nb-Cu 234
quality factor, 3
quench 13, 48, 129, 192, 197, 198, 408
– at defects 192, 194
– at welds 204
– baked 184
– baking effect 48, 138
– BCP baked 155, 165, 171, 198
– EP baked 198, 200
– grain boundary edge 157, 158
– high field Q-drop 200
– HPR accident 259
– increase with EP 154
– large grain 215
– phonon peak 196
– post-purification 132, 142, 218, 238
– single crystal 218
– thermal model 196
– thermo-currents 87
– weld 198
Quarter Wave Resonator (QWR) 26, 28–30, 221
– Nb-Cu 371

r

reaccelerator 378
recrystallization, niobium 208
re-entrant cavity 4, 13, 14, 16
residual resistance 4, 41, 48, 49, 65, 66, 69–71, 165, 183
rf processing 5, 14, 95, 106, 109–111, 127, 128, 382, 383
– couplers 287, 288
– simulation 116

RFQ 4, 30, 371
RHIC-II, see energy recovery linac
RIA 7, 25, 29, 378, 400, 401
RIB 7, 18, 373, 376, 378, 399
Rossendorf, see free electron laser

s

SARAF, see heavy-ion linac
S-DALINAC, see free electron laser
seamless cavities 222
self-amplified-spontaneous-emission (SASE) 362
Shanghai Light Source, see storage ring light source
SIMS, see surface analysis
single-crystal
– cavity 180, 212, 218
– niobium 211, 212, 218
SNS 6, 18, 38, 380, 381, 384, 387, 402, 404
SOLEIL, see storage ring light source
spinning 227
SPIRAL-II, see heavy-ion accelerator
SPL, see high intensity proton accelerator
split ring resonator 26, 369
spoke resonator 9, 22–25, 33, 38, 221
sputtered NbCu, low-β resonators 231
SQUID scanning 210
SRRC (Taiwan), see storage ring light source
stiffeners 35, 36
storage ring light source 342
– BESSY 391
– Canadian Light Source 342
– CHESS 6, 338, 339, 342, 396
– DIAMOND 342, 343
– ELETTRA 348
– Shanghai Light Source 342
– SLS 348
– SOLEIL 343, 346
– SRRC (Taiwan Light Source) 342
storage rings 335
sulfur contamination EP 253
surface analysis
– EBSD 187
– – dislocations 187
– SIMS 188, 189
– TEM 192
– XPS 46, 66, 72, 163, 181, 182, 184, 253
surface roughness 242

t

Taiwan Light Source, see SRRC
Tata Institute of Fundamental Research (TIFR), Mumbai 373
TEM, see surface analysis
temperature map
– field emission 93, 96, 106
– large grain cavity 178
– Q-disease 66, 69
– Q-drop 130
– trapped flux 69
Tesla Test Facility, see free electron laser
thermal breakdown 196
thermal conductivity 52, 55, 57, 196
– medium field Q-slope 51, 60
– post-purification 132
thermal feedback effect 55, 60, 132, 145, 146
thermal feedback model 51, 52, 57
third harmonic SRF 347
TEM, see surface analysis
transverse kick, input coupler 266
trapped dc magnetic flux 65, 69, 87
TRASCO, see high intensity proton linac
TRISTAN, e+e-collider 335
TRIUMF 374, see also ISAC-II
TTF 6, 361, 362, 364, 365, 391
– operating experience 362
tube forming 229
tumbling 237
tuners 5, 29, 35, 313–316, 326, 329, 331, 364, 387
– fast 313, 324, 326
– low-β resonators 329, 330
– magnetostrictive actuator, fast tuner 314, 329
two-fluid model 41, 42
two-point MP 87, 88
– couplers 285

u

ultrafast x-rays, crab cavities for 353
ultrasonic degreasing, EP 251

v

variable reactance, tuners 331
viscous layer, EP 248

w

warm window 270
– input coupler 265
waveguide coupler 268, 269, 280
waveguide couplers, HOM 296
windows 270, 283
– antimultipactor coating 283
– cold 270
– cylindrical 271
– planar disk 271
– planar rectangular 271
– warm 270

x

XADS, see high intensity proton linac
XFEL, see free electron laser
XPS, see surface analysis
X-rays, field emission 96

y

yield strength of Nb 205